SEISMIC SURFACE WAVES IN A LATERALLY INHOMOGENEOUS EARTH

MODERN APPROACHES IN GEOPHYSICS

formerly Seismology and Exploration Geophysics

VOLUME 9

The titles published in this series are listed at the end of this volume.

SEISMIC SURFACE WAVES IN A LATERALLY INHOMOGENEOUS EARTH

Edited by

V. I. KEILIS-BOROK

Institute of Physics of the Earth, Academy of Sciences of the U.S.S.R., Moscow, U.S.S.R.

With contributions by

A. L. Levshin*, T. B. Yanovskaya**, A. V. Lander*, B. G. Bukchin*, M. P. Barmin*, L. I. Ratnikova* and E. N. Its**

* *Institute of Physics of the Earth, Academy of Sciences of the U.S.S.R., Moscow, U.S.S.R.*

** *Faculty of Physics, Leningrad State University, Leningrad, U.S.S.R.*

KLUWER ACADEMIC PUBLISHERS

DORDRECHT / BOSTON / LONDON

Library of Congress Cataloging-in-Publication Data CIP

Poverkhnostnye seĭsmicheskie volny v gorizontal'no-neodnorodnoĭ Zemle.
English.
Seismic surface waves in a laterally inhomogeneous earth / edited by V. I. Keilis-Borok; with contributions by A.L. Levshin ... [et al.].
p. cm. – (Modern approaches in geophysics; 9)
Translation of:
Poverkhnostnye seĭsmicheskie volny v gorizontal 'noneodnorodnoĭ Zemle.
Includes index.

ISBN-13: 978-94-010-6885-7 e-ISBN-13: 978-94-009-0883-3
DOI: 10.1007/978-94-009-0883-3

1. Seismic waves. I. Keĭlis-Borok, Vladimir Isaakovich. II. Levshin, A. L. (Anatoliĭ L'vovich), 1932– . III. Title. IV. Series.
QE538.5.P6813 1989
551.2'2–dc 19 88–39095

Published by Kluwer Academic Publishers,
P.O. Box 17, 3300 AA Dordrecht, The Netherlands.

Kluwer Academic Publishers incorporates the publishing programmes of
D. Reidel, Martinus Nijhoff, Dr W. Junk and MTP Press.

Sold and distributed in the U.S.A. and Canada
by Kluwer Academic Publishers,
101 Philip Drive, Norwell, MA 02061, U.S.A.

In all other countries, sold and distributed
by Kluwer Academic Publishers Group,
P.O. Box 322, 3300 AH Dordrecht, The Netherlands.

Originally published as

Poverkhnostnye seĭsmicheskie volny v gorizontal'no-neodnorodnoĭ Zemle

Translated from the Russian by A. L. Petrosyan

printed on acid free paper

CONTENTS

INTRODUCTION

Surface waves form the longest and strongest portion of a seismic record excited by explosions and shallow earthquakes. Traversing areas with diverse geologic structures, they 'absorb' information on the properties of these areas which is best reflected in dispersion, the dependence of velocity on frequency. The other properties of these waves — polarization, frequency content, attenuation, azimuthal variation of the amplitude and phase — are also controlled by the medium between the source and the recording station; some of these are affected by the properties of the source itself and by the conditions around it.

In recent years surface wave seismology has become an indispensable part of seismological practice. The maximum amplitude in the surface wave train of virtually every earthquake or major explosion is being measured and used by all national and international seismological surveys in the determination of the most important energy parameter of a seismic source, namely, the magnitude M_s. The relationship between M_s and the body wave magnitude m_b is routinely employed in identification of underground nuclear explosions. Surface waves of hundreds of earthquakes recorded every year are being analysed to estimate the seismic moment tensor of earthquake sources, to determine the periods of free oscillations of the Earth, to construct regional dispersion curves from which in turn the crustal and upper mantle structure in various areas is derived, and to evaluate the dissipative parameters of the mantle material. Work has begun on surface wave 'tomography' of the Earth's mantle — identification of lateral inhomogeneities and elastic anisotropy from worldwide seismic observations. Surface waves from explosions and earthquakes are being used to study relatively inaccessible areas, for instance in a search for major sedimentary basins in the Arctic shelf.

These spectacular successes of surface wave seismology have been made possible by recent developments in long-period seismic instrumentation, seismic arrays, and the array-like use of national and international networks of long-period stations.

On the other hand, computerization has helped in the development of methods for the analysis of surface wave observations that could more accurately estimate surface wave dispersion and polarization. Techniques of quantitative data interpretation have been developed that can effectively determine velocity and density models consistent with the observations and evaluate the non-uniqueness and resolution of the data.

All these results were essentially based on an advanced theory and efficient techniques for calculating the spectral characteristics and wave fields of surface waves in vertically and radially varying structures.

The progress in surface wave seismology highlighted the inadequacies in the theoretical model of the medium underlying data interpretation as contrasted with the presently available accuracy and detail in the observations, as well as with the problems that needed to be solved. The structures studied today have considerable near-surface and deep lateral inhomogeneities that manifest themselves both as smooth variations in velocity and density with distance (within major blocks of the continental and oceanic lithosphere) and as sharp contrasts in velocity and density at boundaries between blocks with differing deep structure, in subduction zones, geosynclinal depressions, intra-continental basins, rifts, island arcs etc.

In recent years investigators in the USSR and elsewhere have been conducting a number of theoretical and experimental studies (on models and in the field) of surface wave effects caused by such lateral inhomogeneities. This work has yielded results that are valuable for applications: approaches have been found that have enabled the treatment of observations, under certain assumptions, within the framework of laterally varying models and the derivation of meaningful information on the inhomogeneities using theoretically sound methods (from the standpoint of physics). It has become possible to pose new problems for surface wave seismology that are based on advanced concepts of the properties of seismic fields in inhomogeneous media. This monograph is an attempt to present a unified exposition of these novel theoretical and interpretation approaches, as well as of their application in specific problems.

The book consists of two parts. The first part is devoted to forward problems in the surface wave seismology of inhomogeneous media.

Chapter 1 presents the fundamentals of surface wave theory for vertically and radially varying media. A description of seismic sources is provided based on the theory of stress glut; the main properties of waves excited by some kinds of point sources are considered together with various factors that affect them; examples are discussed for the simplest (classical) sources. Included is an account of efficient methods for incorporating sphericity, anelasticity, and the main properties of seismograms.

Chapter 2 is concerned with smooth lateral inhomogeneities as they affect the surface wave field. Asymptotic approaches are described providing ray representations for the leading term in the displacement transported by harmonic and transient surface waves. Transient wave fields are being studied by spatial ray tracing; the fundamentals of the theory of space-time surface wave rays are presented for media involving smooth lateral inhomogeneities.

Chapter 3 propounds an approximate technique for treating a vertical contact in surface wave propagation. The technique uses Green functions to calculate reflection and transmission coefficients for plane waves incident at an arbitrary angle on a plane vertical contact between two quarter-spaces of differing structure. The accuracy of the method is evaluated by comparison with the results of more exact calculations that include the contribution of diffracted waves. Complications are considered such as inclined contact discontinuities or fault zones with very low rigidity.

Chapter 4 contains algorithms for solving forward problems in surface wave theory. It describes the algorithms the authors used to compute spectral parameters for vertically varying media, reflection and transmission coefficients for vertical contacts. We present ray-geometric algorithms for media involving smooth lateral inhomogeneities. Each technique is illustrated by practical examples.

The second part of the book treats questions arising in data analysis and interpretation.

Chapter 5 describes the modern methods in use for recording and analyzing long-period information. A mathematical model of the dispersed seismic signal is presented and examined. The potential and limitations of spectral and frequency-time methods of signal separation and dispersion measurement are demonstrated. Polarization and spatial analysis are used to provide ways of separating interfering waves and measuring their parameters.

Chapter 6 considers quantitative interpretation of surface wave dispersion. It begins with a review of recent progress. Two different approaches are considered in great detail which allow extraction of local information from observations involving areas with different deep structures. One of these approaches does not require prior information about the area of study and seeks lateral perturbations to phase and group velocities using a two-dimensional analogue of the widely known Backus—Gilbert technique. The other implies a regionalization of the study area and determines regional dispersion curves for individual laterally homogeneous regions using nonlinear programming methods.

Errors are examined that arise in the interpretation of surface wave data because of neglected lateral inhomogeneities, along with some inhomogeneity-related phenomena that might be attributed to anisotropy. Inverse problems for earthquake sources based on surface wave observations are formulated.

Chapter 7 gives an account of some results in the interpretation of surface waves recorded by seismological stations. Crustal velocity structures are presented for a number of areas in Eurasia obtained by successive elimination of paths and maps of group velocity distribution for some continental and oceanic areas. Dispersion and polarization effects due to an extended anomalous zone in northeastern Siberia are studied, the zone being located around a hypothetical continental riftogenesis area.

This monograph is a joint work by several authors. The Introduction, Chapter 1 (except for Section 1.2), Sections 5.1, 5.5, and 6.7 are written by A. L. Levshin; Chapters 2 and 3 (excluding Section 3.4), Sections 4.1, 4.3, 6.1, 6.5, 6.6 and 7.2 are by T. B. Yanovskaya; A. V. Lander is responsible for Sections 5.2 through 5.4, and 7.3, B. G. Bukchin for Sections 1.2, 3.4, and 6.9, M. P. Barmin for Sections 4.2, 6.2 through 6.4, L. I. Ratnikova for Sections 6.8 and 7.1, E. N. Its for Section 4.4.

The materials used were collected in collaboration with K. A. Berteussen, P. Malischewski, T. M. Sabitova, S. E. Kapitanova, O. E. Starovoit, A. N. Nesterov and published in joint papers. We are sincerely grateful to all these colleagues.

PART 1

THEORY

1. SURFACE WAVES IN VERTICALLY INHOMOGENEOUS MEDIA

1.1. EQUATIONS OF MOTION FOR AN ELASTIC MEDIUM

Surface waves are considered within the framework of linear elasticity theory, mainly for perfectly elastic bodies. Internal sources of disturbance are described in Section 1.2 by means of equivalent forces. Small departures from perfect elasticity in wave propagation theory can be treated as shown in Section 1.6.

Equations of motion. Equations of motion for a point $\mathbf{x}$ with coordinates (x_1, x_2, x_3) acted on by body forces have the form (Levshin, 1973; Aki and Richards, 1980)

$$\sigma_{ij,j} + f_i = \rho \ddot{u}_i. \tag{1.1}$$

Here the σ_{ij} are components of the symmetrical stress tensor; the f_is are components of body force per unit volume; ρ is the density; the u_is are components of the displacement vector. The subscripts i, j take on the values 1, 2, 3. A symbol with a dot above it denotes a time derivative, a subscript j when preceded by a comma means differentiation with respect to the spatial coordinates x_j. Summation over double (mute) subscripts is understood everywhere below, unless otherwise indicated.

The stress tensor is linearly related to small strains e_{ij} through Hooke's law

$$\sigma_{ij} = c_{ijpq} e_{pq} \tag{1.2}$$

c_{ijpq} being a tensor of elasticity constants that depends on $\mathbf{x}$. It has the symmetries $c_{ijpq} = c_{jipq} = c_{ijqp} = c_{pqij}$.

Subsequent discussion is mainly confined to isotropic bodies: $c_{ijpq} = \lambda \delta_{ij} \delta_{pq} + \mu(\delta_{ip}\delta_{jq} + \delta_{iq}\delta_{jp})$, δ_{ij} being the Kronecker delta. The Lame constants λ and μ determine the velocities of elastic waves, a and b:

$$a = [(\lambda + 2\mu)/\rho]^{1/2} \quad \text{for compressional (P) waves,}$$

$$b = [\mu/\rho]^{1/2} \quad \text{for shear (S) waves.}$$

Hooke's law for isotropic media has the form

$$\sigma_{ij} = \lambda \delta_{ij} \vartheta + 2\mu e_{ij} \tag{1.2a}$$

where $\vartheta = e_{kk}$ is the dilatation.

Small strains and displacements are related through

$$e_{ij} = \tfrac{1}{2}(u_{i,j} + u_{j,i}). \tag{1.3}$$

Substitution of (1.2a) and (1.3) into (1.1) yields equations of motion in **u** in an invariant form:

$$(\lambda + \mu)\operatorname{div}\mathbf{u} + \mu\Delta\mathbf{u} + \nabla\lambda\operatorname{div}\mathbf{u} + \nabla\mu \times \operatorname{rot}\mathbf{u} + \\ + 2(\mu, \nabla)\,\mathbf{u} + \mathbf{f} = \rho\ddot{\mathbf{u}} \tag{1.4}$$

where ∇ denotes the gradient, Δ is the Laplacian, and $(\mathbf{x}, \mathbf{y})$ a scalar product.

Initial conditions. The medium is assumed to be at rest before the time $t = 0$:

$$u_j \equiv \dot{u}_j \equiv 0 \quad \text{when} \quad t < 0. \tag{1.5}$$

Boundary conditions. Any elastic body that we shall consider below is bounded (wholly or partially) by a free surface S_0. This means that the following relation holds

$$T_i(\mathbf{n}) = \sigma_{ij} n_j \,|_{S_0} = 0 \tag{1.6}$$

where the n_j are components of the outward normal to S_0, $\mathbf{n}$. Here and below, the $T_i(\mathbf{n})$ are components of traction $\mathbf{T}(\mathbf{n})$, i.e., the force on an element of area $\mathrm{d}S_0$.

Conditions at interfaces. The functions $a(\mathbf{x})$, $b(\mathbf{x})$, $\rho(\mathbf{x})$ are assumed to be positive and piecewise continuous. There is welded contact at any surface where these functions are discontinuous, i.e., the displacement vector **u** and stress tensor σ_{ij} are continuous across an interface.

Sources of disturbance. Elastic motion is excited by a vector field of body force **f** that is a function of spatial and time coordinates (its physical meaning is discussed in the next section). It will be assumed in what follows that the field is spatially finite, i.e. $f_j(\mathbf{x}, t) = 0$ outside some closed region Ω, which will be termed the source region.

The time derivative of f_j is assumed to be a finite function of time, i.e. $f_j(\mathbf{x}, t) = \dot{f}_j(\mathbf{x}, t) = 0$ when $t < 0$ and $\dot{f}_j(\mathbf{x}, t) = 0$ when $t > t_e$.

Green function. To solve a forward problem in elastic wave theory is to determine the displacement vector field $u_j(\mathbf{x}, t)$ and the stress tensor field $\sigma_{ij}(\mathbf{x}, t)$ that obey Equations (1.1), (1.2a), (1.3) and the initial and boundary conditions (1.5), (1.6).

When the body force field is concentrated in space at $\mathbf{x}_0$ and in time at t_0, and points along the x_j-axis, i.e.,

$$f_i(\mathbf{x}, t) = \delta_{ij}\delta(\mathbf{x} - \mathbf{x}_0)\delta(t - t_0)$$

δ being the Dirac delta function, the resulting displacement $\mathbf{u}(\mathbf{x}, t)$ is called the fundamental solution, or Green function for the boundary value problem (1.1), (1.5), (1.6) (Aki and Richards, 1980). It will be denoted $\mathbf{G}^{(j)}(\mathbf{x}; \mathbf{x}_0; t - t_0)$, its ith component being $G_{ij}(\mathbf{x}; \mathbf{x}_0; t - t_0)$.

The solution for an arbitrary force $f_j(\mathbf{x}, t)$ that is zero when $t < 0$ can be expressed in terms of the Green function as follows

$$u_i(\mathbf{x}, t) = \int_0^t \mathrm{d}\tau \int_\Omega G_{ij}(\mathbf{x}; \mathbf{y}; t - \tau) f_j(\mathbf{y}, \tau)\, \mathrm{d}V_y \tag{1.7}$$

where $\mathrm{d}V$ is an element of the volume Ω (source region) within which $f_j(\mathbf{x}, t)$ is not identically zero.

In addition to the Green function itself, it is convenient to define its time integral:

$$H_{ij}(\mathbf{x};\mathbf{y};t)=\int_0^t G_{ij}(\mathbf{x};\mathbf{y};\tau)\,\mathrm{d}\tau.$$

It is easy to show that the displacement for any bounded force can also be represented in the form

$$u_i(\mathbf{x},t)=\int_0^{t_e}\mathrm{d}\tau\int_\Omega H_{ij}(\mathbf{x};\mathbf{y};t-\tau)\dot{f}_j(\mathbf{y},\tau)\,\mathrm{d}V_y \tag{1.8}$$

where $(0, t_e)$ is the interval in which $\dot{f}_j(t)$ is not identically zero.

Representation theorem. Consider a three-dimensional region V bounded by a surface S whose outward normal is $\mathbf{n}$. Displacement at any point $\mathbf{x}$ within the region at time t can, according to the representation theorem (Aki and Richards, 1980), be expressed through the displacement $\mathbf{u}$ and traction $\mathbf{T}$ at the surface S and the body forces within V, as follows:

$$\begin{aligned}u_i(\mathbf{x},t)&=\int_0^\infty\mathrm{d}\tau\int_V G_{ji}(\mathbf{y};\mathbf{x};t-\tau)f_j(\mathbf{y},\tau)\,\mathrm{d}V_y+\\&+\int_0^\infty\mathrm{d}\tau\int_S[G_{ji}(\mathbf{y};\mathbf{x};t-\tau)T_j(\mathbf{u}(\mathbf{y},\tau),\mathbf{n})-\\&-(\mathbf{u}(\mathbf{y},\tau)T_j(\mathbf{G}^{(i)}(\mathbf{y};\mathbf{x};t-\tau),\mathbf{n})]\,\mathrm{d}S_y.\end{aligned}\tag{1.9}$$

Here, the Green function $\mathbf{G}^{(i)}$ is the fundamental solution to (1.1) in V with arbitrary boundary conditions at S.

The traction $\mathbf{T}(\mathbf{G}^{(i)}(\mathbf{y};\mathbf{x};t),\mathbf{n})$ is caused by the displacement $\mathbf{G}^{(i)}$ and according to (1.2), (1.3) has the form

$$T_j=c_{jklm}n_kG_{li,m}.$$

Now suppose there are no body forces, but there is a surface Σ with normal $\mathbf{n}$ in the medium across which the displacement vector $\mathbf{u}$ is discontinuous and $\mathbf{T}(\mathbf{u},\mathbf{n})$ continuous. In such a case, choosing as S in (1.9) $S_0+\Sigma^-+\Sigma^+$ where Σ^- and Σ^+ are the two sides of Σ, assuming homogeneous boundary conditions (1.6) for $\mathbf{G}^{(i)}$ and for the displacement $\mathbf{u}$ at S_0, we get

$$u_i(\mathbf{x},t)=\int_0^\infty\mathrm{d}t\int_\Sigma[u_j(\mathbf{y},t)]c_{jklm}(\mathbf{y})n_k(\mathbf{y})G_{il,m}(\mathbf{x};\mathbf{y};t-\tau)\,\mathrm{d}\Sigma_y. \tag{1.9a}$$

The Green function is here differentiated with respect to the y_j. Enclosing a quantity in square brackets means taking the difference of its two values at $\mathbf{y}$ measured on the two sides of Σ, Σ^+ and Σ^-, the normal $\mathbf{n}$ pointing from Σ^- to Σ^+. The reciprocity of the Green function used here (Aki and Richards, 1980)

$$G_{ij}(\mathbf{x};\mathbf{y};t-\tau)=G_{ji}(\mathbf{y};\mathbf{x};t-\tau)$$

was employed in the derivation of (1.9a).

1.2. SOURCES OF SEISMIC DISTURBANCE

Seismic disturbances most frequently arise from the action of internal sources (earthquakes or explosions) in the absence of any external body forces. One must then set $f_i \equiv 0$ in (1.1), so that the only solution that satisfies the homogeneous initial (1.5) and boundary (1.6) conditions, as well as Hooke's law (1.2), will be $u_i \equiv 0$. Non-zero displacements cannot arise in the medium, unless at least one of the above conditions is not true. Following Backus and Mulcahy (1976, 1976a), we assume seismic motion to be caused by a departure from Hooke's law within some volume of the medium at some instant of time $t > 0$.

Let $u_i(\mathbf{x}, t)$ describe the displacements and $\sigma_{ij}(\mathbf{x}, t)$ the stresses that would have existed in the medium had Hooke's law (1.2) been true everywhere in it. Let $s_{ij}(\mathbf{x}, t)$ be the actual stresses. The difference

$$\Gamma_{ij}(\mathbf{x}, t) = \sigma_{ij}(\mathbf{x}, t) - s_{ij}(\mathbf{x}, t), \tag{1.10}$$

called the stress glut tensor, is not identically zero within the three-dimensional region Ω. Within, and only within, that region, the tensor $\dot{\Gamma}_{ij}(\mathbf{x}, t)$ too is not identically zero. We shall assume that Ω lies wholly within the medium (does not come out to the surface) and that, since some instant of time $t_e > 0$, $\dot{\Gamma}_{ij}(\mathbf{x}, t) = 0$ everywhere in the medium. The integral of Γ_{ij} over Ω is called the seismic moment tensor (Kostrov, 1970; Aki and Richards, 1980). As the true motion obeys the equation $s_{ij,j} = \rho \ddot{u}_i$, in accordance with (1.1), one derives from (1.10)

$$\sigma_{ij,j} + g_i = \rho \ddot{u}_i, \tag{1.11}$$

$$g_i = -\Gamma_{ij,j}. \tag{1.12}$$

The equivalent forces $g_i(\mathbf{x}, t)$ that enter (1.11) satisfy all the restrictions that we have imposed on body forces f_i in 1.1, so the resulting displacements are given by the same formulas, (1.7) and (1.8), with f_i replaced by g_i. Using relation (1.12) for g_i and the Gauss–Ostrogradsky theorem, we finally get

$$u_i(\mathbf{x}, t) = \int_0^t \mathrm{d}\tau \int_\Omega G_{ij,k}(\mathbf{x}; \mathbf{y}; t - \tau)\Gamma_{jk}(\mathbf{y}, \tau)\, \mathrm{d}V_y \tag{1.13}$$

or

$$u_i(\mathbf{x}, t) = \int_0^{t_e} \mathrm{d}\tau \int_\Omega H_{ij,k}(\mathbf{x}; \mathbf{y}; t - \tau)\dot{\Gamma}_{jk}(\mathbf{y}, \tau)\, \mathrm{d}V_y. \tag{1.14}$$

The G_{ij}, H_{ij} are here differentiated with respect to y_k.

If the departure from perfect elasticity is confined to some arbitrary finite area at the inner surface Σ, the stress glut tensor becomes $\Gamma_{jk}(\mathbf{x}, t) = m_{jk}(\mathbf{x}, t)\delta_\Sigma(\mathbf{x})$, where $\delta_\Sigma(\mathbf{x})$ is a distribution that satisfies

$$\int_V \delta_\Sigma(\mathbf{x})\varphi(\mathbf{x})\, \mathrm{d}V_x = \int_\Sigma \varphi(\mathbf{x})\, \mathrm{d}\Sigma_x$$

for any function $\varphi(\mathbf{x})$. Integration over the volume V_y in (1.13), (1.14) will then reduce to that over the surface Σ:

$$u_i(\mathbf{x}, t) = \int_0^t \mathrm{d}t \int_\Sigma G_{ij,k}(\mathbf{x}; \mathbf{y}; t-\tau) m_{jk}(\mathbf{y}, \tau)\, \mathrm{d}\Sigma_y$$

where the points $\mathbf{y}$ belong to Σ. If the departure from perfect elasticity is defined as a discontinuity in displacement $\mathbf{u}$ at Σ without a stress discontinuity, then (1.9a) yields

$$m_{jk}(\mathbf{x}, t) = n_q(\mathbf{x})\,[u_p(\mathbf{x}, t)]\,c_{jkpq}(\mathbf{x}),$$

where $\mathbf{n}$ is the normal to Σ (the meaning of the discontinuity $[u_p]$ has been explained above). For an isotropic medium we shall have

$$m_{jk} = \lambda [u_p] n_p \delta_{jk} + \mu(n_j [u_k] + n_k [u_j]);$$

in the case of tangential (shear) dislocation we have

$$n_p [u_p] \equiv 0 \quad \text{and}$$

$$m_{jk} = \mu(n_j [u_k] + n_k [u_j]). \tag{1.15}$$

If the departure from perfect elasticity is confined to a small vicinity of $\mathbf{x}_0$ (the region Ω shrinks to a point), then

$$\Gamma_{jk}(\mathbf{x}, t) = m_{jk}(t)\delta(\mathbf{x} - \mathbf{x}_0)$$

and the equivalent forces g_i take the dipole form

$$g_i = -m_{jk}(t) \frac{\partial \delta(\mathbf{x} - \mathbf{x}_0)}{\partial x_k}. \tag{1.16}$$

Such a source excites a field of the form

$$u_i = \int_0^t m_{jk}(t) G_{ij,k}(\mathbf{x}; \mathbf{x}_0; t - \tau)\, \mathrm{d}t, \tag{1.17}$$

or

$$u_i = \int_0^{t_c} \dot{m}_{jk}(t) H_{ij,k}(\mathbf{x}; \mathbf{x}_0; t - \tau)\, \mathrm{d}t, \tag{1.17a}$$

where the G_{ij}, H_{ij} are differentiated with respect to y_k at the point $\mathbf{y} = \mathbf{x}_0$.

A point center of expansion (an ideally concentrated explosion) in an isotropic medium will produce (Aki and Richards, 1980)

$$m_{jk} = m(t)\delta_{jk}, \tag{1.18}$$

while for a point source of slip we shall have

$$m_{jk} = m(t)(\kappa_j n_k + \kappa_k n_j), \tag{1.19}$$

where the κ_j are unit vector components in the direction of the discontinuity $[\mathbf{u}]$

(slip vector) and $m(t) = \mu|[\mathbf{u}]|$. The quantity $m_0 = \lim_{t\to\infty} m(t)$ is called the seismic moment.

1.3. SURFACE WAVES DUE TO A POINT SOURCE IN A VERTICALLY VARYING HALF-SPACE

We shall now confine ourselves to considering a far more restricted class of models compared with those outlined above, viz., ones in which the properties of the medium are functions of a single coordinate. The theory of surface waves for such media is fairly complete (Levshin, 1973; Aki and Richards, 1980); we shall summarize it below for use in more complex situations.

We shall deal with a vertically varying half-space having the Cartesian coordinates $x = x_1$ $(-\infty < x < \infty)$, $y = x_2$ $(-\infty < y < \infty)$, $z = x_3$ $(0 < z < \infty)$ and bounded by a free surface S_0 $(z = 0)$. The velocities a and b, the density ρ are functions of z only. The medium is homogeneous below $z = Z$, that is, $a(z) = a(Z + 0)$, $b(z) = b(Z + 0)$, $\rho(z) = \rho(Z + 0)$ when $Z < z < \infty$; also, $b(Z+0) = \max b(z)$, $a(Z+0) = \max a(z)$.

It is shown by Levshin (1973), Aki and Richards (1980) that the full solution $\mathbf{u}(\mathbf{x}, t)$ for such a medium can be represented in an integral form. Its principal part at large (compared with the wavelength) distances from the source region Ω to the observation site is formed by surface waves. It is supposed that the depths of both source region and observation site are much smaller than the horizontal distance between them. The surface wave part of the solution separates into two independent fields $\mathbf{u}_R$ and $\mathbf{u}_L$ (Rayleigh and Love waves) with different polarizations. We shall frequently employ the symbol D (D = R or L) in what follows to indicate the relevant wave type.

One-dimensional eigenvalue problems for Rayleigh and Love waves. Consider partial solutions to (1.1) for the above model with no body force having the form of plane harmonic waves propagating along the x-axis

$$\mathbf{u}(\mathbf{x}, t) = \mathbf{V}(z)\exp[i(\omega t - \xi x)],$$

and satisfy the boundary condition (1.6)

$$\sigma_{13} = \sigma_{23} = \sigma_{33} = 0 \quad \text{for} \quad z = 0.$$

Substituting these solutions in (1.1) and (1.6), we are faced with two independent one-dimensional eigenvalue problems for components V_x, V_y, V_z of the amplitude factor $\mathbf{V}(z)$. One of these describes motion in the (x, z)-plane (P, SV) and determines, in particular, the velocity and amplitude of Rayleigh waves. The other describes the motion along the y-axis (SH) and determines the velocity and amplitude of Love waves.

The P—SV problem has the form

$$\begin{aligned}
&\frac{\mathrm{d}}{\mathrm{d}z}\,\bar{\sigma}_{33} - \xi\mu\,\frac{\mathrm{d}V^{(2)}}{\mathrm{d}z} + V^{(1)}(\rho\omega^2 - \xi^2\mu) = 0\\
&\frac{\mathrm{d}}{\mathrm{d}z}\,\bar{\sigma}_{13} + \xi\lambda\,\frac{\mathrm{d}V^{(1)}}{\mathrm{d}z} + V^{(2)}[\rho\omega^2 - \xi^2(\lambda + 2\mu)] = 0
\end{aligned} \tag{1.20}$$

where

$$V^{(1)} = V_z; \qquad V^{(2)} = iV_x;$$

$$\bar{\sigma}_{13} = \mu \left(\frac{\mathrm{d}V^{(2)}}{\mathrm{d}z} + \xi V^{(1)} \right); \tag{1.21}$$

$$\bar{\sigma}_{33} = (\lambda + 2\mu) \frac{\mathrm{d}V^{(1)}}{\mathrm{d}z} - \xi\lambda V^{(2)}$$

$\bar{\sigma}_{13}$ and $\bar{\sigma}_{33}$ are amplitude factors in the expressions for the true stresses σ_{13}, σ_{33}:

$$\sigma_{13} = -i\bar{\sigma}_{13}(z) \exp[i(\omega t - \xi x)],$$
$$\sigma_{33} = \bar{\sigma}_{33}(z) \exp[i(\omega t - \xi x)].$$

The boundary condition at $z = 0$ is

$$\bar{\sigma}_{13} = \bar{\sigma}_{33} = 0. \tag{1.22}$$

The functions σ_{13}, σ_{33}, $V^{(1)}$ and $V^{(2)}$ are continuous and bounded for all values of z.

The problem has both a discrete and a continuous spectrum, the wavenumber ξ being a free parameter. The discrete eigenvalues occur within the range $\xi^2 C_0^2 < \omega_{kR}^2 < \xi^2 b^2\ (Z + 0)$ where $k = 1, 2, \ldots$; C_0 is the smallest of three possible velocities: $b_{\min}$, the smallest shear wave velocity; C_{R}, Rayleigh wave velocity in a homogeneous half-space with parameters $a(0)$, $b(0)$; min C_s^i, smallest velocity of Stoneley waves (Aki and Richards, 1980) associated with velocity or density discontinuities at $z = z_i$, provided these waves can propagate along the interface between two homogeneous half-spaces with parameters $a(z_i \pm 0)$, $b(z_i \pm 0)$, $\rho(z_i \pm 0)$. Corresponding to discrete spectral eigenvalues are real vector eigenfunctions whose components (for $z > Z$) are linear combinations of exponential terms that decrease with z:

$$\begin{aligned} V_k^{(1)} = {} & A_1 \exp[-(\xi^2 - (\omega_{kR}/a(Z+0))^2)^{1/2}] + \\ & + B_1 \exp[-(\xi^2 - (\omega_{kR}/b(Z+0))^2)^{1/2}] \\ V_k^{(2)} = {} & A_2 \exp[-(\xi^2 - (\omega_{kR}/a(Z+0))^2)^{1/2}] + \\ & + B_2 \exp[-(\xi^2 - (\omega_{kR}/b(Z+0))^2)^{1/2}]. \end{aligned}$$

The continuous spectrum lies in the range $\xi^2 b^2\ (Z + 0) < \omega^2 < \infty$. The vector function components are (for $z > Z$) linear combinations of upgoing and downgoing plane waves

$$\exp[\pm i((\omega/a(Z+0))^2 - \xi^2)^{1/2} z], \quad \exp[\pm i((\omega/b(Z+0))^2 - \xi^2)^{1/2} z].$$

The following orthogonality conditions hold for the eigenfunctions when ξ is fixed:

$$\int_0^\infty \rho[V_k^{(1)}V_l^{(1)} + V_k^{(2)}V_l^{(2)}]\,\mathrm{d}z = 0, \qquad k \neq l$$

$$\int_0^\infty \rho[V_k^{(1)}V^{(1)}(\omega^2, z) + V_k^{(2)}V^{(2)}(\omega^2, z)]\,\mathrm{d}z = 0 \tag{1.23}$$

$$\int_0^\infty \rho[V^{(1)}(\omega^2, z)V^{(2)*}(\beta^2, z) + (V^{(2)}(\omega^2, z)V^{(2)*}(\beta^2, z)]\,\mathrm{d}z \sim \delta(\omega - \beta)$$

where * denotes complex conjugation.

Two types of orthogonality condition hold for fixed ω:

$$\begin{aligned}(\xi_k + \xi_l)\left[\int_0^\infty (\lambda + 2\mu)V_k^{(2)}V_l^{(2)}\,\mathrm{d}z + \int_0^\infty \mu V_k^{(1)}V_l^{(1)}\,\mathrm{d}z\right] - \\ - \int_0^\infty \lambda\left(\frac{\mathrm{d}V_k^{(1)}}{\mathrm{d}z}V_l^{(2)} + \frac{\mathrm{d}V_l^{(1)}}{\mathrm{d}z}V_k^{(2)}\right)\mathrm{d}z + \\ + \int_0^\infty \mu\left(\frac{\mathrm{d}V_k^{(2)}}{\mathrm{d}z}V_l^{(1)} + \frac{\mathrm{d}V_l^{(2)}}{\mathrm{d}z}V_k^{(1)}\right)\mathrm{d}z = 0 \qquad k \neq l\end{aligned} \tag{1.23a}$$

$$\begin{aligned}(\xi_k - \xi_l)\left[\int_0^\infty (\lambda + 2\mu)V_k^{(2)}V_l^{(2)}\,\mathrm{d}z - \int_0^\infty \mu V_k^{(1)}V_l^{(1)}\,\mathrm{d}z\right] - \\ - \int_0^\infty \lambda\left(\frac{\mathrm{d}V_k^{(1)}}{\mathrm{d}z}V_l^{(2)} + \frac{\mathrm{d}V_l^{(1)}}{\mathrm{d}z}V_k^{(2)}\right)\mathrm{d}z + \\ - \int_0^\infty \mu\left(\frac{\mathrm{d}V_k^{(2)}}{\mathrm{d}z}V_l^{(1)} + \frac{\mathrm{d}V_l^{(2)}}{\mathrm{d}z}V_k^{(1)}\right)\mathrm{d}z = 0 \qquad \text{for any } k \text{ and } l.\end{aligned} \tag{1.23b}$$

Conditions similar to (1.23a), (1.23b) exist for eigenfunctions of the continuous spectrum and between those of the discrete and continuous spectrum. Given a value of ω, the continuous spectrum exists in the interval $-\infty < \xi^2 < \omega^2/b^2$ $(Z + 0)$.

The problem for SH motion has the form

$$\frac{\mathrm{d}}{\mathrm{d}z}\bar{\sigma}_{23} + V^{(3)}(\rho\omega^2 - \xi^2\mu) = 0 \tag{1.24}$$

where $V^{(3)} = V_y$; $\bar{\sigma}_{23} = \mu\,\mathrm{d}V^{(3)}/\mathrm{d}z$ is an amplitude factor in the expression for the tangential stress $\sigma_{23} = \bar{\sigma}_{23}\exp[i(\omega t - \xi x)]$. The boundary condition at $z = 0$ is

$$\bar{\sigma}_{23} = 0 \tag{1.25}$$

$V^{(3)}(z)$ and $\bar{\sigma}_{23}(z)$ are continuous and bounded for all values of z. This problem too has both a discrete and a continuous spectrum (Figure 1.1). The discrete eigenvalues lie within the interval

$$\xi^2 b_{\min}^2 < \omega_{kl}^2 \leqslant \xi^2 b^2 (Z+0) \qquad \text{where } k = 1, 2, \ldots .$$

Corresponding to this spectrum are real eigenfunctions that fall off exponentially with z for $z > Z$:

$$V_k = A \exp[-(\xi^2 - (\omega_{kL}/b(Z+0))^2)^{1/2}].$$

The continuous spectrum lies in the interval $\xi^2 b^2 (Z+0) < \omega^2 < \infty$ (Figure 1.1). The corresponding eigenfunctions for $z > Z$ are linear combinations of upgoing and downgoing plane waves:

$$\exp[\pm i((\omega/b(Z+0))^2 - \xi^2)z].$$

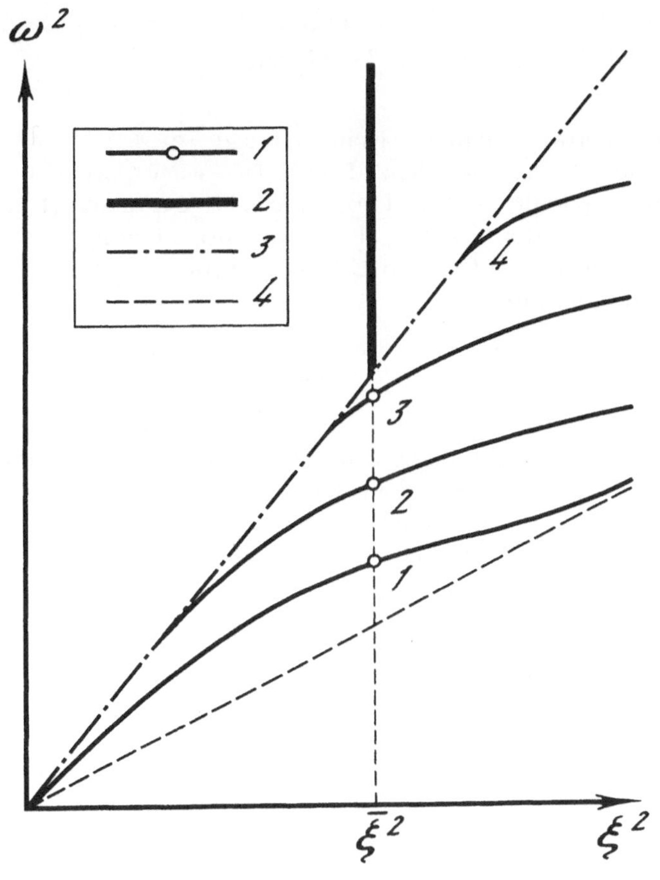

Fig. 1.1. Discrete and continuous spectrum in the one-dimensional boundary-value problem for Love waves. (1) $\omega_k^2(\xi)$ curves, $k = 1, 2, 3$, open circles corresponding to some value $\xi^2 = \bar{\xi}^2$; (2) interval occupied by the continuous spectrum for $\xi^2 = \bar{\xi}^2$; (3) upper bound of the discrete spectrum $\omega/\xi = b(Z+0)$; (4) lower bound of the discrete spectrum $\mu/\xi = \min b(z)$.

When ξ is fixed, the eigenfunctions $V_k^{(3)}$, $V^{(3)}$ obey weighted orthogonality conditions (with weight $\rho(z)$):

$$\int_0^\infty \rho V_k^{(3)}(z)\, V_l^{(3)}(z)\, \mathrm{d}z = 0, \qquad k \neq l \tag{1.26}$$

$$\int_0^\infty \rho V_k^{(3)}(z)\, V^{(3)}(\omega^2, z)\, \mathrm{d}z = 0$$

$$\int_0^\infty \rho V^{(3)}(\omega^2, z)\, V^{(3)*}(\beta^2, z)\, \mathrm{d}z = 0.$$

One may use ω as a free parameter in this problem. In that case (for ω fixed) the orthogonality conditions (1.26) involve a weight $\mu(z)$:

$$\int_0^\infty \mu V_k^{(3)} V_l^{(3)}\, \mathrm{d}z = 0, \qquad k \neq l \quad \text{etc.} \tag{1.26a}$$

The continuous spectrum will then exist in the range $-\infty < \xi^2 < \omega^2/b^2\,(Z+0)$.

Green function; surface-wave part. The surface-wave part of the displacement field excited by a spatially localized source can be expressed (Levshin, 1973) in terms of the eigenfunctions of the above operators belonging to their discrete spectra. The relevant Green function $\mathbf{G}^{\mathrm{D}}$ can be represented as a sum of infinitely many terms (modes, harmonics)

$$G_{ij}^{\mathrm{D}} = \sum_{k=1}^{\infty} G_{ij}^{k\mathrm{D}}(\mathbf{x};\mathbf{x}_0;t), \qquad \mathrm{D} = \mathrm{R}, \mathrm{L}$$

the mode with $k = 1$ usually being called the fundamental and the other, higher modes (overtones). In Part I we shall denote Fourier transforms, spectral transforms of time functions, by the same letters as their originals, but with a ^ above. The contribution of each mode can then be represented as

$$G_{ij}^{k\mathrm{D}} = \frac{1}{\pi} \operatorname{Re} \int_{\bar{\omega}_{k\mathrm{D}}}^{\infty} \hat{G}_{ij}^{k\mathrm{D}}(\mathbf{x};\mathbf{x};\omega)\, e^{i\omega t}\, \mathrm{d}\omega$$

where $\bar{\omega}_{k\mathrm{D}}$ is the lowest frequency (the mode does not exist at frequencies below $\omega_{k\mathrm{D}}$).

We shall employ two sets of coordinates whose origins are at the 'epicenter' — a point at the free surface lying at the same vertical with the source (point $\mathbf{x}_0$): a Cartesian set (x, y, z) and a cylindrical one (z, r, φ) in which the angle φ is measured clockwise from the x-axis (as viewed from above). A force applied at the source will be projected onto the Cartesian unit vectors $\mathbf{e}_1$, $\mathbf{e}_2$, $\mathbf{e}_3$. We also introduce at a receiver point $\mathbf{x}$ a local vector basis $\mathbf{e}_z$, $\mathbf{e}_r$, $\mathbf{e}_\varphi$ for which $\mathbf{e}_z$ is parallel to $\mathbf{e}_3$; $\mathbf{e}_r$ is directed along the radius vector of the cylindrical system and $\mathbf{e}_\varphi$ is in the direction of increasing φ. The corresponding components of the Green function in this basis will be denoted $\hat{G}_{pq}^{k\mathrm{D}}$, where p stands for the z-, r- or φ-component, the index q taking on the values 1, 2, 3. The connection with the ordinary (Cartesian) components of Green function $\hat{G}_{sq}$ $(s, q = 1, 2, 3)$ is given by the obvious relations

$$\hat{G}^{kR}_{1q} = G^{kR}_{rq} \cos\varphi \qquad \hat{G}^{kR}_{2q} = G^{kR}_{rq} \sin\varphi \qquad \hat{G}^{kR}_{3q} = G^{kR}_{zq}$$
$$\hat{G}^{kL}_{1q} = -\hat{G}^{kL}_{\varphi q} \sin\varphi \qquad \hat{G}^{kL}_{2q} = \hat{G}^{kL}_{\varphi q} \cos\varphi \qquad \hat{G}^{kL}_{3q} = \hat{G}^{kL}_{\varphi q} = 0.$$

The components $\hat{G}^{kD}_{pq}$ are found from

$$\hat{G}^{kD}_{pq} = \frac{\exp(-i\pi/4)}{\sqrt{8\pi}} \frac{\exp(-i\xi_{kD} r)}{\sqrt{\xi_{kD} r}} \times \\ \times \frac{\varepsilon_p V^{(i_p)}_k(\omega, z)}{\sqrt{C_{kD} U_{kD} I^{(0)}_{kD}}} \frac{\varepsilon_q V^{(i_q)}_k(\omega, h)}{\sqrt{C_{kD} U_{kD} I^{(0)}_{kD}}}. \tag{1.27}$$

Here the $V^{(i)}_k$ are eigenfunctions belonging to the discrete spectra of the one-dimensional problems discussed above; since an eigenfunction is determined apart from a constant, we shall assume that $V^{(1)}_k(\omega, 0) \equiv V^{(3)}_k(\omega, 0) \equiv 1$; $C_{kD}(\omega) = \omega/\xi_{kD}(\omega)$ is phase velocity, $\xi_{kD}(\omega)$ being the inverse function of $\omega_{kD}(\xi)$;

$$U_{kD}(\omega) = \left(\frac{d\xi_{kD}}{d\omega}\right)^{-1} = C_{kD}\left(1 - \frac{\omega}{C_{kD}} \frac{dC_{kD}}{d\omega}\right)^{-1} \tag{1.28}$$

is group velocity; $I^{(0)}_{kD}$ is proportional to the mean kinetic energy transported by a mode over a cycle of oscillation

$$I^{(0)}_{kR} = \int_0^\infty \rho(z) [(V^{(1)}_k)^2 + (V^{(2)}_k)^2] \, dz; \tag{1.29}$$

$$I^{(0)}_{kL} = \int_0^\infty \rho(z) [(V^{(3)}_k)]^2 \, dz. \tag{1.30}$$

The values of ε_p, ε_q and i_p, i_q are given in Table 1.I.

TABLE 1.I

D	p	ε_p	i_p	q	ε_q	i_q
	z	1	1	1	$i \cos\varphi$	2
R	r	$-i$	2	2	$i \sin\varphi$	2
	φ	0	—	3	1	1
	z	0	—	1	$i \sin\varphi$	3
L	r	0	—	2	$-i \cos\varphi$	3
	φ	i	3	3	0	—

Note that (1.27) is asymptotic, being true when $\xi_{kD} r \gg 1$, $r \gg h$, $r \gg z$. If these conditions are not satisfied, not only is the asymptotic formula (1.27) not true, but the very separation of the field into the 'body' and 'surface' wave parts may turn out to be somewhat forced.

Displacement field due to spatially concentrated forces. When the source is a combination of point forces

$$f_q = K_q(t)\delta(\mathbf{x}-\mathbf{x}_0)$$

the resulting displacement becomes

$$\hat{u}_p^{k\mathrm{D}}(\mathbf{x};\mathbf{x}_0;\omega) = \hat{G}_{pq}^{k\mathrm{D}}\hat{K}_q(\omega).$$

This gives an asymptotic formula for $\hat{u}_{pq}^{k\mathrm{D}}$

$$\hat{u}_p^{k\mathrm{D}} = \frac{\exp(-i\pi/4)}{\sqrt{8\pi}}\ \frac{\exp(-i\xi_{k\mathrm{D}}r)}{\sqrt{\xi_{k\mathrm{D}}r}} \times \times \frac{\varepsilon_p V_k^{(i_p)}(\omega,z)}{\sqrt{C_{k\mathrm{D}}U_{k\mathrm{D}}I_{k\mathrm{D}}^{(0)}}}\ \frac{W^{k\mathrm{D}}(\omega,\varphi,h)}{\sqrt{C_{k\mathrm{D}}U_{k\mathrm{D}}I_{k\mathrm{D}}^{(0)}}} \tag{1.31}$$

where

$$W^{kD} = (\varepsilon_q V_k^{(i_q)})\hat{K}_q(\omega). \tag{1.32}$$

Displacement field due to spatially concentrated dipoles. When the source is a dipole of the form (1.16)

$$f_q = -m_{qs}(t)\,\frac{\partial\delta(\mathbf{x}-\mathbf{x}_0)}{\partial x_s} \qquad (q, s = 1, 2, 3),$$

the resulting displacement is

$$\hat{u}_p^{k\mathrm{D}} = \hat{G}_{pq,s}^{k\mathrm{D}}\hat{m}_{qs}(\omega).$$

Making use of asymptotic estimates for $\hat{G}_{pq,s}^{k\mathrm{D}}$ that do not involve terms that would fall off with distance faster than $r^{-1/2}$, we get an expression for $\hat{u}_p^{k\mathrm{D}}$ of the form (1.31), where

$$W^{k\mathrm{D}} = B_{qs}^{k\mathrm{D}}\hat{m}_{qs}. \tag{1.33}$$

Like $\hat{m}_{qs}$, $B_{qs}^{k\mathrm{D}}$ is symmetric in q and s. The formulas for $B_{qs}^{k\mathrm{D}}$ are listed in Table 1.II.

Formulas (1.31) to (1.33) together with the complementary Tables 1.I, 1.II provide a complete asymptotic description of surface wave fields excited by point forces and dipoles in a vertically varying half-space. Extension to more complex point forces is straightforward.

1.4. PHYSICAL INTERPRETATION; SOME SIMPLE EXAMPLES

Formulas such as (1.27) and (1.31) for spectral displacement amplitudes in surface waves can be given a simple physical interpretation. Apart from the first factor, a complex constant, they involve three more factors, each of these being controlled by certain physical conditions and parameters of the observational procedure. The second factor, $(\xi_{k\mathrm{D}}r)^{-1/2}\exp(-i\xi_{k\mathrm{D}}r)$, describes the effect of cylindrical geometrical spreading affecting the energy flux of a surface wave and its propagation-associated phase delay, $\xi_{k\mathrm{D}}r = \omega r/C_{k\mathrm{D}}$, which steadily increases with the distance and is a nonlinear function of frequency. The functions $C_{k\mathrm{D}}(\omega)$, $U_{k\mathrm{D}}(\omega)$, i.e., the dispersion curves, are determined by the properties of the medium only, namely, by the velocity and density distributions $a(z)$, $b(z)$, $\rho(z)$.

TABLE 1.II

D	q	s	$B_{qs}^{k\mathrm{D}}$
	1	1	$-\xi_{k\mathrm{R}} \cos^2 \varphi \, V_k^{(2)}(\omega, h)$
	2	2	$-\xi_{k\mathrm{R}} \sin^2 \varphi \, V_k^{(2)}(\omega, h)$
	3	3	$\left.\dfrac{\mathrm{d}V_k^{(1)}(\omega, z)}{\mathrm{d}z}\right\|_{z=h}$
R	1	2	$-(1/2)\xi_{k\mathrm{R}} \sin 2\varphi \, V_k^{(2)}(\omega, h)$
	1	3	$(i/2) \cos \varphi \left[\xi_{k\mathrm{R}} V_k^{(1)}(\omega, h) + \left.\dfrac{\mathrm{d}V_k^{(2)}(\omega, z)}{\mathrm{d}z}\right\|_{z=h} \right]$
	2	3	$(i/2) \sin \varphi \left[\xi_{k\mathrm{R}} V_k^{(1)}(\omega, h) + \left.\dfrac{\mathrm{d}V_k^{(2)}(\omega, z)}{\mathrm{d}z}\right\|_{z=h} \right]$
	1	1	$-(1/2)\xi_{k\mathrm{L}} \sin 2\varphi \, V_k^{(3)}(\omega, h)$
	2	2	$(1/2)\xi_{k\mathrm{L}} \sin 2\varphi \, V_k^{(3)}(\omega, h)$
	3	3	0
L	1	2	$(1/2)\xi_{k\mathrm{L}} \cos 2\varphi \, V_k^{(3)}(\omega, h)$
	1	3	$(i/2) \sin \varphi \left.\dfrac{\mathrm{d}V_k^{(3)}(\omega, z)}{\mathrm{d}z}\right\|_{z=h}$
	2	3	$-(i/2) \cos \varphi \left.\dfrac{\mathrm{d}V_k^{(3)}(\omega, z)}{\mathrm{d}z}\right\|_{z=h}$

When weak dissipation is present, the resulting attenuation and the extra dispersion due to it (see Section 1.6 below) will naturally be incorporated in the same factor.

The third factor, $\varepsilon_p V_k^{(i_p)}(\omega, z)(C_{k\mathrm{D}} U_{k\mathrm{D}} I_{k\mathrm{D}}^{(0)})^{-1/2}$, is controlled by the receiver depth z and the recorded component of displacement $p (p = z, r, \varphi)$. Actually in a seismic experiment, the horizontal seismometers are usually oriented east—west and north—south; knowing the epicenter coordinates, however, one can convert the N—S and E—W components of the seismogram to the r- and φ-components by means of a simple linear transformation.

It follows from (1.27), (1.31) that a Rayleigh wave (D = R) is elliptically polarized in a vertical plane that contains the source at $\mathbf{x}_0$ and the receiver at $\mathbf{x}$, that is, its φ-component is equal to zero, while the z- and r-components have a phase difference of $\pi/2$. The direction of particle motion and the form of the ellipse are controlled by the ratio $V_k^{(2)}(\omega, z)/V_k^{(1)}(\omega, z)$ which depends on the frequency, receiver depth, and the properties of the medium. The quantity $\chi_k(\omega) = [V_k^{(2)}(\omega, 0)/V_k^{(1)}(\omega, 0)]$ is called ellipticity; it equals the ratio of horizontal and vertical axes of the ellipse along which particles of the free surface are moving in the process of R-wave propagation. It must be borne in mind that it is only for purely sinusoidal oscillations that one can speak of a strictly elliptic particle motion; particle paths in transient motion may be very unlike ellipses, even though remaining in the vertical plane. Love waves (D = L) have the φ-component of motion alone, i.e., are linearly polarized in a horizontal direction normal to the Rayleigh wave polarization plane.

The variation of displacement amplitude over depth is fully determined by the eigenfunctions $V_k^{(i_p)}(\omega, z)$, i.e. by the properties of the medium, receiver depth, and the frequency, being independent of epicentral distance r.

When the frequency response of the recording instrument is to be included, the relevant complex expression can conveniently be incorporated in the third factor.

The last, fourth factor is $W^{k\mathrm{D}}(\omega, \varphi, h)(C_{k\mathrm{D}} U_{k\mathrm{D}} I_{k\mathrm{D}}^{(0)})^{-1/2}$. This depends both on the medium and source parameters: source depth, the relative locations of source and receiver, and the source mechanism, i.e. the relation between components of the force vector $K_q(t)$ or of the seismic moment tensor $m_{qs}(t)$.

Expressions for $W^{k\mathrm{D}}$ are given below for simple point sources constructed to imitate explosions and earthquakes.

(1) Center of dilatation (1.18)

$$W^{k\mathrm{D}} = \left(\left. \frac{\mathrm{d} V_k^{(1)}(\omega, z)}{\mathrm{d} z} \right|_{z=h} - \xi_{k\mathrm{R}}(\omega) V_k^{(2)}(\omega, h) \right) \hat{m}(\omega) \tag{1.34}$$

$$W^{k\mathrm{L}} = 0.$$

In this case the fourth factor depends on frequency, the medium, source depth and spectrum.

(2) Point shear dislocation (1.19) along a direction $\boldsymbol{\kappa}$ which is tangent to an area with normal $\mathbf{n}$. Denote

$$n_x = \sin\gamma \cos\alpha \qquad n_y = \sin\gamma \sin\alpha \qquad n_z = \cos\gamma$$
$$\kappa_x = \sin\beta \cos\delta \qquad \kappa_y = \sin\beta \sin\delta \qquad \kappa_z = \cos\beta$$

(α and δ are azimuths of the horizontal projections of $\mathbf{n}$ and $\boldsymbol{\kappa}$ as measured from the x-axiz; γ and β are angles which the two vectors make with the vertical z). $W^{k\mathrm{D}}$ can be expressed as follows:

$$\begin{aligned} W^{k\mathrm{R}} = \Bigg[& 2 \cos\beta \cos\gamma \left. \frac{\mathrm{d} V_k^{(1)}}{\mathrm{d} z} \right|_{z=h} - \\ & - 2\xi_{k\mathrm{R}} \sin\beta \sin\gamma \cos(\delta - \varphi) \cos(\alpha - \varphi) V_k^{(2)}(\omega, h) + \\ & + i [\sin\beta \sin\gamma \cos(\alpha - \varphi)] \times \\ & \times \left(\xi_{k\mathrm{R}} V_k^{(1)}(\omega, h) + \left. \frac{\mathrm{d} V_k^{(2)}(\omega, z)}{\mathrm{d} z} \right|_{z=h} \right) \Bigg] \hat{m}(\omega), \end{aligned} \tag{1.35}$$

$$\begin{aligned} W^{k\mathrm{L}} = \Bigg[& \xi_{k\mathrm{L}} \sin\gamma \sin\beta \sin(\alpha + \delta - \varphi) V_k^{(3)}(\omega, h) - \\ & - i [\sin\beta \cos\gamma \sin(\delta - \varphi) + \\ & + \sin\gamma \cos\beta \sin(\alpha - \varphi)] \left. \frac{\mathrm{d} V_k^{0(3)}(\omega, z)}{\mathrm{d} z} \right|_{z=h} \Bigg] \hat{m}(\omega). \end{aligned}$$

In this case the fourth factor is controlled by frequency, the medium, source depth and spectrum, as well as by source geometry, *viz.* orientation of the fault plane and slip vector with respect to the source-receiver pair.

Simple examples. To illustrate the above expressions we now consider some simple examples of elastic media and surface waves arising in them.

Homogeneous half-space. The Rayleigh wave fundamental mode alone can exist in a medium with a, b, ρ constant. It exists for all frequencies and is not subject to dispersion. The phase velocity C_R depends on the ratio $\gamma = b/a$ only (or, which amounts to the same thing, on Poisson's ratio σ):

$$\sigma = \lambda/[2(\lambda + \mu)] = (1 - 2\gamma^2)/[2(1 - \gamma^2)]$$

C_R is a single real root to

$$(2 - \kappa)^2 - 4\sqrt{1 - (\gamma\kappa)^2}\,\sqrt{1 - \kappa^2} = 0$$

where $\kappa = C_R/b$. The functions $C_R(\gamma)$ and $C_R(\sigma)$ are illustrated in Figure 1.2. Group velocity U is equal to the phase velocity. The variation of intensity with depth obeys the law (Figure 1.3)

$$V^{(1)}(\omega, z) = \frac{\kappa^2 - 2}{\kappa^2}\left(\exp(-r_\alpha z) + \frac{2}{\kappa^2 - 2}\exp(-r_\beta z)\right)$$

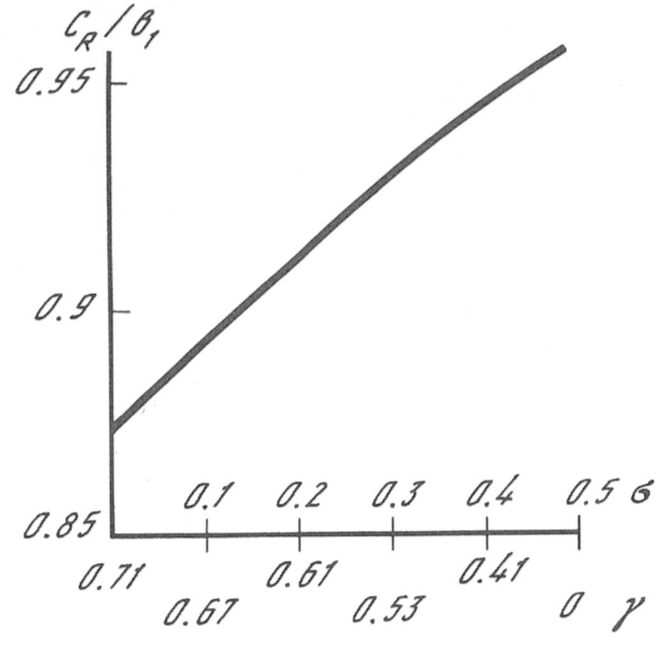

Fig. 1.2. Rayleigh wave phase velocity in a homogeneous half-space as a function of $\gamma = b/a$ and Poisson's ratio.

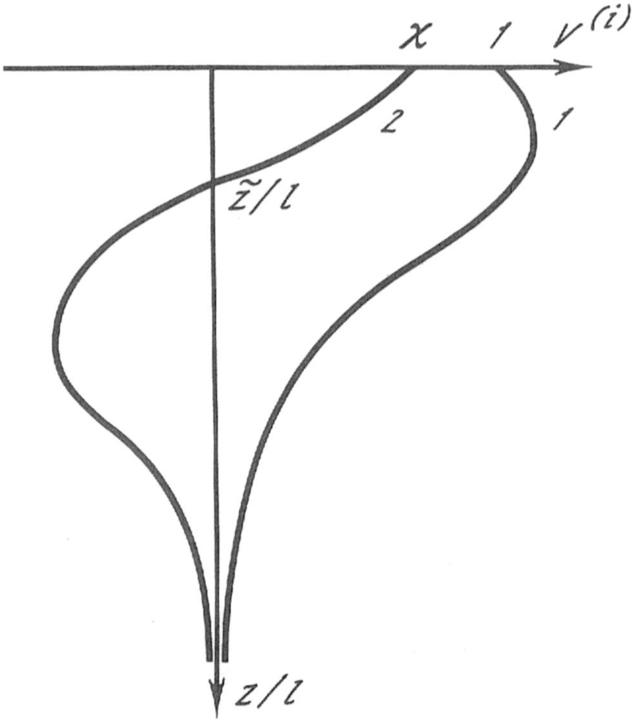

Fig. 1.3. Components of an eigenfunction in the one-dimensional Rayleigh wave boundary-value problem for a homogeneous half-space. Numerals by the curves are component numbers.

for the vertical component, and

$$V^{(2)}(\omega, z) = \frac{2-\kappa^2}{\kappa^2\sqrt{1-\gamma^2\kappa^2}}\left(\exp(-r_\alpha z) + \frac{\kappa^2-2}{2}\exp(-r_\beta z)\right)$$

for the horizontal component. Here

$$r_\alpha = 2\pi(1-\gamma^2\kappa^2)^{1/2}/l, \qquad r_\beta = 2\pi(1-\kappa^2)^{1/2}/l$$

$l = 2\pi C_R/\omega$ being surface wave length.

Particle motion occurs in ellipses in the rz-plane, the ratio of major to minor axis for $z = 0$,

$$\chi = (1-\kappa^2/2)/(1-\gamma^2\kappa^2)^{1/2} = V^{(2)}(\omega, 0)$$

being independent of frequency.

The depth $\tilde{z}$ where $V^{(2)}(z)$ changes sign is

$$\tilde{z} = Kl = \frac{\ln[(2-\kappa^2)/2]}{2\pi[(1-\kappa^2)^{1/2} - (1-\kappa^2\gamma^2)^{1/2}]}\, l.$$

Particle motion is retrograde for $z < \tilde{z}$ (toward the source in the upper portion of particle path) and prograde for $z > \tilde{z}$.

The kinetic energy integral is

$$I_{\mathrm{R}}^{(0)} = \frac{b\rho}{\omega} \frac{8[(1-\gamma^2\kappa^2)^{1/2} - (1-\kappa^2)^{1/2}]^2 + \kappa^4(2-\kappa^2)}{4(1-\kappa^2)^{1/2}(1-\gamma^2\kappa^2)\kappa^3}.$$

The spectral amplitude of vertical displacement at point $\mathbf{x}(0, r, \varphi)$ in a Rayleigh wave excited by a vertical point force at the surface for $r \gg l$ is equal to

$$\hat{G}_{zz}^{\mathrm{R}}(\mathbf{x};0;\omega) = \frac{\exp(-i\pi/4)}{\sqrt{8\pi}} \frac{\exp(-i\omega r/C_{\mathrm{R}})}{\sqrt{r}} \times$$

$$\times \frac{\omega^{1/2}4(1-\gamma^2\kappa^2)(1-\kappa^2)^{1/2}\kappa^{3/2}}{b^{5/2}\rho[8((1-\gamma^2\kappa^2)^{1/2} - (1-\kappa^2)^{1/2})^2 + \kappa^4(2-\kappa^2)^2]}$$

$$\hat{G}_{rz}^{\mathrm{R}}(\mathbf{x};0;\omega) = -i\chi\hat{G}_{zz}^{\mathrm{R}}.$$

In the case of a horizontal point force acting towards the recording site (i.e., one having azimuth φ), the displacements are

$$\hat{u}_z^{\mathrm{R}} = i\chi\hat{G}_{zz}^{\mathrm{R}} \qquad \hat{u}_r^{\mathrm{R}} = \chi^2\hat{G}_{zz}^{\mathrm{R}}.$$

Homogeneous layer on homogeneous half-space. In the simplest case of a vertically varying medium

$$a(z) = a_1, \quad b(z) = b_1, \quad \rho(z) = \rho_1 \quad \text{for} \quad z \leqslant Z;$$
$$a(z) = a_2, \quad b(z) = b_2, \quad \rho(z) = \rho_2 \quad \text{for} \quad z > Z;$$

both Love and Rayleigh waves exist in infinitely many modes.

Love wave. The phase velocity of Love wave obeys the dispersion relation

$$\tan(r_{\beta_1}Z) - \frac{b_2^2\rho_2 r_{\beta_2}}{b_1^2\rho_1 r_{\beta_1}} = 0$$

where

$$r_{\beta_1} = \omega(\kappa^2-1)^{1/2}/(\kappa b_1), \qquad r_{\beta_2} = \omega(1-(b_1\kappa/b_2)^2)^{1/2}/(\kappa b_1),$$
$$\kappa = C_{k\mathrm{L}}/b_1.$$

For fixed ω, the above equation has one or more solutions; the fundamental mode can thus exist at any ω and higher modes at frequencies $\omega_{k\mathrm{L}} < \omega < \infty$, where

$$\omega_{k\mathrm{L}} = \frac{\pi(k-1)b_2}{Z((b_2/b_1)^2-1)^{1/2}} \qquad k = 2, 3, \ldots .$$

The group velocity obeys the relation

$$U_{k\mathrm{L}} = \frac{b_1^2}{C_{k\mathrm{L}}} \frac{\omega Z + \kappa^2\Lambda b_1}{\omega Z + \Lambda b_1}$$

where

$$\Lambda = \frac{\rho_2\kappa\cos^2(r_{\beta_1}Z)(b_2^2/b_1^2-1)}{\rho_1(\kappa^2-1)(1-(b_1\kappa/b_2)^2)^{1/2}}.$$

The behavior of $C_{k\mathrm{L}}(\omega)$ and $U_{k\mathrm{L}}(\omega)$ is illustrated in Figure 1.4.

Intensity as a function of depth is given by (see Figure 1.5)

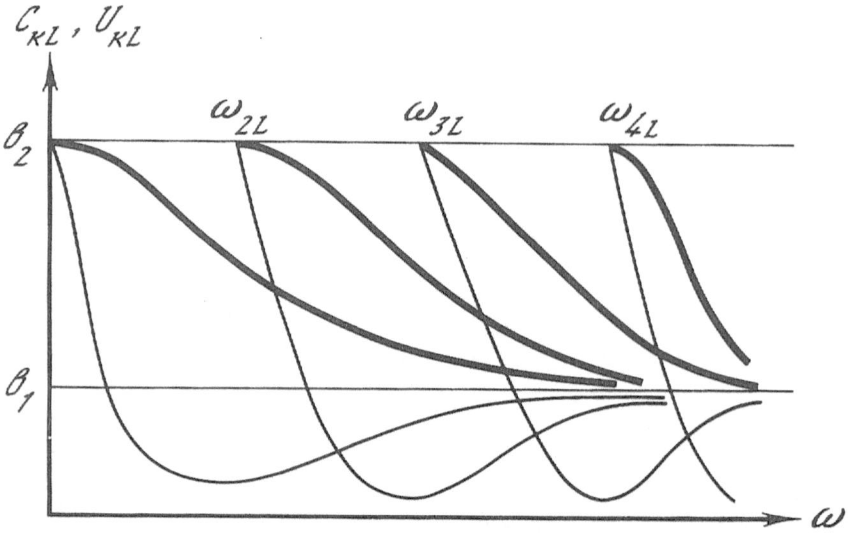

Fig. 1.4. Dispersion of phase (heavy line) and group (light line) Love wave velocities for a homogeneous layer overlying a homogeneous half-space.

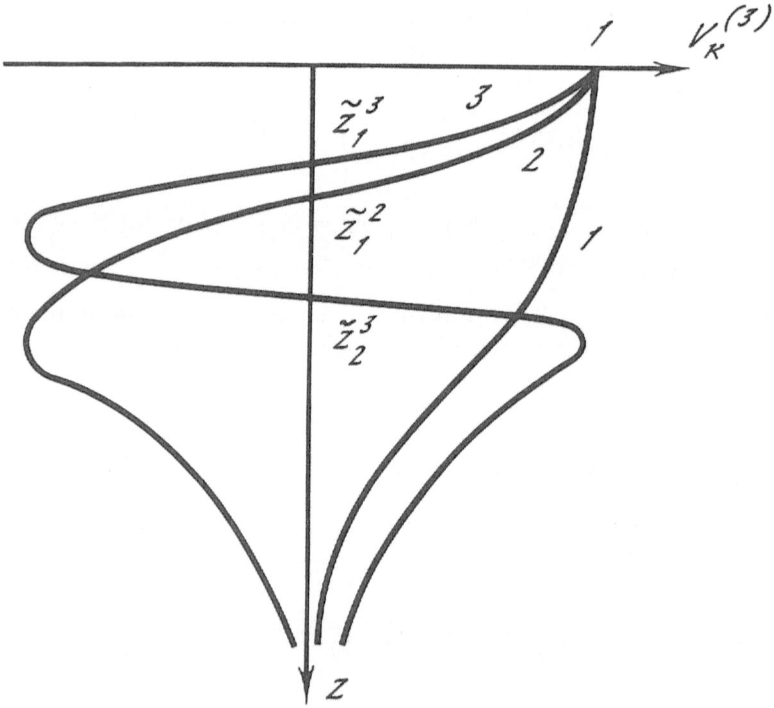

Fig. 1.5. Eigenfunctions of the one-dimensional Love wave problem for a layer overlying a half-space. Numerals by the curves are mode numbers.

$$V_k^{(3)} = \cos(r_{\beta_1} z), \qquad 0 \leqslant z \leqslant Z$$
$$V_k^{(3)} = \cos(r_{\beta_1} Z)\exp[-r_{\beta_2}(z-Z)], \qquad z \geqslant Z.$$

The depths at which the $V_k^{(3)}$ change sign are equal to

$$\tilde{Z}_m^k = \frac{(2m-1)\pi}{2r_{\beta_1}} \qquad m = 1, \ldots, k-1 \quad (k > 1).$$

The kinetic energy integral is

$$I_{k\mathrm{L}}^{(0)} = \frac{\rho_1}{2\omega}[\omega Z + \Lambda b_1].$$

The spectral amplitude of Love wave displacement at the free surface excited by a horizontal point force applied at (0, 0, $h < Z$) and acting perpendicularly to the source-receiver line to the right of that line has the form

$$\hat{u}_\varphi^{k\mathrm{L}} = \frac{e^{-i\pi/4}}{\sqrt{8\pi}} \frac{e^{-i\omega r/C_{k\mathrm{L}}}}{\sqrt{r}} \frac{2(\omega C_{k\mathrm{L}})^{1/2}\cos(r_{\beta_1} h)}{b_1^2 \rho_1(\omega Z + \kappa^2 b_1 \Lambda)}.$$

We shall not present a similar discussion for Rayleigh waves, since the calculations are cumbersome, even for the case of the simple model considered here.

1.5. SPHERICITY CORRECTIONS; VARIATIONAL FORMULAS

Sphericity corrections. Much work in seismology uses surface waves that propagate to such great distances and penetrate to such great depths as to require corrections for the Earth's spherical shape. Indeed, large magnitude earthquakes excite surface waves of so great an amplitude that they can travel several times round the Earth before dying out because of dissipation in the medium. A seismic station then records two sequences of waves that travel in opposite directions.

We define a set of spherical coordinates with the origin at the center of a sphere of radius R_0, putting the receiver at the point $\mathbf{x}(R, \theta, \varphi)$ and the source at $\mathbf{x}_0(R_s, 0, 0)$ where $R_s = R_0 - h$, and shall consider the pth components of displacement corresponding to the local vector basis $\mathbf{e}_p$ ($p = R, \theta, \varphi$). Let $\tilde{\theta}$ be the total distance travelled by the wave:

$$\tilde{\theta} = (-1)^g \theta + 2\pi(l + g),$$

where $l = 0, 1, 2, \ldots$ is the number of the Earth's great circles the wave has travelled; $g = 0$ or 1, depending on whether the wave comes from the epicentre or from the opposite direction. In that case formulas (1.27) become for $\theta \neq k\pi$

$$\hat{G}_{pq}^{k\mathrm{D}}(\mathbf{x}; \mathbf{x}_0; \omega) = \frac{\exp(-i\pi/4)}{\sqrt{8\pi}} \frac{\exp[-i\omega R_0 \tilde{\theta}/C_{k\mathrm{D}} + (i\pi/2)(2l + g)}{\sqrt{\nu_{k\mathrm{D}} \sin\theta}} \times$$
$$\times \frac{\varepsilon_p V_k^{(i_p)}(\omega, R)}{\sqrt{C_{k\mathrm{D}} U_{k\mathrm{D}} I_{k\mathrm{D}}^{(0)}}} \frac{\varepsilon_q V_k^{(i_q)}(\omega, R_s)}{\sqrt{C_{k\mathrm{D}} U_{k\mathrm{D}} I_{k\mathrm{D}}^{(0)}}}; \tag{1.36}$$

phase velocity is defined as

$$C_{k\mathrm{D}}(\omega) = \frac{\omega R_0}{\nu_{k\mathrm{D}} + 1/2} \tag{1.37}$$

and group velocity as

$$U_{k\mathrm{D}}(\omega) = \left[\frac{\mathrm{d}\nu_{k\mathrm{D}}(\omega)}{\mathrm{d}\omega} \right]^{-1} R_0. \tag{1.38}$$

The eigenfunctions $V_k^{(i)}(\omega, R)$ belong to slightly different one-dimensional boundary value problems than those given by (1.20)—(1.25); specifically, for Rayleigh waves, we have

$$\begin{aligned} l_1(V^{(1)}, V^{(2)}) &\equiv \frac{\mathrm{d}\bar{\sigma}_{RR}}{\mathrm{d}R} + \frac{\mu}{R^2}\left[4R\,\frac{\mathrm{d}V^{(1)}}{\mathrm{d}R} - 4V^{(1)} + \right. \\ &\quad \left. + N\left(3V^{(2)} - R\,\frac{\mathrm{d}V^{(2)}}{\mathrm{d}R} - NV^{(1)} \right) \right] + \omega^2 \rho V^{(1)} = 0, \\ l_2(V^{(1)}, V^{(2)}) &\equiv \frac{\mathrm{d}\bar{\sigma}_{R\theta}}{\mathrm{d}R} + \frac{\lambda N}{R}\left[\frac{\mathrm{d}V^{(1)}}{\mathrm{d}R} + \frac{2}{R}\,V^{(1)} - \frac{N}{R}\,V^{(2)} \right] + \\ &\quad + \frac{\mu}{R^2}\left[5NV^{(1)} + 3R\,\frac{\mathrm{d}V^{(2)}}{\mathrm{d}R} - V^{(2)} - 2N^2V^{(2)} \right] + \\ &\quad + \omega^2\rho V^{(2)} = 0 \end{aligned} \tag{1.39}$$

with the boundary conditions

$$\begin{aligned} \bar{\sigma}_{RR} &\equiv (\lambda + 2\mu)\,\frac{\mathrm{d}V^{(1)}}{\mathrm{d}R} + \frac{2\lambda}{R}\,V^{(1)} - \frac{\lambda N}{R}\,V^{(2)} = 0 \qquad R = R_0 \\ \bar{\sigma}_{R\theta} &\equiv \mu\left[\frac{\mathrm{d}V^{(2)}}{\mathrm{d}R} - \frac{V^{(2)}}{R} + \frac{NV^{(1)}}{R} \right] = 0 \qquad R = R_0 \\ V^{(1)} &= V^{(2)} = 0 \qquad R = 0. \end{aligned} \tag{1.40}$$

Here $N = \sqrt{\nu(\nu + 1)}$, the parameter ν being a non-dimensional analogue of the wave number; $\bar{\nu}$ is the lower bound of the wavenumber range under consideration ($\bar{\nu} \gg 1$).

Similarly for Love waves, $V_k^{(3)}(\omega, R)$ are the eigenfunctions of the problem

$$\begin{aligned} l_3(V^{(3)}) &\equiv \frac{\mathrm{d}\bar{\sigma}_{R\varphi}}{\mathrm{d}R} + \frac{3\mu}{R}\,\frac{\mathrm{d}V^{(3)}}{\mathrm{d}R} - \\ &\quad - \frac{\mu}{R^2}(N^2 + 1)V^{(3)} + \rho\omega^2 V^{(3)} = 0 \end{aligned} \tag{1.41}$$

with the boundary conditions

$$\begin{aligned} \bar{\sigma}_{R\varphi} &\equiv \mu\left[\frac{\mathrm{d}V^{(3)}}{\mathrm{d}R} - \frac{V^{(3)}}{R} \right] = 0 \qquad R = R_0 \\ V^{(3)} &= 0 \qquad R = 0. \end{aligned} \tag{1.42}$$

Formulas (1.36) to (1.38) involve $\nu_{k\mathrm{D}}(\omega)$, the inverse functions of $\omega_{k\mathrm{D}}(\nu)$.

The kinetic energy integrals are

$$I_{kR}^{(0)} = \frac{1}{R_0^2} \int_0^{R_0} \rho R^2 \left[(V_k^{(1)})^2 + (V^{(2)})^2 \right] \mathrm{d}R$$

$$I_{kL}^{(0)} = \frac{1}{R_0^2} \int_0^{R_0} \rho R^2 [(V_k^{(3)})^2] \, \mathrm{d}R. \tag{1.43}$$

The expressions for ε_p, ε_q, i_p, i_q are listed in Table 1.III.

TABLE 1.III

D	p	ε_p	i_p	q	ε_q	i_q
R	R	1	1	x	$i(-1)^g \cos\varphi$	2
	θ	$-i(-1)^g$	2	y	$i(-1)^g \sin\varphi$	2
	φ	0	—	z	-1	1
L	R	0	—	x	$i(-1)^g \sin\varphi$	3
	θ	0	—	y	$-i(-1)^g \cos\varphi$	3
	φ	$i(-1)^g$	3	z	0	—

The factor W^{kD} for simple sources is given by (1.34) and (1.35), except that one has to replace ξ_{kD} by $(\nu_{kD} + 1/2)/R_s$, differentiate with respect to z by that with respect to R, and insert the factor $(-1)^{g+1}$ in (1.35) in front of expressions that involve $V_k^{(1)}$, $\mathrm{d}V_k^{(2)}/\mathrm{d}R$ and $\mathrm{d}V_k^{(3)}/\mathrm{d}R$.

Variational formulas. The theory of perturbations yields the following integral formulas for phase and group velocity in a vertically varying half-space:

$$C_{kR} = \left\{ \left[I_{kR}^{(1)} + I_{kR}^{(2)} + \frac{2}{\xi_{kR}} (I_{kR}^{(3)} + I_{kR}^{(4)}) + \right.\right.$$

$$\left.\left. + \frac{1}{\xi_{kR}^2} (I_{kR}^{(5)} + I_{kR}^{(6)}) \right] \Big/ I_{kR}^{(0)} \right\}^{1/2} \tag{1.44}$$

$$C_{kL} = \left[\left(I_{kL}^{(1)} + \frac{1}{\xi_{kL}^2} I_{kL}^{(2)} \right) \Big/ I_{kL}^{(0)} \right]^{1/2}$$

$$U_{kR} = \left[I_{kR}^{(1)} + I_{kR}^{(2)} + \frac{1}{\xi_{kR}} (I_{kR}^{(3)} + I_{kR}^{(4)}) \right] \Big/ (C_{kR} I_{kR}^{(0)}) \tag{1.45}$$

$$U_{kL} = I_{kL}^{(1)} / (C_{kL} I_{kL}^{(0)})$$

where the integrals $I_{kD}^{(j)}$ are

$$I_{kR}^{(1)} = \int_0^\infty b^2\rho[V_k^{(1)}]^2\,\mathrm{d}z$$

$$I_{kR}^{(2)} = \int_0^\infty a^2\rho[V_k^{(2)}]^2\,\mathrm{d}z$$

$$I_{kR}^{(3)} = \int_0^\infty b^2\rho\left[\frac{\mathrm{d}V_k^{(2)}}{\mathrm{d}z}V_k^{(1)} + 2\frac{\mathrm{d}V_k^{(1)}}{\mathrm{d}z}V_k^{(2)}\right]\mathrm{d}z$$

$$I_{kR}^{(4)} = -\int_0^\infty a^2\rho\frac{\mathrm{d}V_k^{(1)}}{\mathrm{d}z}V_k^{(2)}\,\mathrm{d}z \tag{1.46}$$

$$I_{kR}^{(5)} = \int_0^\infty b^2\rho\left[\frac{\mathrm{d}V_k^{(2)}}{\mathrm{d}z}\right]^2\mathrm{d}z$$

$$I_{kR}^{(6)} = \int_0^\infty a^2\rho\left[\frac{\mathrm{d}V_k^{(1)}}{\mathrm{d}z}\right]^2\mathrm{d}z$$

$$I_{kL}^{(1)} = \int_0^\infty b^2\rho[V_k^{(3)}]^2\,\mathrm{d}z$$

$$I_{kL}^{(2)} = \int_0^\infty b^2\rho\left[\frac{\mathrm{d}V_k^{(3)}}{\mathrm{d}z}\right]^2\mathrm{d}z.$$

The formulas for a radially varying sphere are similar:

$$C_{kR} = \left\{\left[I_{kR}^{(1)} + I_{kR}^{(2)} + \frac{2}{N}(I_{kR}^{(3)} + I_{kR}^{(4)}) + \frac{1}{N^2}(I_{kR}^{(5)} + I_{kR}^{(6)})\right]\Big/ I_{kR}^{(0)}\right\}^{1/2}$$

$$C_{kL} = \left[\left(I_{kL}^{(1)} + \frac{1}{N^2}I_{kL}^{(2)}\right)\Big/ I_{kL}^{(0)}\right]^{1/2} \tag{1.47}$$

$$U_{kR} = \left[I_{kR}^{(1)} + I_{kR}^{(2)} + \frac{1}{N}(I_{kR}^{(3)} + I_{kR}^{(4)})\right]\Big/(C_{kR}I_{kR}^{(0)}) \tag{1.48}$$

$$U_{kL} = I_{kL}^{(1)}/(C_{kL}I_{kL}^{(0)})$$

where $I^{(0)}_{kR}$, $I^{(0)}_{kL}$ are given by (1.43) and the $I^{(j)}_{kD}$ have the form

$$I^{(1)}_{kR} = \int_0^{R_0} b^2\rho[V^{(1)}_k]^2\,dR$$

$$I^{(2)}_{kR} = \int_0^{R_0} a^2\rho[V^{(2)}_k]^2\,dR$$

$$I^{(3)}_{kR} = \int_0^{R_0} b^2\rho\left[R\left(\frac{dV^{(2)}_k}{dR}V^{(1)}_k + 2\frac{dV^{(1)}_k}{dR}V^{(2)}_k\right) + V^{(1)}_kV^{(2)}_k\right]dR$$

$$I^{(4)}_{kR} = -\int_0^{R_0} a^2\rho\left[R\frac{dV^{(1)}_k}{dR}V^{(2)}_k + 2V^{(1)}_kV^{(2)}_k\right]dR$$

$$I^{(5)}_{kR} = \int_0^{R_0} b^2\rho\left[\left(R\frac{dV^{(2)}_k}{dR}\right)^2 - 2R\left(4\frac{dV^{(1)}_k}{dR}V^{(1)}_k + \frac{dV^{(2)}_k}{dR}V^{(2)}_k\right) - \right.$$

$$\left. - 4(V^{(1)}_k)^2 - (V^{(2)}_k)^2\right]dR \tag{1.49}$$

$$I^{(6)}_{kR} = \int_0^{R_0} a^2\rho\left[\left(R\frac{dV^{(1)}_k}{dR}\right)^2 + 4R\frac{dV^{(1)}_k}{dR}V^{(1)}_k + 4(V^{(1)}_k)^2\right]dR$$

$$I^{(1)}_{kL} = \int_0^{R_0} b^2\rho[V^{(3)}_k]^2\,dR$$

$$I^{(2)}_{kL} = \int_0^{R_0} b^2\rho\left[\left(R\frac{dV^{(3)}_k}{dR}\right)^2 - 2R\frac{dV^{(3)}_k}{dR}V^{(3)}_k - (V^{(3)}_k)^2\right]dR.$$

Partial derivatives of phase velocity. The solution of inverse problems in surface wave seismology is greatly facilitated by the formalism of partial derivatives which is used to determine phase velocity perturbations due to small perturbations in the velocity and density distributions. Consider a small perturbation $\delta_{\kappa}(z)$ in the parameter $\kappa(z)$ ($\kappa = a$, b, or ρ) that vanishes everywhere except within the interval $z_i < z < z_{i+1}$. The resulting phase velocity perturbation $\delta C_{k\mathrm{D}}(\omega)$ for a fixed frequency ω is

$$\delta C_{k\mathrm{D}} = \int_{z_{i-1}}^{z_i} \frac{\partial C_{k\mathrm{D}}(\omega, z)}{\partial \kappa}\, \delta\kappa(z)\, \mathrm{d}z.$$

The kernel $\partial C_{k\mathrm{D}}/\partial\kappa$ is a partial derivative of $C_{k\mathrm{D}}$ with respect to κ (or, to be more exact, the 'response' of $C_{k\mathrm{D}}$ to a δ-like variation in $\kappa(z)$ at the point z). When $\omega = \text{constant}$, the $\partial C_{k\mathrm{D}}/\partial\kappa$ are given by

$$\frac{\partial C_{k\mathrm{R}}}{\partial a} = \frac{a\rho}{U_{k\mathrm{R}} I^{(0)}_{k\mathrm{R}}} \left[V_k^{(2)} - \frac{1}{\xi_{k\mathrm{R}}} \frac{\mathrm{d}V_k^{(1)}}{\mathrm{d}z} \right]^2$$

$$\frac{\partial C_{k\mathrm{R}}}{\partial b} = \frac{b\rho}{U_{k\mathrm{R}} I^{(0)}_{k\mathrm{R}}} \left[\left(V_k^{(1)} + \frac{1}{\xi_{k\mathrm{R}}} \frac{\mathrm{d}V_k^{(2)}}{\mathrm{d}z} \right) + \frac{4}{\xi_{k\mathrm{R}}} \frac{\mathrm{d}V_k^{(1)}}{\mathrm{d}z} V_k^{(2)} \right]$$

$$\frac{\partial C_{k\mathrm{R}}}{\partial \rho} = \frac{1}{2\rho} \left[\frac{\partial C_{k\mathrm{R}}}{\partial a} a + \frac{\partial C_{k\mathrm{R}}}{\partial b} b \right] - \frac{C^2_{k\mathrm{R}}}{2U_{k\mathrm{R}} I^{(0)}_{k\mathrm{R}}} [(V_k^{(1)})^2 + (V_k^{(2)})^2] \tag{1.50}$$

$$\frac{\partial C_{k\mathrm{L}}}{\partial a} = 0$$

$$\frac{\partial C_{k\mathrm{L}}}{\partial b} = \frac{b\rho}{U_{k\mathrm{L}} I^{(0)}_{k\mathrm{L}}} \left[(V_k^{(1)})^2 + \left(\frac{1}{\xi_{k\mathrm{L}}} \frac{\mathrm{d}V_k^{(2)}}{\mathrm{d}z} \right)^2 \right]$$

$$\frac{\partial C_{k\mathrm{L}}}{\partial \rho} = \frac{1}{2\rho} \frac{\partial C_{k\mathrm{L}}}{\partial b} b - \frac{C^2_{k\mathrm{L}}}{2U_{k\mathrm{L}} I^{(0)}_{k\mathrm{L}}} (V_k^{(3)})^2.$$

For the spherical case these formulas become

$$\frac{\partial C_{kR}}{\partial a} = \frac{a\rho}{U_{kR} I^{(0)}_{kR}} \left[\left(V^{(2)}_k - \frac{R}{N} \frac{dV^{(1)}_k}{dR} \right)^2 - \frac{4}{N} V^{(1)}_k V^{(2)}_k + \right.$$

$$\left. + \frac{4}{N^2} V^{(1)}_k \left(V^{(2)}_k + R \frac{dV^{(1)}_k}{dR} \right) \right]$$

$$\frac{\partial C_{kR}}{\partial b} = \frac{b\rho}{U_{kR} I^{(0)}_{kR}} \left[\left(V^{(1)}_k - \frac{R}{N} \frac{dV^{(2)}_k}{dR} \right)^2 + \right.$$

$$+ \frac{2}{N} V^{(2)}_k \left(V^{(2)}_k + 2R \frac{dV^{(1)}_k}{dR} \right) - \frac{4}{N^2} \left(V^{(1)}_k + 2R \frac{dV^{(1)}_k}{dR} \right) -$$

$$\left. - \frac{1}{N^2} \left(V^{(2)}_k + \frac{R}{N} \frac{dV^{(1)}_k}{dR} \right)^2 \right]$$

$$\frac{\partial C_{kR}}{\partial \rho} = \frac{1}{2\rho} \left[\frac{\partial C_{kR}}{\partial a} a + \frac{\partial C_{kR}}{\partial b} b \right] -$$

$$- \frac{C^2_{kR}}{2 U_{kR} I^{(0)}_{kR}} \frac{R^2}{R_0} [(V^{(1)}_k)^2 + (V^{(2)}_k)^2] \qquad (1.50a)$$

$$\frac{\partial C_{kL}}{\partial a} = 0$$

$$\frac{\partial C_{kL}}{\partial b} = \frac{b\rho}{U_{kL} I^{(0)}_{kL}} \left[(V^{(3)}_k)^2 + \left(\frac{R}{N} \frac{dV^{(3)}_k}{dR} \right)^2 - \right.$$

$$\left. - \frac{2}{N^2} R V^{(3)}_k \frac{dV^{(3)}_k}{dR} - \frac{1}{N^2} (V^{(3)}_k)^2 \right]$$

$$\frac{\partial C_{kL}}{\partial \rho} = \frac{b}{2\rho} \frac{\partial C_{kL}}{\partial b} - \frac{C^2_{kL}}{2 U_{kL} I^{(0)}_{kL}} \frac{R^2}{R^2_0} (V^{(3)}_k)^2.$$

Partial derivatives of group velocity at frequency ω can be found numerically from derivatives of phase velocity at two close points $\omega_1 = \omega e^{\delta}$, $\omega_2 = \omega e^{-\delta}$ (Rodi *et al.* 1975).

$$\left.\frac{\partial U}{\partial \kappa}\right|_{\omega} = \frac{U|_{\omega}}{2C|_{\omega}}\left(2 - \frac{U|_{\omega}}{C|_{\omega}}\right)\left(\left.\frac{\partial C}{\partial \kappa}\right|_{\omega_1} + \left.\frac{\partial C}{\partial \kappa}\right|_{\omega_2}\right) +$$

$$+ \frac{1}{2}\left[\frac{U|_{\omega}}{C|_{\omega}}\right]^2\left(\left.\frac{\partial C}{\partial \kappa}\right|_{\omega_1} - \left.\frac{\partial C}{\partial \kappa}\right|_{\omega_2}\right)\delta^{-1}. \tag{1.51}$$

Dependence of phase velocity on the depth of an interface between layers. The variational relation stated in Woodhouse (1976) may be used to obtain formulas for partial derivatives of phase velocity with respect to interface depth. Let h be the depth of some interface within a vertically varying half-space and $[\chi]_-^+$ be the jump of any function $\chi(z)$ at the interface $z = h$, i.e. $[\chi]_-^+ = \chi(h+0) - \chi(h-0)$. Then the following relation is valid for Rayleigh waves

$$\frac{\partial C_{kR}}{\partial h} = \frac{C_{kR}^3}{2\omega[\omega(I_{kR}^{(1)} + I_{kR}^{(2)}) + C_{kR}(I_{kR}^{(3)} + I_{kR}^{(4)})]} \times$$

$$\times \left\{\omega^2[(V_k^{(2)}(h))^2 + (V_k^{(1)}(h))^2]\,[\rho]_-^+ - \right.$$

$$- \left(\frac{\omega}{C_{kR}}\right)^2 (V_k^{(1)}(h))^2[\mu]_-^+ - \left(\frac{\omega}{C_{kR}}\right)^2 (V_k^{(2)}(h))^2[\lambda + 2\mu]_-^+ +$$

$$\left. + \left[(\lambda + 2\mu)\left(\frac{dV_k^{(1)}}{dz}\right)^2\right]_-^+ + \left[\mu\left(\frac{dV_k^{(2)}}{dz}\right)^2\right]_-^+\right\}.$$

For the particular case of Rayleigh waves in the structure consisting of homogeneous layers this formula was obtained by Kushnir *et al.*, 1988.

For Love waves we have

$$\frac{\partial C_{kL}}{\partial h} = \frac{C_{kL}^3}{2\omega^2 I_{kL}^{(1)}}\left\{\omega^2(V_k^{(3)}(h))^2[\rho]_-^+ - \right.$$

$$\left. - \left(\frac{\omega}{C_{kL}}\right)^2 (V_k^{(3)}(h))^2[\mu]_-^+ + \left[\mu\left(\frac{dV_k^{(3)}}{dz}\right)^2\right]_-^+\right\}.$$

Similar formulas hold in the spherical case. Let $\tilde{R}$ be the radius of an internal interface and $[\chi]_-^+ = \chi(\tilde{R}+0) - \chi(\tilde{R}-0)$. Then

$$\frac{\partial C_{kR}}{\partial \tilde{R}} = \frac{C_{kR}^3 R_0}{2\omega\,[\omega\,[I_{kR}^{(1)} + I_{kR}^{(2)}) + C_{kR}(I_{kR}^{(3)} + I_{kR}^{(4)})]} \times$$

$$\times \left\{ \omega^2 \left(\frac{\tilde{R}}{R_0} \right)^2 [(V_k^{(2)}(\tilde{R}))^2 + (V_k^{(1)}(\tilde{R}))^2]\,[\rho]_-^+ - \right.$$

$$- \left[\left(\frac{\omega R_0 V_k^{(1)}(\tilde{R}) + C_{kR} V_k^{(2)}(\tilde{R})}{C_{kR} R_0} \right)^2 - \right.$$

$$\left. - 2\,\frac{2(V_k^{(1)}(\tilde{R}))^2 + (V_k^{(2)}(\tilde{R}))^2}{R_0^2} \right] [\mu]_-^+ -$$

$$- \left[\frac{\omega}{C_{kR}}\, V_k^{(2)}(\tilde{R}) - \frac{2}{R_0}\, V_k^{(1)}(\tilde{R}) \right]^2 [\lambda + 2\mu]_-^+ +$$

$$+ \left(\frac{\tilde{R}}{R_0} \right)^2 \left[(\lambda + 2\mu) \left(\frac{\mathrm{d}V_k^{(1)}}{\mathrm{d}z} \right)^2 \right]_-^+ +$$

$$\left. + \left(\frac{\tilde{R}}{R_0} \right)^2 \left[\mu \left(\frac{\mathrm{d}V_k^{(2)}}{\mathrm{d}z} \right)^2 \right]_-^+ \right\}$$

$$\frac{\partial C_{kL}}{\partial \tilde{R}} = \frac{C_{kL}^3}{2\omega^2 I_{kL}^{(1)}} \left\{ \omega^2 \left(\frac{\tilde{R}}{R_0} \right)^2 (V_k^{(3)}(\tilde{R}))^2 [\rho]_-^+ - \right.$$

$$\left. - \left[\left(\frac{\omega}{C_{kL}} \right)^2 - \frac{1}{R_0^2} \right] (V_k^{(3)}(\tilde{R}))^2 [\mu]_-^+ + \left[\mu \left(\frac{\mathrm{d}V_k^{(3)}}{\mathrm{d}z} \right)^2 \right]_-^+ \right\}.$$

1.6. EFFECTS OF ANELASTICITY

The stress—strain relations in a weakly dissipating medium become integro-differential expressions (Kogan, 1966; Akopyan, Zharkov and Lyubimov, 1975; Aki and Richards, 1980; Levshin, Ratnikova and Saks, 1980). Propagation of harmonic waves can then be conveniently examined by using complex frequency-dependent elastic moduli $\bar{K}(\omega, \mathbf{x})$, $\bar{\mu}(\omega, \mathbf{x})$ instead of frequency-independent ones $K(\mathbf{x})$, $\mu(\mathbf{x})$ for a perfectly elastic medium. Accordingly, we have complex frequency-dependent P and S wave velocities $a(\omega, \mathbf{x})$ and $b(\omega, \mathbf{x})$, respectively.

Dissipation caused by confining pressure is characterized by the quantity $Q_k = \operatorname{Re} \bar{K}/\operatorname{Im} \bar{K}$, while that caused by pure shear, by $Q_\mu = \operatorname{Re} \bar{\mu}/\operatorname{Im} \bar{\mu}$. Dissipation in propagating elastic waves is controlled by Q_a and Q_b

$$Q_b = Q_\mu = \operatorname{Re} b/\operatorname{Im} b$$

$$Q_a = \left[Q_k^{-1} \left(1 - \frac{4b^2}{3a^2} \right) + Q_\mu^{-1} \frac{4b^2}{3a^2} \right]^{-1} = \frac{\operatorname{Re} a}{2 \operatorname{Im} a} .$$

A frequent assumption is $Q_k^{-1} = 0$, giving $Q_a = Q_b(3a^2/4b^2)$. Denote complex velocity perturbations caused by small dissipation as $\delta a(\omega, z)$, $\delta b(\omega, z)$. The corresponding perturbations in phase velocity $C_{k\mathrm{D}}(\omega)$ can be evaluated from the perturbation theory:

$$\delta C_{k\mathrm{D}}(\omega) = \int_0^\infty \left[\frac{\partial C_{k\mathrm{D}}(\omega, z)}{\partial a} \delta a(\omega, z) + \frac{\partial C_{k\mathrm{D}}(\omega, z)}{\partial b} \delta b(\omega, z) \right] dz.$$

Assuming Q_a, Q_b to be independent of frequency over a broad frequency range, the velocity perturbations δa, δb become (Akopyan, Zharkov and Lyubimov, 1975; Aki and Richards, 1980; Levshin, Ratnikova and Saks, 1980)

$$\delta a(\omega) = \frac{a(\omega_0)}{Q_a} \left(\frac{1}{\pi} \ln \frac{\omega}{\omega_0} + \frac{i}{2} \right)$$

$$\delta b(\omega) = \frac{b(\omega_0)}{Q_b} \left(\frac{1}{\pi} \ln \frac{\omega}{\omega_0} + \frac{i}{2} \right) \tag{1.52}$$

where ω_0 is the reference frequency for which we know the velocity distributions $a(z)$, $b(z)$. In that case

$$\delta C_{k\mathrm{D}}(\omega) = \left(\frac{1}{\pi} \ln \frac{\omega}{\omega_0} + \frac{i}{2} \right) S_{k\mathrm{D}}(\omega) \tag{1.53}$$

where

$$S_{kD} = \int_0^\infty \left[\frac{a(z)}{Q_a(z)} \frac{\partial C_{kD}}{\partial a} + \frac{b(z)}{Q_b(z)} \frac{\partial C_{kD}}{\partial b} \right] dz$$

and, for the particular case $Q_k = 0$,

$$S_{kD} = \int_0^\infty \left[\frac{4b(z)}{3a(z)} \frac{\partial C_{kD}}{\partial a} + \frac{\partial C_{kD}}{\partial b} \right] \frac{b}{Q_b} dz. \tag{1.53a}$$

Knowing $\delta C_{kD}(\omega)$, one can easily find the attenuation coefficient for surface waves at frequency ω:

$$\alpha_{kD}(\omega) = \frac{\omega \operatorname{Im}(\delta C_{kD})}{C_{kD}^2} = \frac{\omega S_{kD}}{2C_{kD}^2} \tag{1.54}$$

and the apparent $Q_{kD}(\omega)$ as determined from surface waves:

$$Q_{kD}(\omega) = \omega/[2\alpha_{kD}(\omega)U_{kD}(\omega)]. \tag{1.55}$$

With dissipation present, the second factor in surface wave displacements, (1.27) and (1.31), becomes

$$(\xi_{kD} r)^{-1/2} \exp\left[-\frac{i\omega r}{C_{kD}} \left(1 - \frac{S_{kD}}{\pi C_{kD}} \ln \frac{\omega}{\omega_0} \right) \right] \exp[-\alpha_{kD}(\omega) r]$$

or, in the spherical case (1.36),

$$(\nu_{kD} \sin\theta)^{-1/2} \exp\left[-\frac{i\omega\tilde{\theta}R_0}{C_{kD}} \left(1 - \frac{S_{kD}}{\pi C_{kD}} \ln \frac{\omega}{\omega_0} \right) + \right.$$

$$\left. + i \frac{\pi}{2} (2l + g) \right] \exp[-\alpha_{kD}(\omega)\tilde{\theta}R_0].$$

1.7. SYNTHETIC SEISMOGRAMS

To pass from spectral to time representation of surface waves, i.e., to synthetic seismograms, one should be able to compute Fourier integrals of rapidly oscillating functions involving the factor $\exp(-i\xi_{kD}\omega r)$, where r is large. Wave forms can conveniently be evaluated by using asymptotic formulas of stationary phase (Copson, 1965; Olver, 1974) or Airy integrals (Olver, 1974; Pekeris, 1948).

We write the spectrum of a surface wave as

$$\hat{u}_q^{kD} = \frac{1}{\sqrt{r}} \Phi_q^{kD}(\omega, h, z, \varphi) \exp[-i\xi_{kD}(\omega) r],$$

where Φ_q^{kD} is a complex-valued function that slowly oscillates as ω is varied. In that case

$$
\begin{aligned}
u_q^{kD}(t) &= \frac{1}{\pi\sqrt{r}} \operatorname{Re} \int_{\omega_{kD}}^{\infty} \Phi_q^{kD}(\omega) \exp[i(\omega t - \xi_{kD}(\omega) r)]\, \mathrm{d}\omega \\
&\approx \sqrt{\frac{2}{\pi}}\, \frac{1}{r} \operatorname{Re} \sum_j \frac{U_{kD}(\omega_j)\Phi_{pq}^{kD}(\omega_j)}{\sqrt{|\mathrm{d}U_{kD}(\omega)/\mathrm{d}\omega|_{\omega=\omega_j}}} \times \qquad (1.56) \\
&\quad \times \exp\left[i\omega t - i\xi_{kD}(\omega) r + i\frac{\pi}{4} \operatorname{sign}\left(\frac{\mathrm{d}U_{kD}(\omega)}{\mathrm{d}\omega}\right)\right] + o(r^{-1}),
\end{aligned}
$$

where stationary phase points $\omega_j(t)$ are roots of the equation $t - r/U_{kD}(\omega) = 0$.

The approximation (1.56) is valid within those frequency ranges where $\mathrm{d}U_{kD}(\omega)/\mathrm{d}\omega$ does not vanish, the conditions being stated more precisely by Pekeris (1948). To evaluate the contribution due to the vicinity of $\bar{\omega}$ where $\mathrm{d}U_{kD}(\omega)/\mathrm{d}\omega = 0$ (called the Airy phase), one needs the Airy integral

$$
\begin{aligned}
u_q^{kD}(t) &= \frac{2^{4/3}}{\sqrt{\pi}}\, \frac{1}{r^{5/6}} \operatorname{Re} \frac{\Phi(\bar{\omega})E(\tau)}{(-\mathrm{d}^3\xi_{kD}/\mathrm{d}\omega^3|_{\omega=\bar{\omega}})^{1/3}} \times \\
&\quad \times \exp[i(\bar{\omega} t - \xi_{kD}(\bar{\omega}) r)] + o(r^{-2/3}) \qquad (1.57)
\end{aligned}
$$

where $E(\tau)$ is an Airy function of τ:

$$
\tau = (t - r/U_{kD}(\bar{\omega}))\left(-\frac{r}{2}\, \frac{\mathrm{d}^3\xi_{kD}}{\mathrm{d}\omega^3}\bigg|_{\omega=\bar{\omega}}\right)^{-1/3}.
$$

It follows from (1.56) and (1.57) that the waveform is controlled by group velocity. If we divide the dispersion curve $U_{kD}(\omega)$ into portions having the sign of $\mathrm{d}U_{kD}(\omega)/\mathrm{d}\omega$ constant, each such portion can be represented by a quasi-sinusoidal oscillation of varying frequency $\hat{\omega}(r/t)$ where $\hat{\omega}(r/t)$ is the inverse function of $U_{kD}(\omega)$ within that portion. The contribution of an Airy phase can be represented by an oscillation of quasi-constant frequency $\bar{\omega}$.

The amplitude of motion with apparent frequency ω_j falls off with increasing distance like r^{-1} ($r^{-1/2}$ is due to geometrical spreading and the other $r^{-1/2}$ is caused by the spreading of a signal with time). The decay of the Airy phase amplitude is slightly slower, as $r^{-5/6}$ ($r^{-1/3}$ instead of $r^{-1/2}$ due to a slower spreading with time). The actual amplitude of a harmonic component depends on several factors; we recall that $\Phi_q^{kD}(\omega)$ depends on source mechanism and depth, source spectrum, and structure of the medium.

The complete seismogram $u_p^{D}(t)$ is obtained by adding all the modes present inside the time interval of interest. One cannot always ascribe a physical meaning to the contribution of a particular mode, because several modes may interfere in such a way that they cannot be separated either in the time or the frequency domain (Levshin, 1973). For this reason one should not attach significance to anomalies in the polarization, dispersion and other spectral characteristics of a mode, unless the mode can be separated from other modes in the time or frequency ranges studied under actual experimental conditions.

The formulas for evaluation of theoretical seismograms discussed above are not always convenient for actual computation, owing to the narrow intervals of frequency where asymptotic approximations are valid and the difficulties in joining the intervals. Numerical methods are usually employed for calculation of synthetic seismograms (Aki and Richards, 1980; Schwab *et al.*, 1984).

2. SURFACE WAVES IN MEDIA WITH WEAK LATERAL INHOMOGENEITY

2.1. FORMULATION OF PROBLEMS OF SURFACE WAVE PROPAGATION IN A HALF-SPACE WITH WEAK LATERAL INHOMOGENEITY

Surface wave propagation in a vertically varying half-space was treated in the preceding chapter by evaluating the leading term in the exact integral solution of the equations of motion, which describes the surface wave field far from the source (see (1.27), (1.32), (1.33)). Exact formulas for a structure with both vertical and lateral variation cannot be derived, so the above approach to the analysis of surface waves is inapplicable. An arbitrary lateral inhomogeneity can be handled only by numerical methods. Nevertheless, when the lateral variation is much smaller than the vertical, a procedure similar to the ray method can be used to construct an approximate solution corresponding to the surface wave field (Woodhouse, 1974; Babich and Chikhachev, 1975; Babich *et al.*, 1976; Babich and Grigoryeva, 1980). Similarly to the ray method (Alekseev *et al.*, 1961; Babich and Buldyrev, 1972), the solution is expanded in powers of a small parameter (this is the inverse of frequency ω in the ray method and a parameter describing the weak inhomogeneity in our case).

It is assumed that the elastic moduli λ, μ and density ρ in the half-space can be represented in the form

$$\lambda = \lambda(\varepsilon x, \varepsilon y, z), \quad \mu = \mu(\varepsilon x, \varepsilon y, z), \quad \rho = \rho(\varepsilon x, \varepsilon y, z),$$

where ε is a small parameter. Obviously, in such case the derivatives of elastic moduli along a horizontal direction are of order ε compared with those along the vertical. All interfaces are assumed to be smooth and nearly horizontal. This assumption is implied in our representation of interface depth as

$$z_i = \zeta_i(\varepsilon x, \varepsilon y).$$

The free boundary of a half-space can be defined in a similar manner: $z_0 = \zeta_0(\varepsilon x, \varepsilon y)$.

The parameter ε must be so small that the relative lateral variation of elastic moduli should be small over distances the order of a wavelength. Denoting phase velocity as C (obviously, $C = C(\varepsilon x, \varepsilon y)$), the conditions imposed on the lateral variation of λ, μ, ρ can be written in the form

$$|\nabla_\perp \lambda| \ll \frac{\omega}{C}\lambda, \; |\nabla_\perp \mu| \ll \frac{\omega}{C}\mu, \; |\nabla_\perp \rho| \ll \frac{\omega}{C}\rho$$

where $\nabla_\perp$ is the horizontal component of the gradient. The functions that give interface depths obey the requirement $|\nabla_\perp \zeta_i| \ll 1$.

In case there are transition layers having large gradients of the elastic constants instead of interfaces, it will be required that the gradient should be nearly vertical. It should also be remembered that we deal with the surface wave field which is described by the principal part of the solution to the equations of motion at great distances r from the source. The amplitude of a surface wave decays with distance from the source as $r^{-1/2}$ (for harmonic motion) and as r^{-1} (transient motion). At large r this results in the amplitude varying much more rapidly vertically than horizontally, so that the relative variation of amplitude over distances the order of a wavelength is small ($\omega r/C \gg 1$). When solving for surface wave displacements in a medium with weak lateral variation, we shall assume the derivative of amplitude along any horizontal direction to be small, of the order of $0(\varepsilon)$.

We should also remember that surface waves propagate horizontally, that is, the phase factor describing wave propagation should not depend on z. The amplitude is written in the form of a series in powers of ε. A solution in this form is substituted into the equation of motion for an inhomogeneous elastic medium and into the boundary conditions, both the equation and the boundary conditions being written as series in powers of ε. Then the coefficients of successive powers of ε in the equation and boundary conditions are equated to zero, resulting in a sequence of boundary value problems for successive approximations for the desired solution.

2.2. PROPAGATION OF HARMONIC SURFACE WAVES

The problem in the propagation of harmonic surface waves in a half-space with weak lateral variation has been solved by Woodhouse (1974); Babich and Chikhachev (1975); Babich *et al.* (1976); Babich and Grigoryeva (1980). The field of a harmonic surface wave can be represented in the form

$$\mathbf{u}(x, y, z, t, \varepsilon) = \mathbf{V}(\varepsilon x, \varepsilon y, z, \varepsilon) \exp[i\omega(t - \tau(x, y, \varepsilon))]. \tag{2.1}$$

Similarly to the ray theory, the solution is sought in the form of waves travelling along rays at the velocity $C = |\nabla\tau|^{-1}$, the difference being that the rays are horizontal now, because surface waves propagate horizontally. A ray is understood to be a line normal to the front $\tau =$ constant. The principal difference from the ray theory, however, is that the solution for $\mathbf{V}$ is represented as an asymptotic expansion in powers of ε.

Substituting (2.1) into the equation of motion (1.4), we get

$$\begin{aligned}&(\lambda + 2\mu)\nabla \operatorname{div} \mathbf{V} - \mu \operatorname{rot}\operatorname{rot} \mathbf{V} + \nabla\lambda \operatorname{div} \mathbf{V} + \nabla\mu \times \operatorname{rot} \mathbf{V} +\\ &+ 2(\nabla\mu, \nabla)\mathbf{V} - i\omega \left\{ (\lambda + \mu)\,[\operatorname{div} \mathbf{V}\nabla\tau + \nabla(\mathbf{V}, \nabla\tau)] + \mu \left[\mathbf{V}\Delta\tau + \right.\right.\\ &\left.\left. + 2\frac{\partial \mathbf{V}}{\partial\tau}(\nabla\tau)^2 \right] + \nabla\lambda(\mathbf{V}, \nabla\tau) + (\nabla\mu, \mathbf{V})\nabla\tau + (\nabla\mu, \nabla\tau)\mathbf{V} \right\} -\\ &- \omega^2\{(\lambda + \mu)(\mathbf{V}, \nabla\tau) + (\mu(\nabla\tau)^2 - \rho)\mathbf{V}\} = 0.\end{aligned} \tag{2.2}$$

Substitution of (2.1) into the boundary condition (1.6) gives

$$\lambda \mathbf{n} \operatorname{div} \mathbf{V} + 2\mu \left[\frac{\partial \mathbf{V}}{\partial n} + \frac{1}{2} (\mathbf{n} \times \operatorname{rot} \mathbf{V}) \right] -$$

$$- i\omega \left[\lambda \mathbf{n}(\nabla \tau, \mathbf{V}) + \mu (\mathbf{n} \times (\nabla \tau \times \mathbf{V})) + 2\mu \mathbf{V} \frac{\partial \tau}{\partial n} \right] = 0 \qquad (2.3)$$

where **n** is the unit vector normal to the surface:

$$\mathbf{n} = \left\{ \frac{1}{B} \frac{\partial \zeta_0}{\partial x}, \frac{1}{B} \frac{\partial \zeta_0}{\partial y}, -\frac{1}{B} \right\}$$

$$B = \sqrt{1 + \left(\frac{\partial \zeta_0}{\partial x} \right)^2 + \left(\frac{\partial \zeta_0}{\partial y} \right)^2}.$$

We now define an orthogonal set of curvilinear coordinates: the 'ray' coordinates τ, φ in the xy-plane and the vertical coordinate z. The coordinate φ characterizes a ray, while the coordinate τ defines a point on a ray. The direction of $\mathbf{e}_\varphi$ is fixed $\mathbf{e}_\varphi = \mathbf{e}_z \times \mathbf{e}_\tau$. Geometrical spreading in the xy-plane is denoted as

$$J = \left[\left(\frac{\partial x}{\partial \varphi} \right)^2 + \left(\frac{\partial y}{\partial \varphi} \right)^2 \right]^{1/2}.$$

We now represent the desired solution **V** as a series in powers of ε:

$$\mathbf{V} = \sum_{j=0}^{\infty} \varepsilon^j \mathbf{V}_j.$$

The components of $\mathbf{V}_j$ along the coordinates z, τ, φ are denoted $V_j^{(z)}$, $V_j^{(\tau)}$, $V_j^{(\varphi)}$, respectively. We substitute the expansion for **V** into (2.2) and (2.3) and write out the leading terms. We shall use the remark in Section 2.1 about the derivative of amplitude along any horizontal direction being of the order of $0(\varepsilon)$. This is also true for

$$\Delta \tau = \frac{|\nabla \tau|}{J} \frac{\mathrm{d}}{\mathrm{d}\tau} (J |\nabla \tau|)$$

(Alekseev *et al.*, 1961), because the relative variation of geometrical spreading is small at great distances from the source. The result is a set of equations in $V_0^{(z)}$, $V_0^{(\tau)}$, $V_0^{(\varphi)}$:

$$l_1(V_0^{(z)}, V_0^{(\tau)}) = \frac{\partial}{\partial z}\left(\mu \frac{\partial V_0^{(\tau)}}{\partial z}\right) - i\omega|\nabla\tau|\left[\lambda \frac{\partial V_0^{(z)}}{\partial z} + \right.$$

$$\left. + \frac{\partial(\mu V_0^{(z)})}{\partial z}\right] + \omega^2[\rho - (\lambda + 2\mu)(\nabla\tau)^2]V_0^{(\tau)} = 0 \quad (2.4)$$

$$l_2(V_0^{(z)}, V_0^{(\tau)}) = \frac{\partial}{\partial z}\left((\lambda + 2\mu)\frac{\partial V_0^{(z)}}{\partial z}\right) - i\omega|\nabla\tau|\left[\mu \frac{\partial V_0^{(\tau)}}{\partial z} + \right.$$

$$\left. + \frac{\partial(\lambda V_0^{(\tau)})}{\partial z}\right] + \omega^2[\rho - \mu(\nabla\tau)^2]V_0^{(z)} = 0 \quad (2.5)$$

$$l_3(V_0^{(\varphi)}) = \frac{\partial}{\partial z}\left(\mu \frac{\partial V_0^{(\varphi)}}{\partial z}\right) + \omega^2[\rho - \mu(\nabla\tau)^2]V_0^{(\varphi)} = 0. \quad (2.6)$$

The boundary conditions are as follows. When $z = \zeta_0$

$$\frac{\partial V_0^{(\tau)}}{\partial z} - i\omega|\nabla\tau|V_0^{(z)} = 0 \quad (2.7)$$

$$(\lambda + 2\mu)\frac{\partial V_0^{(z)}}{\partial z} - i\omega\lambda V_0^{(\tau)} = 0 \quad (2.8)$$

$$\frac{\partial V_0^{(\varphi)}}{\partial z} = 0. \quad (2.9)$$

When $z \to \infty$

$$V_0^{(\tau)} \to 0, \qquad V_0^{(z)} \to 0 \quad (2.10a)$$

$$V_0^{(\varphi)} \to 0. \quad (2.10b)$$

This set of equations and boundary conditions is completely analogous to that for a laterally homogeneous half-space (see formulas (1.20)–(1.21), (1.23), (1.25)). However, the coefficients in the equations and boundary conditions are no longer constants, but depend on εx, εy, and (2.7) to (2.9) are satisfied, not when $z = 0$, but when $z = \zeta_0(\varepsilon x, \varepsilon y)$, corresponding to the position of the free surface at the point (x, y).

As in the case of a laterally homogeneous half-space, the problem (2.4)–(2.10) splits into two independent parts, corresponding to Rayleigh and Love waves. We have $V_0^{(\varphi)}(z) = 0$ for a Rayleigh wave, the eigenvalues $\omega|\nabla\tau|$ and eigenfunctions $V_0^{(\tau)}(z)$, $V_0^{(z)}(z)$ being determined from (2.4) and (2.5) together with the boundary conditions (2.7), (2.8), (2.10a); for Love waves $V_0^{(\tau)} = V_0^{(z)} = 0$, the eigenfunctions $V_0^{(\varphi)}(z)$ and the eigenvalues being determined from the solution of the boundary value problem (2.6), (2.9), (2.10b). The eigenvalues and the relevant eigenfunctions are thus determined locally for each point (x, y) and turns out to be the same as for a laterally homogeneous half-space with a vertical structure that is identical with that at the point (x, y). An eigenvalue $\omega|\nabla\tau| \equiv \xi$ determines phase velocity $C(x, y) = |\nabla\tau|^{-1}$ which is, as was to be expected, a function of (x, y). Hence $\tau(x, y, \varepsilon)$ is given by

$$\tau(x, y, \varepsilon) = \int_L \frac{ds}{C(x, y)} \tag{2.11}$$

where the integral is taken along a ray in the x, y-plane that is given by the well known ray equation (Cerveny *et al.*, 1977)

$$\frac{d\mathbf{r}}{ds} = C\mathbf{k}, \qquad \frac{d\mathbf{k}}{ds} = -\frac{\nabla C}{C^2}. \tag{2.12}$$

Here $\mathbf{r}$ is the radius vector of a point on the ray, $\mathbf{k}$ the slowness vector equal to $\nabla\tau = \mathbf{e}_\tau/C$. Ray tracing schemes for surface waves were developed in Backus (1964), Gjevik (1974), Gregersen (1974).

Examining the leading terms (of order ε) in the expansions of the equation of motion and boundary conditions, one notes an analogy with the ray expansion: an analysis of the first term separates the field into two wave types whose polarizations differ in the zero-order approximation and determines the wave velocities. Also, at each point (x, y) one can define the amplitude of a wave to be a function of depth z, because it is the eigenfunction of the relevant one-dimensional problems. However, the amplitude can be determined apart from a factor which depends on (x, y) and governs the variation of wave intensity during propagation. This factor is found by examining the second term in the expansion of the equation and boundary conditions. Similarly to the ray theory, the amplitude in the zero-order approximation is determined from the equation corresponding to the next term in the expansion of the equation of motion.

The analysis of the next higher approximation in the expansion of (2.2) with boundary conditions (2.3) should proceed separately for Love and Rayleigh waves.

Love waves. As has been mentioned above, when x and y are fixed, the problem corresponding to the zero-order approximation has the form

$$l_3(V_0^{(\varphi)}) \equiv \frac{\partial}{\partial z}\left(\mu\frac{\partial V_0^{(\varphi)}}{\partial z}\right) + \omega^2[\rho - \mu(\nabla\tau)^2]V_0^{(\varphi)} = 0 \tag{2.13}$$

$$\left.\frac{\partial V_0^{(\varphi)}}{\partial z}\right|_{z=\zeta_0} = 0 \tag{2.14}$$

and $V_0^{(\varphi)} \to 0$ when $z \to \infty$.

In the next higher approximation, all the displacement components are different from zero, so that one should have written a set of three equations in the unknowns $V_1^{(z)}$, $V_1^{(\tau)}$, $V_1^{(\varphi)}$ with three boundary conditions. However, the dependence of the zero-order amplitude $V_0^{(\varphi)}$ on (x, y) can be found by examining a single equation, that corresponding to the principal, φ-component. That equation and the relevant boundary condition have the form

$$\varepsilon l_3(V_1^{(\varphi)}) - i\omega \left[\mu V_0^{(\varphi)} \Delta \tau + 2\mu \frac{\partial V_0^{(\varphi)}}{\partial \tau} (\nabla \tau)^2 + \right.$$

$$\left. + \frac{\partial \mu}{\partial \tau} (\nabla \tau)^2 V_0^{(\varphi)} \right] = 0 \tag{2.15}$$

$$\begin{aligned} &\varepsilon \frac{\partial V_1^{(\varphi)}}{\partial z} + \frac{\partial \zeta_0}{\partial \tau} (\nabla \tau)^2 i\omega V_0^{(\varphi)}|_{z=\zeta_0} = 0 \\ &V_1^{(\varphi)} \to 0 \quad (z \to \infty). \end{aligned} \tag{2.16}$$

It is easy to see that the second term is of order ε both in the equation and in the boundary condition, hence both terms in (2.15) and (2.16) are of the same order.

Equation (2.15) together with the boundary conditions (2.16) form an inhomogeneous boundary-value problem for $V_1^{(\varphi)}$. The corresponding homogeneous boundary-value problem is identical with that for $V_1^{(\varphi)}$, and hence has a nontrivial solution. In that case the inhomogeneous problem cannot be solved, unless the right-hand side satisfies a condition of solvability. The condition can be found by multiplying (2.15) by $V_0^{(\varphi)}$ and integrating the product with respect to z between the limits ζ_0 and ∞, taking (2.14) and (2.16) into account. The result is an equation for $V_0^{(\varphi)}$:

$$\mu (V_0^{(\varphi)})^2 \frac{\partial \zeta_0}{\partial \tau} (\nabla \tau)^2|_{z=\zeta_0} + \Delta \tau \int_{\zeta}^{\infty} \mu (V_0^{(\varphi)})^2 \, dz +$$

$$+ 2(\nabla \tau)^2 \int_{\zeta_0}^{\infty} \mu V_0^{(\varphi)} \frac{\partial V_0^{(\varphi)}}{\partial z} \, dz + (\nabla \tau)^2 \int_{\zeta_0}^{\infty} \frac{\partial \mu}{\partial \tau} (V_0^{(\varphi)})^2 \, dz = 0 \tag{2.17}$$

Using the relations $\Delta \tau = (|\nabla \tau|/J)(\partial/\partial \tau)(J|\nabla \tau|)$ and $|\nabla \tau| = C^{-1}$ we get

$$\frac{1}{CJ} \frac{\partial}{\partial \tau} \left(\frac{J}{C} \int_{\zeta_0}^{\infty} \mu (V_0^{(\varphi)})^2 \, dz \right) = \operatorname{div} \left(\nabla \tau \int_{\zeta_0}^{\infty} \mu (V_0^{(\varphi)})^2 \, dz \right) = 0. \tag{2.18}$$

From this equation it follows that the amplitude at an arbitrary point of the ray at depth z is given by

$$V_0^{(\varphi)}(x, y, z) = A_L(x, y) V^{(3)}(z - \zeta_0) \tag{2.19}$$

$$A_L(x, y) = K_L(\varphi) \sqrt{\frac{C}{J}} \left(\int_0^{\infty} \mu (V^{(3)})^2 \, dz \right)^{-1/2} \tag{2.20}$$

where $V^{(3)}(z)$ is the eigenfunction of the one-dimensional problem for Love waves at point (x, y); $K_L(\varphi)$ is a quantity that is constant along a ray and is determined by the source radiation pattern. The relation gives wave amplitude along a ray, provided it is known at a single point of the ray.

It can easily be demonstrated that (2.18) expresses the conservation of energy in the zero-order approximation, within a region that is in our case an analogue of the ray tube, i.e., a region bounded by cylindrical surfaces whose generatrices are semi-infinite vertical lines passing through rays in the (x, y)-plane (Figure 2.1).

To see this, we express the energy flux of a Love wave passing across an infinite vertical strip normal to the ray and bounded by rays of φ and $\varphi + \mathrm{d}\varphi$ as follows

$$\mathrm{d}E = \frac{U\omega^2}{2} J\,\mathrm{d}\varphi \int_{\zeta_0}^{\infty} \rho(V_0^{(\varphi)})^2\,\mathrm{d}z \tag{2.21}$$

where U is the group velocity. Now recall the integral formula for group velocity in Chapter 1, (1.45):

$$U = \int_0^{\infty} \mu(V^{(3)})^2\,\mathrm{d}z \Big/ C \int_0^{\infty} \rho(V^{(3)})^2\,\mathrm{d}z.$$

Substituting (2.19), (2.20) and the above expression for U into (2.21), we get

$$\mathrm{d}E = \omega^2 K_{\mathrm{L}}^2(\varphi)\,\mathrm{d}\varphi/2. \tag{2.22}$$

We thus see that the energy flux across any vertical strip bounded by two rays is indeed conserved.

Rayleigh waves. When a point (x, y) is fixed, Equations (2.4), (2.5) combined with boundary conditions (2.7), (2.8) and (2.10a) define the same one-dimensional problem for Rayleigh waves as that discussed in Section 1.3. Hence

$$\begin{aligned} V_0^{(z)}(x, y, z) &= A_{\mathrm{R}}(x, y)\, V^{(1)}(z - \zeta_0) \\ V_0^{(\tau)}(x, y, z) &= iA_{\mathrm{R}}(x, y)\, V^{(2)}(z - \zeta_0). \end{aligned} \tag{2.23}$$

To determine the variation of $V_0^{(z)}$, $V_0^{(\tau)}$ along a ray of φ = constant we must

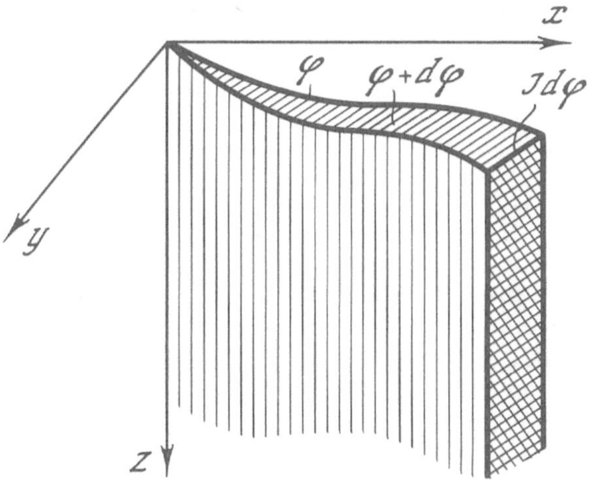

Fig. 2.1. Schematic representation of a ray cylinder, analogue of a ray tube.

examine the next higher approximation. The first-order equations for τ- and z-components of the displacement are

$$\varepsilon l_1(V_1^{(z)}, V_1^{(\tau)}) + |\nabla\tau| \left[\frac{\partial}{\partial\tau}\left(\lambda \frac{\partial V_0^{(z)}}{\partial z}\right) + \frac{\partial}{\partial\tau}\left(\lambda \frac{\partial V_0^{(z)}}{\partial z}\right)\right] -$$

$$- i\omega \left[(\lambda + 2\mu) V_0^{(\tau)} \frac{|\nabla\tau|\partial(J|\nabla\tau|)}{J\,\partial\tau} + 2(\lambda+2\mu)(\nabla\tau)^2 \frac{\partial V_0^{(\tau)}}{\partial\tau} + \right.$$

$$\left. + \frac{\partial(\lambda+2\mu)}{\partial\tau}(\nabla\tau)^2 V_0^{(\tau)} \right] = 0$$

$$\varepsilon l_2(V_1^{(z)}, V_1^{(\tau)}) + \frac{|\nabla\tau|}{J}\left[\frac{\partial}{\partial z}\left(\lambda \frac{\partial(J V_0^{(\tau)})}{\partial\tau}\right) + \frac{\partial}{\partial\tau}\left(\mu J \frac{\partial V_0^{(\tau)}}{\partial z}\right)\right] - \tag{2.24}$$

$$- i\omega \left[\mu V_0^{(z)} \frac{|\nabla\tau|}{J} \frac{\partial}{\partial\tau}(J|\nabla\tau|) + 2\mu(\nabla\tau)^2 \frac{\partial V_0^{(z)}}{\partial\tau} + \right.$$

$$\left. + \frac{\partial\mu}{\partial\tau}(\nabla\tau)^2 V_0^{(z)} \right] = 0$$

where l_1 and l_2 are given by (2.4), (2.5), the relevant boundary conditions being

$$\mu\left(\frac{\partial V_1^{(\tau)}}{\partial z} + |\nabla\tau| \frac{\partial V_0^{(z)}}{\partial\tau} - i\omega|\nabla\tau| V_1^{(z)}\right) -$$

$$- \frac{\partial\zeta_0}{\partial\tau}|\nabla\tau|\left(\lambda \frac{\partial V_0^{(z)}}{\partial z} - i(\lambda+2\mu)\omega|\nabla\tau| V_0^{(\tau)}\right) = 0$$

$$(\lambda+2\mu)\frac{\partial V_1^{(z)}}{\partial z} + \lambda \frac{|\nabla\tau|}{J}\frac{\partial}{\partial\tau}(J V_0^{(\tau)}) - i\omega|\nabla\tau|\lambda V_1^{(z)} - \tag{2.25}$$

$$- \frac{\partial\zeta_0}{\partial\tau}|\nabla\tau|\left(\mu \frac{\partial V_0^{(\tau)}}{\partial z} - i\mu\omega|\nabla\tau| V_0^{(z)}\right) = 0.$$

The subsequent argument proceeds on the lines of the Love wave case. Equations (2.24) together with boundary conditions (2.25) form an inhomogeneous boundary-value problem, the $\omega|\nabla\tau|$ being the eigenvalue of the corresponding homogeneous problem. Therefore, the problem (2.24), (2.25) has a solution when and only when right-hand sides of (2.24) satisfy the following condition. This is obtained by multiplying the first of (2.24) by $V_0^{(\tau)*}$ (the asterisk denotes complex conjugation) and the second by $V_0^{(z)*}$, adding these together, and integrating the result over z between the limits ζ_0 and ∞, taking (2.4), (2.5) and boundary conditions (2.25) into account. After some manipulation of the resulting expressions for fixed φ, we obtain a relation analogous to (2.18)

$$\frac{1}{C}\frac{\partial}{J\,\partial\tau}\left\{\frac{JA_R^2}{C}\left[I_R^{(1)}+I_R^{(2)}+\frac{C}{\omega}(I_R^{(3)}+I_R^{(4)})\right]\right\}$$

$$=\operatorname{div}\left\{\nabla\tau A_R^2\left[I_R^{(1)}+I_R^{(2)}+\frac{C}{\omega}(I_R^{(3)}+I_R^{(4)})\right]\right\}=0 \qquad (2.26)$$

where the integrals $I_R^{(j)}$ are identical with those defined by (1.46). Using (1.45), we get the following expression for A_R:

$$A_R(x,y)=K_R(\varphi)\sqrt{\frac{C}{J}}\left[I_R^{(1)}+I_R^{(2)}+\frac{C}{\omega}(I_R^{(3)}+I_R^{(4)})\right]^{-1/2}$$

$$=K_R(\varphi)\sqrt{\frac{C}{J}}\frac{1}{\sqrt{CUI_R^{(0)}}} \qquad (2.27)$$

where $K_R(\varphi)$ is constant along a ray.

Just as before, (2.27) means that the energy flux across any vertical section of a semi-infinite ray cylinder is constant. A Rayleigh wave travelling across the section carries the energy per unit time given by

$$\mathrm{d}E=\frac{U\omega^2}{2}J\,\mathrm{d}\varphi\int_{\xi_0}^{\infty}\rho(|V_0^{(r)}|^2+|V_0^{(z)}|^2)\,\mathrm{d}z. \qquad (2.28)$$

Using (2.23), (2.27), we get

$$\mathrm{d}E=\omega^2K_R^2(\varphi)\,\mathrm{d}\varphi/2$$

so that wave energy inside a ray cylinder is indeed conserved in the zero-order approximation.

Field of a point source. The above formulas for the components of V_0 merely describe the variation of spectral displacement amplitude along a fixed ray of φ = constant. To derive general expressions for spectral surface wave amplitude due to a point source in the structure under consideration, the factors $K_D(\varphi)$ in (2.20), (2.27) must be specified.

Burridge and Weinberg (1977) have shown for similar scalar problems in wave acoustics that, in the zero-order approximation, the surface wave radiated by a source is locally the same as that for the source in a laterally homogeneous medium with the same distribution of elastic constants both beneath and above the source. Analogous assumptions of local homogeneity around the source (in the zero-order approximation) are typical for the ray geometric theory of body waves. Proceeding on that analogy and using the results of Section 1.3, we arrive at the following expressions for spectral displacement amplitude in a surface wave (kth mode of D-type that is excited at $\mathbf{x}_0\,(0,0,h)$ and recorded at $\mathbf{x}\,(x,y,z)$):

$$\hat{u}^{kD}_{pq}(\omega, x, y, z) = \frac{\exp(-i\pi/4)}{\sqrt{8\pi}} \frac{\exp\left[-i\omega \int_L \mathrm{d}s/C_{kD}(s)\right]}{\sqrt{\xi_{kD} J_{kD}(x, y)}} \times$$

$$\times \left. \frac{\varepsilon_p V_k^{(i_p)}(\omega, z - \zeta_0)}{\sqrt{C_{kD} U_{kD} I^{(0)}_{kD}}} \right|_{(x, y)} \times$$

$$\times \left. \frac{W^{kD}(\omega, \varphi, h - \zeta_0)}{\sqrt{C_{kD} U_{kD} I^{(0)}_{kD}}} \right|_{(0, 0)} + O(\varepsilon). \qquad (2.29)$$

The integral in the exponent of the second factor must be taken along the ray between points (0, 0) and (x, y) in the x, y-plane. The third and fourth factors express the effects of structure around the receiver and the source, respectively; the subscript p takes on the values τ, φ, z ($i_\tau = i_r$, $\varepsilon_\tau = \varepsilon_r$); φ is the azimuth of the ray at the source. It is easy to see that when we deal with a laterally homogeneous structure, (2.29) becomes (1.31).

Summing up, spectral amplitude is evaluated by calculating the scalar field of phase velocity $C_{kD}(x, y)$, by tracing the ray connecting the projections of the source and recording site on to the x, y-plane, finding the geometrical spreading $J(x, y)$ at the receiving site, and calculating the spectral characteristics (group velocity, eigenfunctions, energy integrals at the source and the recording site, the function W^{kD} for the relevant source mechanism).

2.3. PROPAGATION OF TRANSIENT SURFACE WAVES

Knowing the solution for harmonic surface waves, one can construct a solution for transient waves using the procedure outlined in Chapter 1; this consists in representing the field as a superposition of harmonic waves, i.e., as a Fourier integral, and evaluating the integral by the method of stationary phase. The procedure is however not valid in the general case of a laterally varying structure. This can be seen as follows.

As has been shown in 2.2, the field of a harmonic surface wave in a structure with weak lateral variation can be written in the form

$$\mathbf{u}(x, y, z, t, \varepsilon) = \sum_{j=0} \varepsilon^j \mathbf{V}_j(\varepsilon x, \varepsilon y, z, \varepsilon) \exp[i\omega(t - \tau(x, y, \varepsilon))]. \qquad (2.30)$$

The field of a transient wave can be represented in the form

$$\mathbf{u}(x, y, z, t, \varepsilon) = \sum_{j=0} \varepsilon^j \frac{1}{2\pi} \int_{-\infty}^{+\infty} V_j(\varepsilon x, \varepsilon y, z, \varepsilon) \times$$

$$\times \exp[i\omega(t - \tau(x, y, \varepsilon))] \, \mathrm{d}\omega \qquad (2.31)$$

that is, again as an expansion in powers of a small parameter ε. Let us try to evaluate each of the integrals on the right-hand side of (2.31) using the method of stationary phase. The time t can be treated as a large parameter. The phase factor in the integrands on the right-hand side of (2.31) can then be represented in the form

$$\exp\left[it\left(\omega - \frac{\omega\tau}{t}\right)\right] = \exp[it\Psi(\omega, t)]. \tag{2.32}$$

The point of the stationary phase is then given by

$$\frac{\partial\Psi}{\partial\omega} = 1 - \frac{\partial(\omega\tau)}{t\,\partial\omega} = 0 \tag{2.33}$$

or

$$t = \tau + \omega\frac{\partial\tau}{\partial\omega}.$$

Substituting the expression (2.11) for $\tau(x, y, \varepsilon, \omega)$, we obtain

$$t = \int_{L_\omega}\frac{\mathrm{d}s}{C(\varepsilon x, \varepsilon y, \omega)} - \omega\frac{\mathrm{d}}{\mathrm{d}\omega}\int_{L_\omega}\frac{\mathrm{d}s}{C} \approx \int_{L_\omega}\frac{\mathrm{d}s}{U(\varepsilon x, \varepsilon y, \omega)} \tag{2.34}$$

where U is the group velocity given by

$$\frac{1}{U(\omega)} = \frac{1}{C(\omega)} - \frac{\omega}{C^2}\frac{\mathrm{d}C}{\mathrm{d}\omega} \tag{2.35}$$

L_ω being the ray corresponding to frequency ω.

Strictly speaking, when one differentiates the expression (2.11) for $\tau(x, y, \omega)$, one should also have remembered that the integration path L_ω too has been changed with ω, not just the integrand. However, as has been shown by Babich and Buldyrev (1972), the contribution into the variation in τ due to the variation in the path of integration compared with what is caused by the change in C, is a small quantity of the next order; therefore when $\mathrm{d}\tau/\mathrm{d}\omega$ is computed, the variation in the ray due to the change in ω can be neglected. Each of the integrals on the right-hand side of (2.31) is thus treated as a quasiharmonic wave of a frequency that varies slowly in time and space (x, y) and is to be found from (2.34):

$$\mathbf{u}(x, y, z, t, \varepsilon) \approx \sum_{j=0}^{\infty} \varepsilon^j R_j(\varepsilon x, \varepsilon y, z, \omega(t, x, y)) \exp\{i[\omega(t)t - \tau(x, y, \varepsilon, \omega(t, x, y))]\}. \tag{2.36}$$

Babich and Grigoryeva (1980) call a wave of this sort a harmonic wavepacket. From the above it follows that a component of frequency ω in the wavepacket travels with group velocity $U(x, y, \omega)$ along a ray L_ω in the x, y-plane that is determined by the distribution of phase velocity $C(x, y, \omega)$ in accordance with the ray equation (2.12). (Here and below, we write phase and group velocity as functions of the coordinates in the form $C(x, y, \omega)$, $U(x, y, \omega)$, with the understanding that the derivatives of the velocities with respect to x, y are of order ε.) However, this result is the utmost that can be achieved for the field of a transient surface wave within the framework of the Fourier integral. The amplitude of a component in a harmonic wavepacket cannot be evaluated by this procedure. The difficulty lies in the fact that evaluation of the amplitude by stationary phase must be based on the second derivative of the phase function, $d^2\Psi/d\omega^2$ (see 1.7), whereas this depends on the variation in the ray L_ω with varying ω. Such a derivative can only be computed in some particular cases in which $\int_{L_\omega} ds/C(x, y, \omega)$ is expressible in closed form and is represented as a function of ω. This is easy to do for a horizontally stratified structure (one then has $\Psi = \omega - \omega r/Ct$) or, for example, when C varies linearly along some direction in the horizontal plane. Such a procedure is impracticable for an arbitrary function $C = C(x, y)$. The general case of harmonic wavepacket propagation is best treated by the space-time ray method (Whitham, 1974; Babich *et al.*, 1979, 1985) which represents the solution, not as a Fourier integral, but directly as a harmonic wavepacket

$$\mathbf{u}(x, y, z, t, \varepsilon) = \sum_{j=0} \varepsilon^j \mathbf{u}_j(\varepsilon x, \varepsilon y, z, \varepsilon t) \exp[i\Theta(x, y, t)]. \tag{2.37}$$

This form of representation is exactly that resulting from the Fourier integral as evaluated by stationary phase: comparison of (2.36) and (2.37) yields

$$\Theta = \omega(t, x, y)\,(t - \tau(x, y, \omega(t, x, y))). \tag{2.38}$$

Hence the amplitude is a function of time, because it depends on the frequency ω as given by (2.34), both ω and, hence, wave amplitude slowly varying with time. This is easy to see from (2.34), considering that t is large (the solution is constructed far from the source, hence for large values of t).

From (2.38) and (2.34) it follows that

$$\begin{aligned} \frac{\partial\Theta}{\partial t} &= \frac{\partial\omega}{\partial t}\left(t - \frac{\partial(\omega\tau)}{\partial\omega}\right) = \omega \\ \frac{\partial\Theta}{\partial x} &= \frac{\partial\omega}{\partial x}\left(t - \frac{\partial(\omega\tau)}{\partial\omega}\right) - \omega\,\frac{\partial\tau}{\partial x} = -\omega\,\frac{\partial\tau}{\partial x} = -\omega k_x \end{aligned} \tag{2.39}$$

$$\frac{\partial\Theta}{\partial y} = -\omega k_y \tag{2.40}$$

where k_x, k_y are components of $\mathbf{k} = \nabla\tau$ as given by (2.12), so that $\nabla\Theta = \omega\mathbf{k}$ is the wave vector. Analysis of transient surface wave propagation using the space-time ray method can be found in Woodhouse (1974), Babich and Grigoryeva (1980).

The solution is constructed as before: the displacement vector (2.37) is substituted into the equation of motion (1.4) and the boundary conditions (1.6), the left-hand sides being expanded in powers of ε. The coefficients of successive powers of ε are then equated to zero, yielding a sequence of boundary-value problems that determine the travel times and amplitudes of successive approximations. The difference from the harmonic case is that both phase and amplitude are supposed to be unknown functions of t, so that the solution is sought in the four-dimensional space of x, y, z, t rather than in the x, y, z space.

The zero-order equation of motion and the boundary conditions assume a form similar to (2.4)–(2.10). The relationships $d\Theta/dt = \omega$ and $|\nabla\Theta| = \xi$ have been used in the derivation. The resulting equations are

$$l_1(V_0^{(z)}, V_0^{(\tau)}) = \frac{\partial}{\partial z}\left(\mu \frac{\partial V_0^{(\tau)}}{\partial z}\right) - $$

$$- i\xi\left[\lambda \frac{\partial V_0^{(z)}}{\partial z} + \frac{\partial(\mu V_0^{(z)})}{\partial z}\right] +$$

$$+ (\omega^2\rho - (\lambda + 2\mu)\xi^2)V^{(\tau)} = 0 \tag{2.41}$$

$$l_2(V_0^{(z)}, V_0^{(\tau)}) = \frac{\partial}{\partial z}\left((\lambda + 2\mu) \frac{\partial V_0^{(z)}}{\partial z}\right) - $$

$$- i\xi\left[\mu \frac{\partial V_0^{(\tau)}}{\partial z} + \frac{\partial(\lambda V_0^{(\tau)})}{\partial z}\right] +$$

$$+ (\omega^2\rho - \mu\xi^2)V_0^{(z)} = 0 \tag{2.42}$$

$$l_3(V_0^{(\varphi)}) = \frac{\partial}{\partial z}\left(\mu \frac{\partial V_0^{(\varphi)}}{\partial z}\right) + (\omega^2\rho - \mu\xi^2)V_0^{(\varphi)} = 0 \tag{2.43}$$

$$\frac{\partial V_0^{(\tau)}}{\partial z} - i\xi V_0^{(z)}\Big|_{z=\xi_0} = 0, \qquad V_0^{(z)} \to 0\Big|_{z\to\infty} \tag{2.44}$$

$$(\lambda + 2\mu)\frac{\partial V_0^{(z)}}{\partial z} - i\xi\lambda V_0^{(\tau)}\Big|_{z=\xi_0} = 0, \qquad V_0^{(\tau)} \to 0\Big|_{z\to\infty} \tag{2.45}$$

$$\left.\frac{\partial V_0^{(\varphi)}}{\partial z}\right|_{z=\xi_0} = 0, \qquad V_0^{(\varphi)} \to 0\Big|_{z\to\infty}. \tag{2.46}$$

Here, as before, the components of V_0 have the following meanings: $V_0^{(\tau)}$, $V_0^{(\varphi)}$ are the horizontal components along the direction of propagation (along the vector **k**) and that normal to it, $V_0^{(z)}$ is the vertical component. Consequently, Equations (2.41)–(2.43) and the boundary conditions (2.44)–(2.46) are for the components of the equation of motion and boundary conditions along the wave vector, along the vertical and normal to the wave vector in the horizontal plane.

Similarly to the harmonic wave case, this boundary-value problem splits into two, corresponding to Rayleigh and Love waves. Each of the two boundary-value problems defines its own eigenvalues

$$\omega = \omega_k(\xi, x, y) \tag{2.47}$$

and the corresponding eigenfunctions ($V_0^{(\varphi)}(x, y, z)$ for Love waves and $V_0^{(z)}(x, y, z)$, $V_0^{(\tau)}(x, y, z)$ for Rayleigh waves), which turn out to be the same as for a laterally homogeneous medium in which that the vertical distribution of elastic parameters is identical with the vertical cross-section at a given point x, y. Below we confine ourselves to considering a single mode, so will omit the subscript k when writing eigenvalues and eigenfunctions.

The phase function $\Theta(x, y, t)$ can most conveniently be found by the method of characteristics (note that in the harmonic case the method of characteristics when used to determine $\tau(x, y)$ leads to the ray equations (2.12)). With this end in view, the dispersion equation (2.47) is represented as a Hamilton—Jacobi equation

$$\frac{\partial\Theta}{\partial t} - \omega(x, y, \xi_x, \xi_y) = 0 \tag{2.48}$$

where the dependence of ω on x, y, $\xi_x = -\mathrm{d}\Theta/\mathrm{d}x$, $\xi_y = -\mathrm{d}\Theta/\mathrm{d}y$ is defined by the relevant boundary-value problem.

The characteristics of (2.48) are the curves $x = x(s)$, $y = y(s)$, $t = t(s)$ in the x, y, t space. They are determined from the canonical system corresponding to the Hamilton—Jacobi equation (Babich and Grigoryeva, 1980):

$$\begin{gathered}\frac{\mathrm{d}t}{\mathrm{d}s} = 1, \quad \frac{\mathrm{d}x}{\mathrm{d}s} = \frac{\partial\omega}{\partial\xi_x}, \quad \frac{\mathrm{d}y}{\mathrm{d}s} = \frac{\partial\omega}{\partial\xi_y},\\ \frac{\mathrm{d}\xi_x}{\mathrm{d}s} = \frac{\partial\omega}{\partial x}, \quad \frac{\mathrm{d}\xi_y}{\mathrm{d}s} = \frac{\partial\omega}{\partial y}, \quad \frac{\partial\omega}{\partial s} = 0.\end{gathered} \tag{2.49}$$

These curves are called space-time rays (as distinguished from ordinary rays found by the ray method in a real physical space, either x, y or x, y, z).

The vector $\mathbf{U} = \{\mathrm{d}\omega/\mathrm{d}\xi_x, \mathrm{d}\omega/\mathrm{d}\xi_y\}$ is called the group velocity vector. In our case, as one can see from (2.47), ω does not depend separately on ξ_x, ξ_y, but on the absolute value of the wave vector $\xi = \sqrt{\xi_x^2 + \xi_y^2}$. Hence

$$\begin{gathered}U_x = \frac{\mathrm{d}\omega}{\mathrm{d}\xi}\frac{\mathrm{d}\xi}{\mathrm{d}\xi} = U\frac{\xi_x}{\xi}, \qquad U_y = U\frac{\xi_y}{\xi}, \quad \text{so that}\\ \mathbf{U} = U\mathbf{e}_\tau.\end{gathered} \tag{2.50}$$

We shall now determine the derivatives $\mathrm{d}\omega/\mathrm{d}x$, $\mathrm{d}\omega/\mathrm{d}y$. Differentiating the relation $\omega(x, y, z) = \xi C(\omega, x, y)$ with respect to x, the value of ξ being kept constant, we get

$$\frac{\mathrm{d}\omega}{\mathrm{d}x} = \xi\frac{\partial C}{\partial x} + \xi\frac{\partial C}{\partial\omega}\frac{\partial\omega}{\partial x}$$

whence

$$\frac{\mathrm{d}\omega}{\mathrm{d}x} = \left(\xi\frac{\partial C}{\partial x}\right)\bigg/\left(1 - \frac{\omega}{C}\frac{\partial C}{\partial\omega}\right).$$

Taking into account the definition of group velocity (2.35) we have

$$\frac{\mathrm{d}\omega}{\mathrm{d}x} = \frac{\xi U}{C}\frac{\partial C}{\partial x},$$

and similarly

$$\frac{\mathrm{d}\omega}{\mathrm{d}y} = \frac{\xi U}{C}\frac{\partial C}{\partial y},$$

so that Equations (2.49) can now be written as

$$\frac{d\mathbf{r}}{ds} = \mathbf{U}, \quad \frac{dt}{ds} = 1, \quad \frac{d\boldsymbol{\xi}}{ds} = -U\xi\nabla C/C, \quad \frac{d\omega}{ds} = 0, \tag{2.51}$$

where $\mathbf{r} = (x, y)$.

From (2.51) one can see that ω remains constant along a space-time ray, so that each ray is characterized by a definite value of ω.

Let us express the wave vector in terms of the slowness vector, $\mathbf{k} = C^{-1}\mathbf{e}_\tau$ bearing in mind that $ds = dt$. We then arrive at equations that determine the path in the x, y-plane along which a harmonic wavepacket component of frequency ω propagates:

$$\frac{d\mathbf{r}}{dt} = UC\mathbf{k}, \qquad \frac{d\mathbf{k}}{dt} = -U\nabla C/C^2. \tag{2.52}$$

Comparison of (2.52) and (2.12) shows that a wavepacket component travels along the same path as a harmonic wave of the same frequency does, but at group velocity U, not at phase velocity C.

To solve (2.52) one must have some initial values. For propagation from a point source these are $x = y = 0$ at $t = 0$ and $k_x = k_x^0$, $k_y = k_y^0$. The last two conditions imply fixed frequency and take-off azimuth at the source φ, since $k_x^0 = \cos\varphi/C_0(\omega)$, $k_y^0 = \sin\varphi/C_0(\omega)$. When (2.52) is integrated from some surface rather than from the source, for example, from an interface that gives rise to reflected and transmitted surface waves, the initial values for computation of a reflected (transmitted) ray are $x = x_0$, $y = y_0$, $k_x = k_x^0$, $k_y = k_y^0$ at $t = t_0$, where x_0, y_0 is a point at the interface; k_x^0, k_y^0 are components of $\mathbf{k}$ for the relevant wave, to be found from the boundary condition for the ray. The boundary conditions for a ray will be discussed in Section 2.5.

We now discuss the variation of wavepacket amplitude during propagation. To do this, as in the case of a harmonic wave, one should examine the next term in the equations of motion and boundary conditions. Rayleigh and Love waves should be treated separately, exactly as has been done in the preceding subsection.

Love wave. Analogously to the above, the equation that defines the dependence of $V_0^{(\varphi)}$ on the horizontal coordinates can be found by examining only that horizontal component in the equation and boundary condition which is orthogonal to the wave vector. The resulting boundary-value problem for the component $V_1^{(\varphi)}$ has the form

$$\varepsilon l_3(V_1^{(\varphi)}) - i\left[2\rho\omega\frac{\partial V_0^{(\varphi)}}{\partial t} + \rho\frac{\partial\omega}{\partial t}V_0^{(\varphi)} + \mu V_0^{(\varphi)}\operatorname{div}(\omega\mathbf{k}) + \right.$$
$$\left. + 2\mu\omega(\nabla_\perp V_0^{(\varphi)}, \mathbf{k}) + \omega(\nabla_\perp\mu, \mathbf{k})V_0^{(\varphi)}\right] = 0 \tag{2.53}$$

$$\varepsilon\frac{\partial V_1^{(\varphi)}}{\partial z} - i(\nabla\zeta_0, \omega\mathbf{k})V_0^{(\varphi)}|_{z=\zeta_0} = 0$$
$$V_1^{(\varphi)} \to \infty, \qquad z \to \infty \tag{2.54}$$

where $\nabla_\perp = (\mathbf{e}_x\partial/\partial x + \mathbf{e}_y\partial/\partial y)$ is the horizontal gradient.

A comparison of (2.53) and (2.15) readily shows that the last three terms within the square brackets in (2.53) exactly correspond to the second term in (2.15). The first two terms have emerged, because in the transient case, as has been mentioned above, both frequency and amplitude are functions of time. They are slowly varying in time, the variation being of the same order as that of amplitude and elastic parameters with distance, i.e., $0(\varepsilon)$. Thus, the expression enclosed in square brackets in (2.53) is indeed of the same order as the first term, i.e., $0(\varepsilon)$. As to the boundary conditions, (2.54) and (2.16) are completely identical.

Just as has been the case for harmonic Love waves, the condition of solvability for the inhomogeneous boundary-value problem (2.53), (2.54) can be found by multiplying (2.53) by $V_0^{(\varphi)}$ and integrating the result over z between the limits ζ_0 and ∞, taking into account the boundary condition (2.54) and equation (2.43), (2.46) for $V_0^{(\varphi)}$. As (2.53) includes terms that involve time derivatives, the result will of course be different from that obtained for a harmonic wave. The equation for $V_0^{(\varphi)}$ is

$$-\mu(V_0^{(\varphi)})^2(\nabla\zeta_0, \omega\mathbf{k}) + \operatorname{div}\left[\omega\mathbf{k}\int_{\zeta_0}^{\infty}\mu(V_0^{(\varphi)})^2\,dz\right] +$$

$$+ 2\omega\int_{\zeta_0}^{\infty}\mu(\nabla_\perp V_0^{(\varphi)}, \omega\mathbf{k})V_0^{(\varphi)}\,dz + \omega\int_{\zeta_0}^{\infty}(\nabla_\perp\mu, \mathbf{k})(V_0^{(\varphi)})^2\,dz +$$

$$+ 2\omega\int_{\zeta_0}^{\infty}\rho V_0^{(\varphi)}\frac{\partial V_0^{(\varphi)}}{\partial t}\,dz + \frac{\partial\omega}{\partial t}\int_{\zeta_0}^{\infty}\rho(V_0^{(\varphi)})^2\,dz = 0. \qquad (2.55)$$

This can be written, on analogy with (2.18), in the form

$$\operatorname{div}\left[\omega\mathbf{k}\int_{\zeta_0}^{\infty}\mu(V_0^{(\varphi)})^2\,dz\right] + \frac{\partial}{\partial t}\left(\omega\int_{\zeta_0}^{\infty}\rho(V_0^{(\varphi)})^2\,dz\right) = 0 \qquad (2.56)$$

the divergence being defined in the x, y space, i.e.,

$$\operatorname{div}\mathbf{a} = \frac{\partial a_x}{\partial x} + \frac{\partial a_y}{\partial y}.$$

The quantity $\rho\omega(V_0^{(\varphi)})^2/2$ is the density of energy in unit time at point (x, y, z). The integral of this over z can be regarded as the density of surface wave energy at (x, y). Denote it as S and recall the expression (1.45) for group velocity and (2.50) for group velocity vector. One can then rewrite (2.56) as

$$\frac{\partial}{\partial t}\left(\frac{S}{\omega}\right) + \operatorname{div}\left(\frac{S\mathbf{U}}{\omega}\right) = 0 \qquad (2.57)$$

$S\mathbf{U}$ is the energy flux density vector. Since the frequency remains constant along a space-time ray, one gets

$$\frac{\partial}{\partial t}\left(\frac{1}{\omega}\right) + \left(\nabla_\perp\left(\frac{1}{\omega}\right), \mathbf{U}\right) = \frac{d}{ds}\left(\frac{1}{\omega}\right) = 0.$$

Consequently, (2.57) may be expressed as a law of conservation of energy

$$\frac{\mathrm{d}S}{\mathrm{d}s} = \frac{\partial S}{\partial t} + \operatorname{div}(S\mathbf{U}) = 0. \tag{2.58}$$

This equation is completely identical in form with the well-known equation of continuity in fluid mechanics

$$\frac{\partial \rho}{\partial t} + \operatorname{div}(\rho \mathbf{U}) = 0$$

which defines the change in fluid density along lines of flow and expresses the law of conservation of mass within a volume. Quite similarly, (2.58) expresses the law of conservation of energy and yields the variation in wave energy density along a space-time ray.

As $\operatorname{div}(S\mathbf{U}) = (\nabla_{\perp} S, \mathbf{U}) + S \operatorname{div} \mathbf{U}$, (2.58) can be rewritten in the form

$$\frac{\mathrm{d}S}{\mathrm{d}s} + S \operatorname{div} \mathbf{U} = 0. \tag{2.59}$$

We proceed to find $\operatorname{div} \mathbf{U} = \partial U_x/\partial x + \partial U_y/\partial y$. First, we notice that it is equal to the divergence of $\mathbf{w} = (U_x, U_y, 1)$ in the three-dimensional x, y, t space. We shall distinguish the three-dimensional divergence by writing it as Div. By definition

$$\operatorname{Div} \mathbf{w} = \lim_{\Omega \to 0} \frac{1}{\Omega} \iint_{\Sigma_\Omega} w_n \, \mathrm{d}S \tag{2.60}$$

where Σ_Ω is the surface that encloses the volume Ω. This volume is chosen to be that portion of a ray tube (formed by space-time rays) bounded by the planes t = constant and $t + \mathrm{d}t$ = constant. The integral over its lateral surface vanishes, since $\mathbf{w}$ is tangent to a space-time ray. $|w_n| = 1$ at the top and bottom of the tube, because the group velocity vector $\mathbf{U}$ lies in the x, y-plane. Equation (2.60) is thus equal to

$$\operatorname{Div} \mathbf{w} = \frac{1}{\mathrm{d}\Omega} [\mathrm{d}\Sigma_1 - \mathrm{d}\Sigma_2] \tag{2.61}$$

where $\mathrm{d}\Sigma_1$, $\mathrm{d}\Sigma_2$ are the areas of the top and bottom of an element of the tube and $\mathrm{d}\Omega$ is its volume.

We now define the geometrical spreading of a space-time ray tube. With this end in view, we introduce the ray coordinates: two that define a ray in the (x, y, t)-space and the time $t = s$ defining a point along a ray. As has been pointed out, ω remains constant along a space-time ray, so it is natural to choose ω as one of the ray coordinates and the take-off azimuth at the source as the other (denoted by φ). If a ray tube is bounded by the rays φ, $\varphi + \mathrm{d}\varphi$ and ω, $\omega + \mathrm{d}\omega$, then the area that a plane of t = constant cuts within the ray is $J \, \mathrm{d}\varphi \, \mathrm{d}\omega$, where

$$J = \frac{D(x, y)}{D(x, y)} = \left(\frac{\partial x}{\partial \varphi} \frac{\partial y}{\partial \omega} - \frac{\partial x}{\partial \omega} \frac{\partial y}{\partial \varphi} \right) \tag{2.62}$$

is the Jacobian transforming x, y to the ray coordinates φ, ω. This can be written as

$$J = \left| \frac{\partial \mathbf{r}}{\partial \varphi} \times \frac{\partial \mathbf{r}}{\partial \omega} \right|. \tag{2.63}$$

Now consider (2.61). Obviously, the volume of an element of the tube is $J\,\mathrm{d}t$, and the difference in area between the top and the bottom of an element of tube is $\mathrm{d}\Sigma_1 - \mathrm{d}\Sigma_2 = (\mathrm{d}J/\mathrm{d}s)\,\mathrm{d}t$. Hence

$$\operatorname{Div} \mathbf{U} = \operatorname{div} \mathbf{w} = \frac{1}{J}\,\frac{\mathrm{d}J}{\mathrm{d}s} \tag{2.64}$$

so that (2.59) finally becomes

$$\frac{1}{J}\,\frac{\mathrm{d}}{\mathrm{d}s}(SJ) = 0. \tag{2.65}$$

Since $S = \int_{\zeta_0}^{\infty} \rho(V_0^{(\varphi)}(z))^2\,\mathrm{d}z$, the expression for Love wave amplitude can be written as

$$V_0^{(\varphi)}(x, y, z, t) = \frac{K_{\mathrm{L}}(\varphi, \omega)\, V^{(3)}(z - \zeta_0(x, y))}{\sqrt{J}\sqrt{\int_0^{\infty} \rho[V^{(3)}(z)]^2\,\mathrm{d}z}} \tag{2.66}$$

where $K_{\mathrm{L}}(\varphi, \omega)$ is a factor that depends on the source spectrum and radiation pattern.

Rayleigh wave. The equations that determine Rayleigh wave amplitude in the zero-order approximation can be derived quite similarly to the Love wave case. The following equations and boundary conditions are obtained for $V_1^{(z)}$ and $V_1^{(\tau)}$:

$$\begin{aligned}
&\varepsilon l_1(V_1^{(z)}, V_1^{(\tau)}) - i\Bigg[2\rho\omega\,\frac{\partial V_0^{(\tau)}}{\partial t} + \rho\,\frac{\partial \omega}{\partial t}\,V_0^{(\tau)} + \\
&\quad + 2\omega(\lambda + 2\mu)(\nabla_{\perp} V_0^{(\tau)}, \mathbf{k}) + (\lambda + 2\mu) V_0^{(\tau)} \operatorname{div}(\omega\mathbf{k}) + \\
&\quad + V_0^{(\tau)}\omega(\mathbf{k}, \nabla_{\perp}(\lambda + 2\mu))\Bigg] + C\Bigg[\frac{\partial}{\partial z}(\mu\nabla_{\perp} V_0^{(z)}, \mathbf{k}) + \\
&\quad + \left(\nabla_{\perp}\left(\lambda\,\frac{\partial V_0^{(z)}}{\partial z}\right), \mathbf{k}\right)\Bigg] = 0
\end{aligned} \tag{2.67}$$

$$\varepsilon l_2(V_1^{(z)}, V_1^{(\tau)}) - i\left[2\rho\omega\,\frac{\partial V_0^{(z)}}{\partial t} + \rho\,\frac{\partial \omega}{\partial t}\,V_0^{(z)} + \right.$$

$$\left. + 2\omega\mu(\nabla_\perp V_0^{(z)}, \mathbf{k}) + \mu V_0^{(z)}\,\mathrm{div}(\omega\mathbf{k}) + V_0^{(z)}\omega(\mathbf{k}, \nabla_\perp\mu)\right] +$$

$$+ \frac{\partial}{\partial z}(\lambda\,\mathrm{div}(V_0^{(\tau)}C\mathbf{k})) + \mathrm{div}\left(\mu\,\frac{\partial V_0^{(\tau)}}{\partial \mathbf{z}}\cdot C\mathbf{k}\right) = 0, \tag{2.68}$$

$$\mu\left[\frac{\partial V_1^{(\tau)}}{\partial z} - i\,\frac{\omega}{C}\,V_1^{(z)} + (\nabla_\perp V_0^{(z)}, C\mathbf{k})\right] -$$

$$- (\nabla_\perp \zeta_0, \mathbf{k})\left[\lambda C\,\frac{\partial V_0^{(z)}}{\partial z} - i\omega(\lambda + 2\mu)V_0^{(\tau)}\right] = 0 \quad \text{at} \quad z = \zeta_0 \tag{2.69}$$

$$(\lambda + 2\mu)\,\frac{\partial V_1^{(z)}}{\partial z} - i\lambda\,\frac{\omega}{C}\,V_1^{(\tau)} + \lambda\,\mathrm{div}(V_0^{(\tau)}C\mathbf{k}) -$$

$$- (\nabla_\perp \zeta_0, \mathbf{k})\left(\mu C\,\frac{\partial V_0^{(\tau)}}{\partial z} - i\omega\mu V_0^{(\tau)}\right) = 0 \quad \text{at} \quad z = \zeta_0 \tag{2.70}$$

$$V_1^{(z)} \to 0, \quad V_1^{(\tau)} \to 0, \quad z \to \infty. \tag{2.71}$$

As in the preceding subsection, the condition of solvability for this boundary-value problem is obtained by multiplying the first equation by $V_0^{(\tau)*}$ and the second by $V_0^{(z)*}$, integrating over z from ζ_0 to ∞, adding the results and using the boundary conditions (2.69)—(2.71), as well as (2.41), (2.42), (2.44), (2.45). We obtain

$$\frac{\partial}{\partial t}\left[\omega\int_{\zeta_0}^{\infty}\rho(|V_0^{(z)}|^2 + |V_0^{(\tau)}|^2)\,\mathrm{d}z\right] + \mathrm{div}\{\mathbf{k}[\omega(I_R^{(1)} + I_R^{(2)}) +$$

$$+ C(I_R^{(3)} + I_R^{(4)})]\} = 0. \tag{2.72}$$

Denoting the Rayleigh wave energy density at (x, y) by S

$$S \equiv \frac{\omega^2}{2}\int_{\zeta_0}^{\infty}\rho(|V_0^{(z)}|^2 + |V_0^{(\tau)}|^2)\,\mathrm{d}z \tag{2.73}$$

and using the integral expression for group velocity (1.45), one easily obtains that (2.72) can be written in the form (2.57) or (2.58). The subsequent argument proceeds on the lines of Love wave derivation. The final expression describing the variation of displacement vector in a Rayleigh wave along a space-time ray has a form similar to (2.66):

$$(V^{(z)}, V^{(\tau)}) = \frac{K_R(\varphi, \omega)\,(V^{(1)}(z-\zeta_0), iV^{(2)}(z-\zeta_0))}{\sqrt{J}\left\{\int_0^\infty \rho[(V^{(1)}(z))^2 + (V^{(2)}(z)]^2\right\}} \tag{2.74}$$

where $V^{(1)}$, $V^{(2)}$ are components of the vector eigenfunction for Rayleigh waves in the 1-D case corresponding to the distribution of elastic constants over depth at point (x, y).

2.4. CALCULATION OF GEOMETRICAL SPREADING FOR SPACE-TIME RAYS

The above method for calculating the geometrical spreading of space-time rays has much in common with that for space rays of non-dispersive body waves (Popov and Pšenčik, 1978; Azbel, Dmitrieva and Yanovskaya, 1980). The geometrical spreading J is given by (2.63) in Section 2.3. As in the ordinary ray method, the vectors $\mathbf{Q}^{\varphi} = \partial\mathbf{r}/\partial\varphi$ and $\mathbf{Q}^{\omega} = \partial\mathbf{r}/\partial\omega$ are found by integrating the relevant set of differential equations numerically along a ray. To derive the equations, we define two additional vectors $\mathbf{P}^{\varphi} = \partial\mathbf{k}/\partial\varphi$ and $\mathbf{P}^{\omega} = \partial\mathbf{k}/\partial\omega$.

We differentiate the ray equations (2.52) separately with respect to φ and ω:

$$\begin{aligned}
\frac{d\mathbf{Q}^{\varphi}}{dt} &= \left(C\frac{\partial U}{\partial\varphi} + U\frac{\partial C}{\partial\varphi}\right)\mathbf{k} + UC\mathbf{P}^{\varphi} \\
\frac{d\mathbf{P}^{\varphi}}{dt} &= \frac{\partial U}{\partial\varphi}\nabla\left(\frac{1}{C}\right) + U\frac{\partial}{\partial\varphi}\left(\nabla\frac{1}{C}\right)
\end{aligned} \tag{2.75}$$

$$\begin{aligned}
\frac{d\mathbf{Q}^{\omega}}{dt} &= \left(C\frac{\partial U}{\partial\omega} + U\frac{\partial C}{\partial\omega}\right)\mathbf{k} + UC\mathbf{P}^{\omega} \\
\frac{d\mathbf{P}^{\omega}}{dt} &= \frac{\partial U}{\partial\omega}\nabla\left(\frac{1}{C}\right) + U\frac{\partial}{\partial\omega}\left(\nabla\frac{1}{C}\right).
\end{aligned} \tag{2.76}$$

We then define a set of coordinates in the x, y-plane related to the ray: the q_2-axis will be along the ray and q_1 normal to it. Since the vectors $\mathbf{r} = \{x, y\}$ and $\mathbf{k} = \{k_x, k_y\}$ lie in the x, y-plane, then $\mathbf{Q}^{\varphi(\omega)}$ and $\mathbf{P}^{\varphi(\omega)}$ are obviously in the same plane too. Denote

$$\begin{aligned}
\mathbf{Q}^{\varphi(\omega)} &= Q_1^{\varphi(\omega)}\mathbf{e}_1 + Q_2^{\varphi(\omega)}\mathbf{e}_2 \\
\mathbf{P}^{\varphi(\omega)} &= P_1^{\varphi(\omega)}\mathbf{e}_1 + P_2^{\varphi(\omega)}\mathbf{e}_2
\end{aligned} \tag{2.77}$$

where $\mathbf{e}_1$ and $\mathbf{e}_2$ are the unit vectors along the q_1 and q_2 axes.

We shall now set up a set of equations for the components of **Q**, **P** (the superscripts $\varphi(\omega)$ are omitted for the moment). We recall that

$$\begin{aligned} Q_1 &= (\mathbf{Q}, \mathbf{e}_1), \qquad Q_2 = (\mathbf{Q}, \mathbf{e}_2) \\ \mathbf{P}_1 &= (\mathbf{P}, \mathbf{e}_1), \qquad \mathbf{P}_2 = (\mathbf{P}, \mathbf{e}_2). \end{aligned} \tag{2.78}$$

The relation $(\mathbf{k}, \mathbf{e}_1) = 0$ readily yields the derivatives of $\mathbf{e}_1$ and $\mathbf{e}_2$ with respect to t, i.e., along the ray:

$$\begin{aligned} &\left(\frac{\mathrm{d}\mathbf{k}}{\mathrm{d}t}, \mathbf{e}_1\right) + \left(\mathbf{k}, \frac{\mathrm{d}\mathbf{e}_1}{\mathrm{d}t}\right) = 0, \qquad \frac{\mathrm{d}\mathbf{k}}{\mathrm{d}t} = -\frac{U}{C^2} \nabla C \\ &\frac{\mathrm{d}\mathbf{e}_1}{\mathrm{d}t} = \frac{U}{C} \frac{\partial C}{\partial q_1} \mathbf{e}_2, \qquad \frac{\mathrm{d}\mathbf{e}_2}{\mathrm{d}t} = -\frac{U}{C} \frac{\partial C}{\partial q_1} \mathbf{e}_1. \end{aligned} \tag{2.79}$$

We now differentiate (2.78) with respect to t and use (2.75), (2.76), (2.79):

$$\frac{\mathrm{d}Q_1}{\mathrm{d}t} = UCP_1 + \frac{U}{C} \frac{\partial C}{\partial q_1} Q_2 \tag{2.80a}$$

$$\frac{\mathrm{d}Q_2}{\mathrm{d}t} = \left(C \frac{\partial U}{\partial \beta} + U \frac{\partial C}{\partial \beta}\right) \frac{1}{C} + UCP_2 - \frac{U}{C} \frac{\partial C}{\partial q_1} Q_1 \tag{2.80b}$$

$$\frac{\mathrm{d}P_1}{\mathrm{d}t} = -\frac{\partial}{\partial \beta}\left(\frac{U}{C^2}\right) \frac{\partial C}{\partial q_1} - \frac{U}{C^2}\left(\frac{\partial \nabla C}{\partial \beta}, \mathbf{e}_1\right) + P_2 \frac{\partial C}{\partial q_1} \frac{U}{C} \tag{2.80c}$$

$$\frac{\mathrm{d}P_2}{\mathrm{d}t} = -\frac{\partial}{\partial \beta}\left(\frac{U}{C^2}\right) \frac{\partial C}{\partial q_2} - \frac{U}{C^2}\left(\frac{\partial \nabla C}{\partial \beta}, \mathbf{e}_2\right) + P_1 \frac{\partial C}{\partial q_2} \frac{U}{C}. \tag{2.80d}$$

The symbol $\partial/\partial\beta$ denotes differentiation with respect either to φ or to ω. First, consider the set of equations for $\mathbf{P}^{\varphi}$, $\mathbf{Q}^{\varphi}$. Note that a derivative with respect to φ can be written as

$$\frac{\partial}{\partial \varphi} = Q_1 \frac{\partial}{\partial q_1} + Q_2 \frac{\partial}{\partial q_2}.$$

Equations (2.80a) and (2.80b) will then be

$$\frac{\mathrm{d}Q_1^{\varphi}}{\mathrm{d}t} = UCP_1^{\varphi} + \frac{U}{C} \frac{\partial C}{\partial q_1} Q_2^{\varphi} \tag{2.81a}$$

$$\frac{\mathrm{d}Q_2^{\varphi}}{\mathrm{d}t} = Q_1^{\varphi} \frac{\partial U}{\partial q_1} + Q_2^{\varphi}\left(\frac{\partial U}{\partial q_2} + \frac{U}{C} \frac{\partial C}{\partial q_2}\right) + UCP_2^{\varphi}. \tag{2.81b}$$

The other two equations are transformed by substituting

$$\frac{\partial \nabla C}{\partial \varphi} = Q_1^\varphi \frac{\partial \nabla C}{\partial q_1} + Q_2^\varphi \frac{\partial \nabla C}{\partial q_2}$$

$$\left(\frac{\partial \nabla C}{\partial \varphi}, \mathbf{e}_1\right) = Q_1^\varphi \frac{\partial^2 \nabla C}{\partial q_1^2} + Q_2^\varphi \frac{\partial^2 \nabla C}{\partial q_1 \partial q_2}$$

$$\left(\frac{\partial \nabla C}{\partial \varphi}, \mathbf{e}_2\right) = Q_1^\varphi \frac{\partial^2 \nabla C}{\partial q_1 \partial q_2} + Q_2^\varphi \frac{\partial^2 \nabla C}{\partial q_2^2}$$

into them, yielding

$$\frac{\mathrm{d}P_1^\varphi}{\mathrm{d}t} = Q_1^\varphi \frac{\partial}{\partial q_1}\left(U \frac{\partial}{\partial q_1}\left(\frac{1}{C}\right)\right) + Q_2^\varphi \frac{\partial}{\partial q_2}\left(U \frac{\partial}{\partial q_1}\left(\frac{1}{C}\right)\right) + {}$$

$$+ \frac{U}{C} \cdot \frac{\partial C}{\partial q_1} P_2^\varphi \tag{2.81c}$$

$$\frac{\mathrm{d}P_2^\varphi}{\mathrm{d}t} = Q_1^\varphi \frac{\partial}{\partial q_1}\left(U \frac{\partial}{\partial q_2}\left(\frac{1}{C}\right)\right) + Q_2^\varphi \frac{\partial}{\partial q_2}\left(U \frac{\partial}{\partial q_2}\left(\frac{1}{C}\right)\right) + {}$$

$$+ \frac{U}{C} \frac{\partial C}{\partial q_1} P_1^\varphi. \tag{2.81d}$$

Equations (2.81a—d) form a set of equations in the unknowns Q_1^φ, Q_2^φ, P_1^φ, P_2^φ. It can however be shown that one of the equations is superfluous, since there is a relation connecting P_2^φ, Q_1^φ, Q_2^φ similar to that available for ordinary spatial rays (Azbel *et al.*, 1980)

$$P_2 = \frac{1}{C^2} \frac{\partial C}{\partial \varphi} = \frac{1}{C^2}\left(Q_1 \frac{\partial C}{\partial q_1} + Q_2 \frac{\partial C}{\partial q}\right) \tag{2.82}$$

ω remains constant in the differentiation with respect to φ, i.e., $\mathrm{d}\omega/\mathrm{d}\varphi = 0$. Since $\omega = \omega(\xi, x, y)$, we have

$$\frac{\partial \omega}{\partial \varphi} = \frac{\partial \omega}{\partial \xi} \frac{\partial \xi}{\partial \varphi} + \left(\frac{\partial \omega}{\partial x} \frac{\partial x}{\partial \varphi} + \frac{\partial \omega}{\partial y} \frac{\partial y}{\partial \varphi}\right) = 0. \tag{2.83}$$

As $\mathrm{d}\omega/\mathrm{d}\xi = U$, $\mathrm{d}\xi/\mathrm{d}\varphi = \omega \mathrm{d}k/\mathrm{d}\varphi = \omega P_2$, and $\omega = \xi C(\omega, x, y)$, we get

$$U\omega P_2^\varphi + \omega k \left(\frac{U}{C} \frac{\partial C}{\partial x} \frac{\partial x}{\partial \varphi} + \frac{U}{C} \frac{\partial C}{\partial y} \frac{\partial y}{\partial \varphi}\right) = 0$$

or

$$P_2^\varphi = \frac{1}{C^2} \frac{\partial C}{\partial \varphi} = -\frac{1}{C^2}\left(Q_1^\varphi \frac{\partial C}{\partial q_1} + Q_2^\varphi \frac{\partial C}{\partial q_2}\right). \tag{2.84}$$

Substitution of (2.84) into (2.81b), (2.81c) yields a set of equations for determining $Q_1^\varphi, Q_2^\varphi, P_1^\varphi$

$$\frac{dQ_1^\varphi}{dt} = UCP_1^\varphi + \frac{U}{C}\frac{\partial C}{\partial q_1} Q_2^\varphi$$

$$\frac{dQ_2^\varphi}{dt} = Q_1^\varphi C \frac{\partial}{\partial q_1}\left(\frac{U}{C}\right) + Q_2^\varphi \frac{\partial U}{\partial q_2} \tag{2.85}$$

$$\frac{dP_1^\varphi}{dt} = -Q_1^\varphi\left[\frac{1}{C}\frac{\partial}{\partial q_1}\left(\frac{U}{C}\frac{\partial C}{\partial q_1}\right)\right] -$$

$$- Q_2^\varphi\left[\frac{1}{C}\frac{\partial}{\partial q_2}\left(\frac{U}{C}\frac{\partial C}{\partial q_1}\right)\right]$$

to be solved under the following initial conditions at the source:

$$Q_1^\varphi = Q_2^\varphi = 0, \qquad P_1^\varphi = C_0^{-1}. \tag{2.86}$$

We now proceed to construct a set of equations to determine $\mathbf{Q}^\omega$ and $\mathbf{P}^\omega$. The right-hand side of these equations will differ from those for $\mathbf{Q}^\varphi$, $\mathbf{P}^\varphi$ because C, U and ∇C will change from a ray for ω to that for $\omega + d\omega$, not only owing to the passage to the point $x + dx$, $y + dy$, but also to the fact that velocity is directly dependent on the frequency. Equations (2.76) then become:

$$\frac{dQ_1^\omega}{dt} = UCP_1^\omega + \frac{U}{C}\frac{\partial C}{\partial q_1} Q_2^\omega \tag{2.87a}$$

$$\frac{dQ_2^\omega}{dt} = Q_1^\omega \frac{\partial U}{\partial q_1} + Q_2^\omega\left(\frac{\partial U}{\partial q_2} + \frac{U}{C}\frac{\partial C}{\partial q_2}\right) +$$

$$+ UCP_2^\omega + \frac{\partial U}{\partial \omega} + \frac{U}{C}\frac{\partial C}{\partial \omega} \tag{2.87b}$$

$$\frac{dP_1^\omega}{dt} = Q_1^\omega \frac{\partial}{\partial q_1}\left(U\frac{\partial}{\partial q_1}\left(\frac{1}{C}\right)\right) + Q_2^\omega \frac{\partial}{\partial q_2}\left(U\frac{\partial}{\partial q_1}\left(\frac{1}{C}\right)\right) +$$

$$+ \frac{U}{C}\frac{\partial C}{\partial q_1} P_2^\omega - \frac{1}{C^2}\frac{\partial C}{\partial q_1}\frac{\partial U}{\partial \omega} -$$

$$- \frac{U}{C^2}\left(\frac{\partial \nabla C}{\partial \omega}, \mathbf{e}_1\right) + \frac{2U}{C^3}\frac{\partial C}{\partial \omega}\frac{\partial C}{\partial q_1} \tag{2.87c}$$

$$\frac{\mathrm{d}P_2^{\omega}}{\mathrm{d}t} = Q_1^{\omega}\frac{\partial}{\partial q_1}\left(U\frac{\partial}{\partial q_2}\left(\frac{1}{C}\right)\right) + Q_2^{\omega}\frac{\partial}{\partial q_2}\left(U\frac{\partial}{\partial q_2}\left(\frac{1}{C}\right)\right) -$$

$$-\frac{U}{C}\frac{\partial C}{\partial q_1}P_1^{\omega} - \frac{1}{C^2}\frac{\partial C}{\partial q_2}\frac{\partial U}{\partial \omega} -$$

$$-\frac{U}{C^2}\left(\frac{\partial \nabla C}{\partial \omega}, \mathbf{e}_2\right) + \frac{2U}{C^3}\frac{\partial C}{\partial \omega}\frac{\partial C}{\partial q_1}. \tag{2.87d}$$

Then P_2^{ω} too can be expressed in terms of Q_1^{ω}, Q_2^{ω}.

Differentiation of $\omega(\xi, x, y)$ with respect to ω yields

$$\frac{\partial \omega}{\partial \xi}\frac{\partial \xi}{\partial \omega} + \frac{\xi U}{C}\frac{\partial C}{\partial \omega'} = 1 \tag{2.88}$$

where $\partial/\partial\omega'$ means the differentiation with respect to ω that is associated with the change in coordinates only:

$$\frac{\partial}{\partial \omega'} = Q_1^{\omega}\frac{\partial}{\partial q_1} + Q_1^{\omega}\frac{\partial}{\partial q_2}.$$

Since $\xi = \omega k$, (2.88) can be rewritten as

$$U(\omega P_2^{\omega} + k) + \frac{U}{C}\omega k\frac{\partial C}{\partial \omega'} = 1 \tag{2.89}$$

or

$$UP_2^{\omega} + \frac{U}{C^2}\frac{\partial C}{\partial \omega} = -\frac{U}{C^2}\frac{\partial C}{\partial \omega'}. \tag{2.90}$$

We thus obtain an expression for P_2^{ω} similar to (2.84):

$$P_2^{\omega} = -\frac{1}{C^2}\left(Q_1^{\omega}\frac{\partial C}{\partial q_1} + Q_2^{\omega}\frac{\partial C}{\partial q_2}\right) - \frac{1}{C^2}\frac{\partial C}{\partial \omega}. \tag{2.91}$$

Substitution of (2.91) in (2.87b), (2.87c) yields the desired result, a set of three equations in Q_1^{ω}, Q_2^{ω}, P_1^{ω}

$$\frac{\mathrm{d}Q_1^{\omega}}{\mathrm{d}t} = UCP_1^{\omega} + \frac{U}{C}\frac{\partial C}{\partial q_1}Q_2^{\omega} \tag{2.92a}$$

$$\frac{\mathrm{d}Q_2^{\omega}}{\mathrm{d}t} = Q_1^{\omega}C\frac{\partial}{\partial q_1}\left(\frac{U}{C}\right) + Q_2^{\omega}\frac{\partial U}{\partial q_2} + \frac{\partial U}{\partial \omega} \tag{2.92b}$$

$$\frac{\mathrm{d}P_1^\omega}{\mathrm{d}t} = -Q_1^\omega\left[\frac{1}{C}\frac{\partial}{\partial q_1}\left(\frac{U}{C}\frac{\partial C}{\partial q_1}\right)\right] -$$

$$- Q_2^\omega\left[\frac{1}{C}\frac{\partial}{\partial q_2}\left(\frac{U}{C}\frac{\partial C}{\partial q_1}\right)\right] -$$

$$-\frac{1}{C^2}\frac{\partial C}{\partial q}\frac{\partial U}{\partial \omega} - \frac{U}{C}\frac{\partial}{\partial q_1}\left(\frac{1}{C}\frac{\partial C}{\partial \omega}\right) \tag{2.92c}$$

to be solved under zero initial conditions at the source: $Q_1^\omega = Q_2^\omega = P_1^\omega = 0$ at $x = 0$, $y = 0$, $t = 0$.

Note that from these relations one can easily get (apart from a constant factor) the formula for the field of a transient surface wave in a laterally homogeneous half-space derived by the method of stationary phase in Chapter 1 (see (1.56)). In that case (C = constant) one obtains $P_1^\varphi = C^{-1}$, $Q_1^\varphi = Ut = r$, $Q_2^\varphi = 0$ from (2.85), while (2.92) yields $P_1^\omega = 0$, $Q_1^\omega = 0$, $Q_2^\omega = (\mathrm{d}U/\mathrm{d}\omega)t = (\mathrm{d}U/\mathrm{d}\omega)/(r/U)$. The geometrical spreading of a space-time ray is thus given by a product of two factors, one of these (Q_1^φ) expressing the divergence of surface wave paths in the x, y-plane and the other, (Q_2^ω), the dispersive lengthening of a wave packet in time.

2.5. REFLECTION AND REFRACTION OF SPACE-TIME RAYS AT A DISCONTINUITY

When there is a discontinuity (possibly curvilinear in plan), it will cause reflection and transmission of surface waves, as is the case for body waves. In addition, a discontinuity will redirect some of the incoming energy to reflected and transmitted surface waves, as well as to body waves (this is associated with the continuity of stress and displacement across the discontinuity to be discussed in Chapter 3). A ray tube is also subject to reflection and transmission at a discontinuity, hence geometrical spreading and the functions that control it, Q_1^φ, Q_2^φ, Q_1^ω, Q_2^ω, will be discontinuous too. We now derive relations that connect these quantities for reflected and transmitted waves with those for the incident wave at a discontinuity.

We must first derive boundary conditions for a space-time ray. We define a discontinuity S by specifying the equation $f(x, y) = 0$. In the x, y, t space this will be a cylindrical surface with generatrices parallel to the t-axis. The function $\Theta(x, y, t)$ must obviously be continuous at the surface. From this condition one can deduce that r and t are continuous along a ray and the derivative of Θ along any direction on S is continuous too. The latter condition means continuity of the vector $\mathbf{g} = \nabla\Theta - \mathbf{n}(\nabla\Theta, \mathbf{n})$, where $\mathbf{n}(n_x, n_y)$ is the unit vector normal to the surface, and

$$\nabla\Theta = \left(\frac{\partial\Theta}{\partial x}, \frac{\partial\Theta}{\partial y}, \frac{\partial\Theta}{\partial t}\right) = (-\xi_x, -\xi_y, \omega).$$

Define the coordinates n, η in the x, y-plane at each point on S: n is along the normal to the discontinuity $f(x, y) = 0$, η is along the tangent to it. The direction of the η-axis is determined by the condition

$$(\mathbf{k}, \mathbf{e}_\eta) = \frac{\sin\theta}{C} > 0$$

so that the η-axis points towards wave propagation (Figure 2.2). It follows from the condition of continuity of $\mathbf{g}$ that, firstly, ω is continuous and, secondly, the vector $\mathbf{k} - \mathbf{n}(\mathbf{k}, \mathbf{n}) = \mathbf{e}_\eta(\mathbf{k}, \mathbf{e}_\eta)$ is continuous. Since $(\mathbf{k}, \mathbf{e}_\eta) = \sin\theta/C$, this means that, as in the case of nondispersive body-wave rays, Snell's law holds:

$$\frac{\sin\theta^\nu}{C^\nu(\omega)} = \frac{\sin\theta^0}{C^0(\omega)}. \tag{2.93}$$

Here and below, the superscript ν refers to a secondary wave (either reflected or transmitted), and 0 to the incident wave. The inclusion of the time coordinate leads to the extra condition

$$\omega^\nu = \omega^0. \tag{2.94}$$

Now we shall consider boundary value conditions for the functions necessary for determining the geometrical spreading of a space-time ray. First, we derive the conditions for Q_1^φ, Q_2^φ, P_1^φ. Obviously, $(\partial\mathbf{r}/\partial\varphi)_S$ must be continuous, because $\mathbf{r}$ is continuous on S. One can easily see that

$$\left(\frac{\partial\mathbf{r}}{\partial\varphi}\right)_S = \mathbf{e}_\eta \frac{Q_1}{(\mathbf{e}_1, \mathbf{e}_\eta)} \tag{2.95}$$

so that the continuity of $(\partial\mathbf{r}/\partial\varphi)_S$ reduces to

$$\frac{(Q_1^\varphi)^\nu}{(\mathbf{e}_1^\nu, \mathbf{e}_\eta)} = \frac{(Q_1^\varphi)^0}{(\mathbf{e}_1^0, \mathbf{e}_\eta)}. \tag{2.96}$$

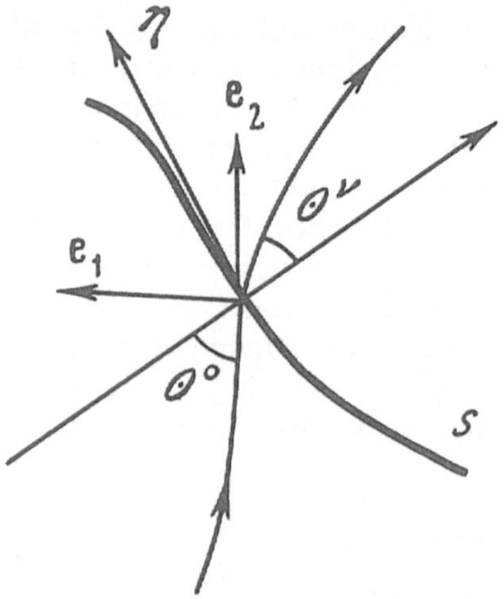

Fig. 2.2. The reference frame with the origin at the discontinuity $S(n, \eta)$ and ray-related unit vectors $\mathbf{e}_1, \mathbf{e}_2$.

The condition for Q_2^φ can be derived from that of continuity of t, yielding the continuity of $(\mathrm{d}t/\mathrm{d}\varphi)_S$. The expression for $(\mathrm{d}t/\mathrm{d}\varphi)_S$ is easily found from Figure 2.3:

$$\left(\frac{\partial t}{\partial \varphi}\right)_S = \frac{1}{U}\frac{\partial l}{\partial \varphi} = \frac{1}{U}\left(\frac{\partial l_1}{\partial \varphi} - \frac{\partial l_2}{\partial \varphi}\right)$$

$$= \frac{1}{U}\left(\frac{Q_1^\varphi \sin\theta}{(\mathbf{e}_1, \mathbf{e}_\eta)} - Q_2^\varphi\right). \tag{2.97}$$

The other boundary condition is thus

$$(Q_1^\varphi)^\nu \frac{\sin\theta^\nu}{U^\nu(\mathbf{e}_1^\nu, \mathbf{e}_\nu)} - \frac{(Q_2^\varphi)^\nu}{U^\nu} = \frac{(Q_1^\varphi)^0 \sin\theta^0}{U^0(\mathbf{e}_1^0, \mathbf{e}_\eta)} - \frac{(Q_2^\varphi)^0}{U^0}. \tag{2.98}$$

The condition that relates the functions P_1^φ for the incident and secondary wave is found from the continuity of $\mathbf{g}$ which also implies the continuity of $(\partial\mathbf{g}/\partial\varphi)_S$. Remembering that

$$\left(\frac{\partial}{\partial \varphi}\right)_S = \left(\frac{\partial}{\partial \varphi}\right)_t + \left(\frac{\partial t}{\partial \varphi}\right)_S \left(\frac{\partial}{\partial t}\right)_\varphi ,$$

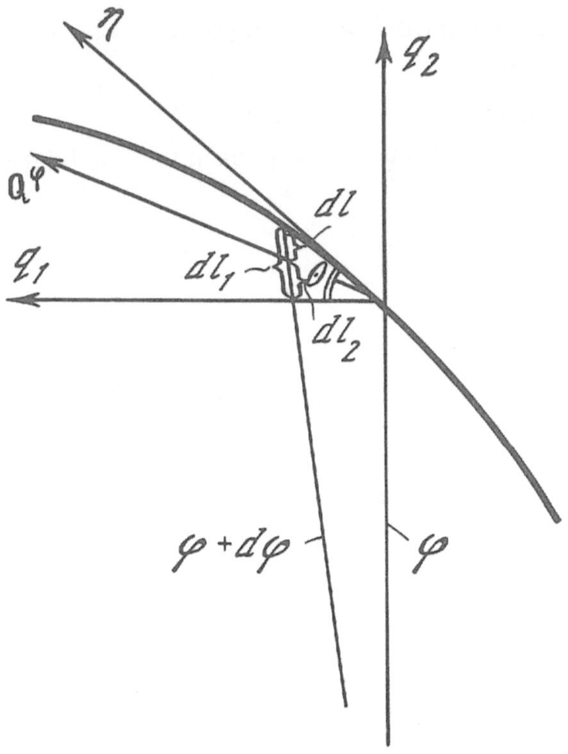

Fig. 2.3. Ray diagram illustrating the derivation of (2.97).

We consider the component of $(\partial \mathbf{g}/\partial\varphi)_S$ along the η-axis:

$$\left(\left(\frac{\partial \mathbf{g}}{\partial\varphi}\right)_S, \mathbf{e}_\eta\right) = \omega\left\{\left(\left(\frac{\partial \mathbf{k}}{\partial\varphi}\right)_t, \mathbf{e}_\eta\right) + \left(\frac{\partial t}{d\varphi}\right)_S\left(\left(\frac{\partial \mathbf{k}}{\partial t}\right)_\varphi, \mathbf{e}_\eta\right) - \right.$$

$$- \left(\left(\frac{\partial \mathbf{n}}{\partial\varphi}\right)_S, \mathbf{e}_\eta\right)(\mathbf{k}, \mathbf{n}) - (\mathbf{n}, \mathbf{e}_\eta)\left[\left(\left(\frac{\partial \mathbf{k}}{\partial\varphi}\right)_S, \mathbf{n}\right) + \right.$$

$$\left.\left. + \left(\left(\frac{\partial \mathbf{n}}{\partial\varphi}\right)_S, \mathbf{k}\right)\right]\right\}. \tag{2.99}$$

After some manipulations similar to those given by Azbel *et al.* (1980) we get

$$\left(\left(\frac{\partial g}{\partial\varphi}\right)_S, \mathbf{e}_\eta\right) = P_1^\varphi \cdot (\mathbf{e}_1, \mathbf{e}_\eta) + P_2^\varphi \sin\theta -$$

$$- \frac{1}{C^2}\left(\frac{Q_1^\varphi \sin\theta}{(\mathbf{e}_1, \mathbf{e}_\eta)} - Q_2^\varphi\right)\frac{\partial C}{\partial\eta} +$$

$$+ \frac{2(\mathbf{e}_2, \mathbf{n}) D Q_1^\varphi}{C(\mathbf{e}_1, \mathbf{e}_\eta)} \tag{2.100}$$

where D is proportional to the discontinuity curvature at the point of incidence:

$$D = \frac{1}{2}\frac{\partial^2 n}{\partial\eta^2}\,. \tag{2.101}$$

Recalling the expression (2.84) for P_2^φ, we obtain that the following condition must hold at the discontinuity

$$(Q_1^\varphi)^\nu\left[-\frac{\sin\theta^\nu}{(C^\nu)^2}\left(\frac{\partial C^\nu}{\partial q_1} + \frac{1}{(\mathbf{e}_1^\nu, \mathbf{e}_\eta)}\frac{\partial C^\nu}{\partial\eta}\right) + \frac{2D(\mathbf{e}_2^\nu, \mathbf{n})}{C^\nu(\mathbf{e}_1^\nu, \mathbf{e}_\eta)}\right] +$$

$$+ \frac{(Q_2^\varphi)^\nu}{(C^\nu)^2}\left[-\sin\theta^\nu\frac{\partial C^\nu}{\partial q_2} + \frac{\partial C^\nu}{\partial\eta}\right] + (P_1^\varphi)^\nu(\mathbf{e}_1^\nu, \mathbf{e}_\eta)$$

$$= (Q_1^\varphi)^0\left[-\frac{\sin\theta^0}{(C^0)^2}\left(\frac{\partial C^0}{\partial q_1} + \frac{1}{(\mathbf{e}_1^0, \mathbf{e}_\eta)}\frac{\partial C^0}{\partial\eta}\right) + \frac{2D(\mathbf{e}_2^0, \mathbf{n})}{C^0(\mathbf{e}_1^0, \mathbf{e}_\eta)}\right] +$$

$$+ \frac{(Q_2^\varphi)^0}{(C^0)^2}\left[-\sin\theta^0\frac{\partial C^0}{\partial q_2} + \frac{\partial C^0}{\partial\eta}\right] + (P_1^\varphi)^0(\mathbf{e}_1^0, \mathbf{e}_\eta). \tag{2.102}$$

The conditions (2.96), (2.98) and (2.102) can conveniently be written in matrix form

$$\mathbf{M}^\nu\mathbf{R}_\varphi^\nu = \mathbf{M}^0\mathbf{R}_\varphi^0 \tag{2.103}$$

where $\mathbf{R}_\varphi$ has the components $Q_1^\varphi, Q_2^\varphi, P_1^\varphi$; $\mathbf{M}$ is the matrix

$$\left\| \begin{matrix} \dfrac{1}{(\mathbf{e}_1, \mathbf{e}_\eta)} & 0 & 0 \\ \dfrac{\sin\theta}{U(\mathbf{e}_1, \mathbf{e}_\eta)} & -\dfrac{1}{U} & 0 \\ -\dfrac{\sin\theta}{C^2}\left(\dfrac{\partial C}{\partial q_1} + \dfrac{1}{(\mathbf{e}_1, \mathbf{e}_\eta)}\dfrac{\partial C}{\partial \eta}\right) + \dfrac{2D(\mathbf{e}_2, \mathbf{n})}{C(\mathbf{e}_1, \mathbf{e}_\eta)} & \dfrac{1}{C^2}\left(-\sin\theta\dfrac{\partial C}{\partial q_2} + \dfrac{\partial C}{\partial \eta}\right) & (\mathbf{e}_1, \mathbf{e}_\eta) \end{matrix} \right\|. \tag{2.104}$$

Now consider the conditions for Q_1^ω, Q_2^ω, P_1^ω. The first two conditions will obviously be identical with those for Q_1^φ, Q_2^φ. The expression for $((\partial\mathbf{g}/\partial\omega)_S, \mathbf{e}_\eta)$ will also be like (2.100) in form. It should be borne in mind, however, that the expression for P_2^ω (see (2.91)) will have an extra term involving a derivative with respect to ω. For this reason the third condition similar to (2.102) will contain extra terms in the right- and left-hand side of the form

$$-\frac{\sin\theta}{C^2}\frac{\partial C}{\partial\omega} = \left(\frac{1}{U} - \frac{1}{C}\right)\sin\theta.$$

Since $\sin\theta/C$ is continuous, it is enough to add $\sin\theta^{\nu(0)}/U^{\nu(0)}$ to the right- and left-hand side of (2.102).

2.6. SPACE-TIME RAYS AND GEOMETRICAL SPREADING IN THE SPHERICAL CASE

It was shown in Section 1.5 that the approximate solution corresponding to the surface wave field in a sphere with radial variation of elastic parameters describes a set of waves propagating along great circles (formula (1.36)), i.e., the solution for a harmonic wave is described by expressions of the form

$$V(R, \theta, t) = \frac{1}{\sqrt{R_0 \sin\theta}}\Phi(R, \omega)\exp\left[i\omega\left(t - \frac{R_0\theta}{C(\omega)}\right)\right] \tag{2.105}$$

where θ is the angular epicentral distance, R_0 the radius of the sphere. The function $\Phi(R, \omega)$ describes the distribution of wave amplitude along a radius; for sufficiently high frequencies it is identical with the eigenfunction of the relevant mode in a half-space. Then, the dispersion equation too is identical with that which holds for a half-space. However, the attenuation due to geometrical spreading is significantly different in a sphere and in a half-space: a harmonic wave decays in a half-space as $1/\sqrt{r}$ and in a sphere as $1/\sqrt{R_0 \sin\theta}$.

Here we shall not construct a complete solution for a laterally varying sphere, but confine ourselves to a consideration of the effect associated with the geometrical spreading of space-time rays. When the wavelength of a surface wave is small compared with the radius, the eigenvalues and eigenfunctions at each point of the spherical surface are the same as for a laterally homogeneous sphere having the radial distribution of elastic parameters as at a given point with coordinates θ, φ.

The analysis will proceed within the framework of the space-time ray method, similarly to the preceding case. The field will be described in a form analogous to (2.37):

$$\mathbf{u}(R, \theta, \varphi, t) = \sum_j \varepsilon^j \mathbf{u}_j(R, \varepsilon\theta, \varepsilon\varphi, \varepsilon t)\, e^{i\Theta(\theta, \varphi, t)}. \tag{2.106}$$

The phase function $\Theta(\theta, \varphi, t)$ obeys the relations

$$\frac{\partial \Theta}{\partial t} = \omega(\theta, \varphi, t), \quad \frac{1}{R}\frac{\partial \Theta}{\partial \theta} = -\xi_\theta, \quad \frac{1}{R \sin\theta}\frac{\partial \Theta}{\partial \varphi} = -\xi_\varphi .$$

A space-time ray is constructed similarly to the case of a half-space. The dispersion relation $\omega = \dot{\omega}(\theta, \varphi, \xi)$ should, generally speaking, correspond to the spherical case, and it is only at sufficiently high frequencies that the relation can be taken to be that for a half-space; it is represented in the form

$$\frac{\partial \Theta}{\partial t} - \omega(\theta, \varphi, \xi_\theta, \xi_\varphi) = 0. \tag{2.107}$$

When ξ_θ, ξ_φ are expressed in terms of partial derivatives of Θ with respect to θ, φ, Equation (2.107) in the Hamilton—Jacobi form becomes

$$\frac{\partial \Theta}{\partial t} - \omega\left(\theta, \varphi, \frac{1}{R}\frac{\partial \Theta}{\partial \theta}, \frac{1}{R \sin\theta}\frac{\partial \Theta}{\partial \varphi}\right) = 0. \tag{2.108}$$

Because this equation is different from (2.48), the canonical system which determines its characteristics will be different from that in Cartesian coordinates. The canonical system in spherical coordinates is

$$\begin{aligned}
&\frac{d\theta}{ds} = \frac{1}{R}\frac{\partial \omega}{\partial \xi_\theta}, \\
&\frac{d(R\xi_\theta)}{ds} = -\left(\frac{\partial \omega}{\partial \theta} + \frac{\partial \omega}{\partial \xi_\varphi}\frac{\partial}{\partial \theta}\left(\frac{1}{R \sin\theta}\frac{\partial \Theta}{\partial \varphi}\right)\right) \\
&\frac{d\varphi}{ds} = \frac{1}{R \sin\theta}\frac{\partial \omega}{\partial \xi_\varphi}, \quad \frac{d(R \sin\theta\, \xi_\varphi)}{ds} = -\frac{\partial \omega}{\partial \varphi}, \\
&\frac{dt}{ds} = 1, \quad \frac{d\omega}{ds} = 0.
\end{aligned} \tag{2.109}$$

After some simple manipulations the system becomes

$$\frac{d\theta}{dt} = \frac{U\xi_\theta}{R\xi}$$

$$\frac{d\varphi}{dt} = \frac{U\xi_\varphi}{R\xi \sin\theta}$$

$$\frac{d\xi_\theta}{dt} = -\frac{U\xi}{RC}\frac{\partial C}{\partial \theta} - \frac{U\xi_\varphi^2}{\xi R \tan\theta} \tag{2.110}$$

$$\frac{d\xi_\varphi}{dt} = -\frac{U\xi}{RC \sin\theta}\frac{\partial C}{\partial \varphi} - \frac{U\xi_\theta\xi_\varphi}{R\xi \tan\theta}.$$

The set of equations (2.110) describes rays on a spherical surface. The set of equations which determines geometrical spreading is conveniently obtained by transforming (2.110) to vectorial form, because one can then perform manipulations similar to those described in Section 2.4.

We again define the slowness vector $\mathbf{k} = k_\theta \mathbf{e}_\theta + k_\varphi \mathbf{e}_\varphi$. It is easy to show that (2.110) can be reduced to a system that is completely analogous to (2.52) in form:

$$\frac{d\mathbf{r}}{dt} = UC\mathbf{k}, \qquad \frac{d\mathbf{k}}{dt} = U\nabla\left(\frac{1}{C}\right) \tag{2.111}$$

where $\mathbf{r}$ is the radius vector of a point on a spherical surface of radius R. It should be emphasized that, in contrast to the preceding case, the slowness vector $\mathbf{k}$ is now three-dimensional, i.e., it does not remain on a plane when moving along a ray, hence $d\mathbf{k}/dt$ will have all the three components. Accordingly, it is to be assumed that the phase velocity C depends not only on θ, φ, but also on the radius R. For, at all points along a radius the wave travels with the same angular velocity $C(\theta, \varphi)/R_0$, where $C(\theta, \varphi)$ is the linear wave velocity on the sphere $R = R_0$ (see (2.105)), so that the linear wave velocity at different points along the same radius will be different and equal to $C(\theta, \varphi)R/R_0$. The right-hand side of the second of (2.111) can be written in a different way:

$$U\nabla_{\theta,\varphi}\left(\frac{1}{C}\right) - \frac{U\mathbf{e}_R}{CR_0}$$

where $U = U(\theta, \varphi)$; $C = C(\theta, \varphi)$. We now use the set of equations for space-time rays in the form (2.111) to derive equations for determining the geometrical spreading. As before, we define coordinates q_1, q_2 associated with a ray on the plane tangent to the sphere at a point of the ray. The parameters that define a ray will be the take-off azimuth at the source, φ_0, the source being placed at the pole of the sphere, and the frequency ω. Geometrical spreading is given by (2.63), as before.

Now put $R = R_0$ and define $\mathbf{Q}^{\varphi(\omega)}$, $\mathbf{P}^{\varphi(\omega)}$, similarly to the preceding. Differentiation of (2.111) with respect to φ_0 and ω yields equations for these vectors. This case is different from the preceding in that $\mathbf{P}^{\varphi(\omega)}$ has components, not only along the coordinates q_1, q_2 on the plane tangent to the sphere, but also along the radius R. The equations for $\mathbf{Q}^\varphi, \mathbf{P}^\varphi$ are

$$\frac{d\mathbf{Q}^\varphi}{dt} = UC\mathbf{P}^\varphi + \mathbf{k}\frac{\partial}{\partial\varphi_0}(UC)$$

$$\frac{d\mathbf{P}^\varphi}{dt} = \frac{\partial U}{\partial\varphi_0}\nabla\left(\frac{1}{C}\right) + U\frac{\partial}{\partial\varphi_0}\nabla\left(\frac{1}{C}\right) - \tag{2.112}$$

$$-\frac{U}{CR_0}\frac{\partial \mathbf{e}_R}{\partial\varphi_0} - \frac{\partial}{\partial\varphi_0}\left(\frac{U}{C}\right)\frac{\mathbf{e}_R}{R_0}.$$

The equations for $\mathbf{Q}^\omega$, $\mathbf{P}^\omega$ are completely analogous, except that the derivatives with respect to φ_0 are to be replaced by those with respect to ω. Two extra terms appear in the second of these equations as compared with (2.75), (2.76). The expression for $\partial\mathbf{e}_R/\partial\varphi_0$ is transformed as follows:

$$\frac{\partial\mathbf{e}_R}{\partial\varphi_0} = Q_1^\varphi\frac{\partial\mathbf{e}_R}{\partial q_1} + Q_2^\varphi\frac{\partial\mathbf{e}_R}{\partial q_2} = \frac{1}{R_0}(Q_1^\varphi\mathbf{e}_1 + Q_2^\varphi\mathbf{e}_2) = \frac{\mathbf{Q}^\varphi}{R_0}. \tag{2.113}$$

Now, as in the preceding case, we construct equations for the components of $\mathbf{Q}$ and $\mathbf{P}$ along the q_1 and q_2 axes. Evidently, the equations for Q_1 and Q_2 would be exactly identical with (2.80a), (2.80b), while those for P_1, P_2 are

$$\frac{dP_1}{dt} = -\frac{\partial}{\partial\varphi_0}\left(\frac{U}{C^2}\right)\frac{\partial C}{\partial q_1} - \frac{U}{C^2}\left(\frac{\partial\nabla C}{\partial\varphi_0}, \mathbf{e}_1\right) +$$

$$+\frac{U}{C}\frac{\partial C}{\partial q_1}P_2 - \frac{U}{CR_0^2}Q_1$$

$$\frac{dP_2}{dt} = -\frac{\partial}{\partial\varphi_0}\left(\frac{U}{C^2}\right)\frac{\partial C}{\partial q_2} - \frac{U}{C^2}\left(\frac{\partial\nabla C}{\partial\varphi_0}, \mathbf{e}_2\right) - \tag{2.114}$$

$$-\frac{U}{C}\frac{\partial C}{\partial q_1}P_1 - \frac{U}{CR_0^2}Q_2.$$

Subsequent transformations are analogous to those described in 2.4. We obtain the following sets of equations for Q_1^φ, Q_2^φ, P_1^φ and Q_1^ω, Q_2^ω, P_1^ω:

$$\frac{dQ_1^\varphi}{dt} = UCP_1^\varphi + \frac{U}{C}\frac{\partial C}{\partial q_1}Q_2^\varphi$$

$$\frac{dQ_2^\varphi}{dt} = Q_1^\varphi C\frac{\partial}{\partial q_1}\left(\frac{U}{C}\right) + Q_2^\varphi\frac{\partial U}{\partial q_2} \tag{2.115}$$

$$\frac{dP_1^\varphi}{dt} = -Q_1^\varphi\left[\frac{1}{C}\frac{\partial}{\partial q_1}\left(\frac{U}{C}\frac{\partial C}{\partial q_1}\right) + \frac{U}{CR_0^2}\right] -$$

$$-Q_2^\varphi\frac{1}{C}\frac{\partial}{\partial q_2}\left(\frac{U}{C}\frac{\partial C}{\partial q_1}\right)$$

$$\frac{dQ_1^{\omega}}{dt} = UCP_1^{\omega} + \frac{U}{C}\frac{\partial C}{\partial q_1}Q_2^{\omega}$$

$$\frac{dQ_2^{\omega}}{dt} = Q_1^{\omega}C\frac{\partial}{\partial q_1}\left(\frac{U}{C}\right) + Q_2^{\omega}\frac{\partial U}{\partial q_2} + \frac{\partial U}{\partial \omega} \tag{2.116}$$

$$\frac{dP_1^{\omega}}{dt} = -Q_1^{\omega}\left[\frac{1}{C}\frac{\partial}{\partial q_1}\left(\frac{U}{C}\frac{\partial C}{\partial q_1}\right) + \frac{U}{CR_0^2}\right] -$$

$$- Q_2^{\omega}\frac{1}{C}\frac{\partial}{\partial q_2}\left(\frac{U}{C}\frac{\partial C}{\partial q_1}\right) -$$

$$- \frac{1}{C^2}\frac{\partial U}{\partial \omega}\frac{\partial C}{\partial q_1} - \frac{U}{C}\frac{\partial}{\partial q_1}\left(\frac{1}{C}\frac{\partial C}{\partial \omega}\right).$$

The initial data for (2.115) are $Q_1^{\varphi} = Q_2^{\varphi} = 0$, $P_1^{\varphi} = C_0^{-1}$, while those for (2.116) are zero: $Q_1^{\omega} = Q_2^{\omega} = P_1^{\omega} = 0$.

For example, take the case of a laterally homogeneous sphere. Equations (2.115) then become

$$\frac{dQ_1^{\varphi}}{dt} = UCP_1^{\varphi}, \quad \frac{dP_1^{\varphi}}{dt} = -Q_1^{\varphi}\frac{U}{CR_0^2}, \quad Q_2^{\varphi} = 0. \tag{2.117}$$

Since, according to the ray tracing equation, we have $dt = R_0\, d\theta/U$, the equations can be written as

$$\frac{dQ_1^{\varphi}}{d\theta} = R_0CP_1^{\varphi}, \qquad \frac{dP_1^{\varphi}}{d\theta} = -\frac{Q_1^{\varphi}}{CR_0}$$

so that, with the initial data $Q_1^{\varphi} = 0$, $P_1^{\varphi} = C^{-1}$, the solution of that system is

$$Q_1^{\varphi} = R \sin\theta, \qquad P_1^{\varphi} = \cos\theta/C. \tag{2.118}$$

The system (2.116) for the case of a laterally homogeneous sphere is

$$\frac{dQ_1^{\omega}}{dt} = UCP_1^{\omega}, \qquad \frac{dQ_2^{\omega}}{dt} = \frac{\partial U}{\partial \omega}, \qquad \frac{dP_1^{\omega}}{dt} = -Q_1^{\omega}\frac{U}{CR_0^2} \tag{2.119}$$

and, with zero initial data, the solution will be

$$Q_1^{\omega} = P_1^{\omega} = 0, \qquad Q_2^{\omega} = \frac{R_0\theta}{U}\frac{\partial U}{\partial \omega}. \tag{2.120}$$

The geometrical spreading of space-time rays will thus be

$$J = \frac{R_0^2\theta}{U}\frac{\partial U}{\partial \omega}\sin\theta \tag{2.121}$$

so that the amplitude of a transient surface wave decays with epicentral distance Δ ($\Delta = \theta$) from the source as

$$\left(R_0\sqrt{\frac{1}{U}\frac{\partial U}{\partial \omega}\Delta\sin\Delta}\right)^{-1}.$$

2.7. ON THE CONFIGURATION OF SPACE-TIME RAYS

The space-time ray field is best examined in a laterally homogeneous structure with the simplest group velocity dispersion law, namely, when the dispersion curve of group velocity has a single minimum (Figure 2.4). In that case a space-time ray is a straight line in the x, y, t space. From the first of (2.52) it follows that the angle the line makes with the t-axis is controlled by group velocity U. Thus, for any given frequency ω, space-time rays form a circular cone, the cone angle being different for different frequencies (Figure 2.5). In a vertical section containing the t-axis, the rays form a fan of straight lines (Figure 2.6). Because in our case the group velocity dispersion curve has two branches, downward ($\omega < \omega_0$) and upward ($\omega > \omega_0$), the rays corresponding to the downward branch (i.e., those for frequencies $\omega < \omega_0$) form a fan of straight lines whose slope increases with ω increasing (Figure 2.6a), whereas the slope of rays corresponding to the upward branch $\omega > \omega_0$ decreases with ω increasing (Figure 2.6b). ω_0 is a critical frequency: a shadow zone appears in the region of the x, y, t space bounded by a cone of $\omega = \omega_0$ rays. This ray cone is in some sense similar to a caustic in the space, even though the caustic is degenerate (in the non-degenerate case, rays are tangent to the caustic at a single point, whereas here the rays themselves form a caustic).

Rays may have more complicated configurations in a laterally varying medium; they may be curved, say, resulting in features of the ray field in the x, y, t space similar to those for the case of rays in a three-dimensional inhomogeneous medium, i.e., the caustic and the shadow zone. The geometrical spreading $J = |\mathbf{Q}^{\varphi} \times \mathbf{Q}^{\omega}|$ vanishes on a space-time caustic, which can occur both when one of the $\mathbf{Q}^{\varphi}$ and $\mathbf{Q}^{\omega}$ vectors vanishes and when these vectors are collinear. Babich *et al.* (1985) present an analysis of space-time caustics for the simplest case of the wave

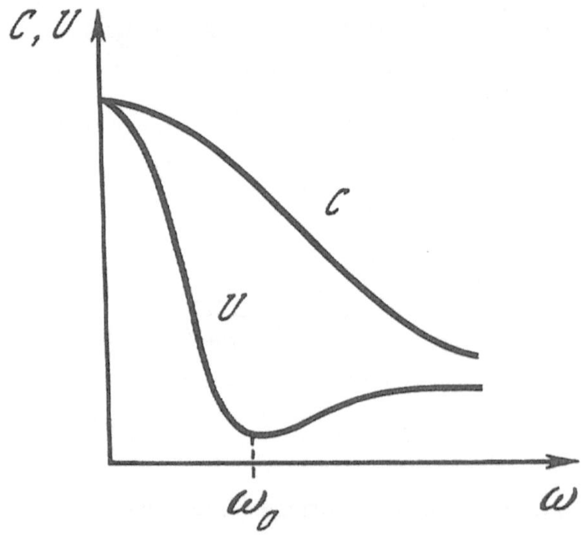

Fig. 2.4. Phase and group velocity as function of frequency.

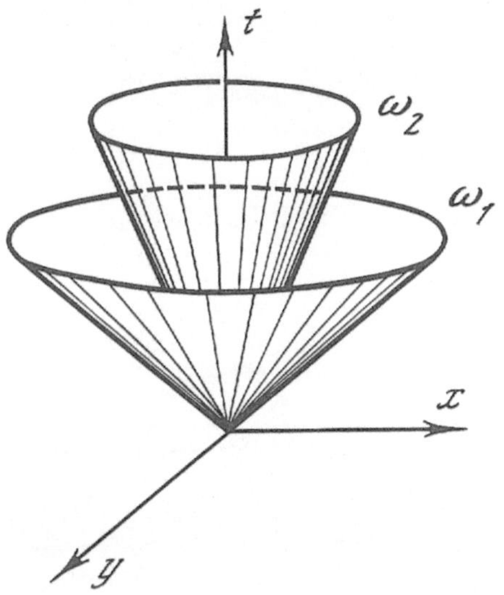

Fig. 2.5. Space-time rays of surface waves in a laterally homogeneous medium.

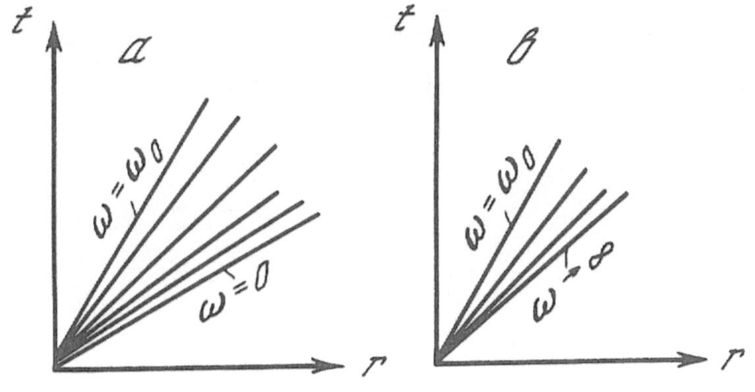

Fig. 2.6. Space-time rays corresponding to frequencies $0 < \omega < \omega_0$ (a) and $\omega > \omega_0$ (b) in the section containing the t-axis.

equation. The analysis for surface waves should be done in a manner similar to the case of the regular ray field, namely, by considering boundary-value problems for successive approximations and including the condition of solvability in the boundary-value problem for the first approximation.

Similarly to the analysis for rays in the physical x, y, z space, a surface wave field can be computed within the framework of the space-time ray method by summing Gaussian beams (Popov, 1983). In our case a beam is a solution that is concentrated around a space-time ray.

3. SURFACE WAVES IN MEDIA INVOLVING VERTICAL CONTACTS

The problems discussed in the preceding chapter can be applied to surface wave propagation in regions with weak lateral variation. There are also well-defined zones within the Earth that exhibit sharp changes in structure. Such are, for example, deep-seated faults and the ocean—continent transition zones. These are a few to tens of kilometers in width, so their transverse dimensions may be less than the wavelength of long-period surface waves. Their effect on surface wave propagation will thus be the same as that of contacts between blocks with different structures. Surface waves are reflected and transmitted at such interfaces. The geometry of reflected and transmitted surface wave rays (changes in propagation direction and geometrical spreading) was discussed in 2.6. Here we shall examine the amplitude characteristics of reflected and transmitted waves, specifically their reflection and transmission coefficients as functions of frequency.

3.1. STATEMENT OF THE PROBLEM

The problem of surface wave propagation across a vertical contact between two structures is formulated as follows. Two vertically varying quarter-spaces, 1 and 2, are in welded contact at the $x = 0$ plane (Figure 3.1). Horizontal discontinuities may exist in each of the quarter-spaces. A plane harmonic surface wave is propagating in structure 1 ($x < 0$). For the sake of simplicity the wave is assumed to be travelling in the direction of increasing x, so there is normal incidence at the interface $x = 0$. We can assume without loss of generality that the incident wave contains just one mode (of Rayleigh or Love waves), so that the displacement is described by

$$\hat{\mathbf{u}}_{\text{inc}}(x, z, \omega)e^{i\omega t} = \mathbf{V}_s(z)\exp[i(\omega t - \xi_{s\text{D}}x)], \quad \text{where} \tag{3.1}$$

$$\mathbf{V}_s(z) = iV_s^{(3)}(z)\mathbf{e}_y, \quad \xi_{s\text{D}} = \xi_{s\text{L}} \quad \text{for a Love wave, and}$$

$$\mathbf{V}_s(z) = V_s^{(1)}\mathbf{e}_z - iV_s^{(2)}(z)\mathbf{e}_x, \quad \xi_{s\text{D}} = \xi_{s\text{R}} \quad \text{for a Rayleigh wave;}$$

$\mathbf{e}_x$, $\mathbf{e}_y$, $\mathbf{e}_z$ are unit vectors along the coordinate axes. The wave field in structure 1 satisfies the following boundary conditions: no stress at the $z = 0$ interface, the displacement vanishes as $z \to \infty$, and there is welded contact (continuity of displacement and stress) at all horizontal discontinuities.

The incidence at the $x = 0$ interface gives rise to harmonic wave fields in

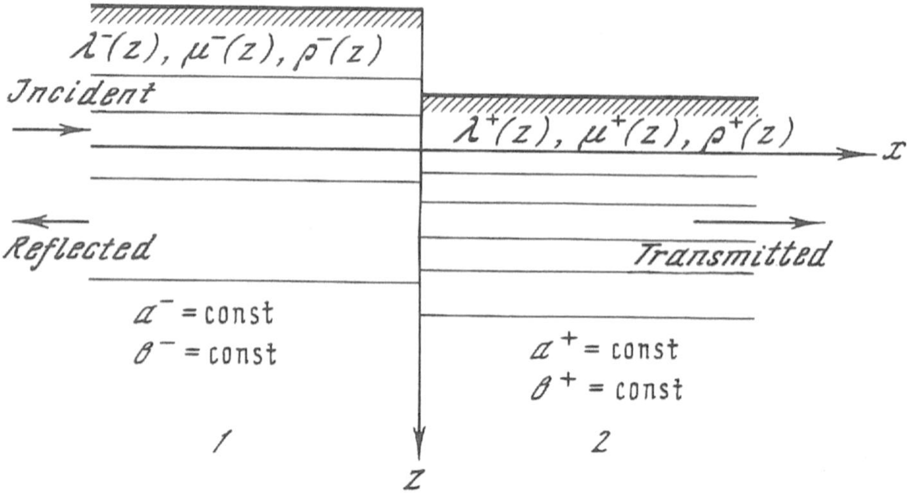

Fig. 3.1. Model of a contact between media 1 and 2 and wave propagation.

structures 1 and 2, denoted $\hat{\mathbf{u}}_{\text{refl}}(x, z, \omega)e^{i\omega t}$ and $\hat{\mathbf{u}}_{\text{trans}}(x, z, \omega)e^{i\omega t}$, respectively. The fields must satisfy the equation of motion (1.4) and the following conditions:

(i) continuity of stress and displacement at $x = 0$

$$\hat{\mathbf{u}}_{\text{inc}}(0, z, \omega) + \hat{\mathbf{u}}_{\text{refl}}(0, z, \omega) = \hat{\mathbf{u}}_{\text{trans}}(0, z, \omega) \tag{3.2}$$

$$\mathbf{T}(\hat{\mathbf{u}}_{\text{inc}} + \hat{\mathbf{u}}_{\text{refl}}, \mathbf{e}_x)|_{x=0} = \mathbf{T}(\hat{\mathbf{u}}_{\text{trans}}, \mathbf{e}_x)|_{x=0} \tag{3.3}$$

(ii) no stress at $z = 0$

$$\mathbf{T}(\hat{\mathbf{u}}_{\text{refl}}, \mathbf{e}_z)|_{z=0} = 0, \quad x < 0 \tag{3.4}$$

$$\mathbf{T}(\hat{\mathbf{u}}_{\text{trans}}, \mathbf{e}_z)|_{z=0} = 0, \quad x > 0 \tag{3.5}$$

(iii) the condition of limiting attenuation for $\hat{\mathbf{u}}_{\text{refl}}(x, z, \omega)$ as $x \to \infty$ and for $\hat{\mathbf{u}}_{\text{trans}}(x, z, \omega)$ as $x \to \infty$.

The $\hat{\mathbf{u}}_{\text{refl}}$ and $\hat{\mathbf{u}}_{\text{trans}}$ fields must satisfy the condition of welded contact at horizontal discontinuities.

These fields involve surface and body waves, including those diffracted at corners. We are seeking the surface wave fields. Obviously, a surface wave at normal incidence will give rise to a surface wave of the parent type (Rayleigh or Love). Oblique incidence will produce waves of both types.

In the case of a Love wave at normal incidence, the resulting displacement field on both sides of the contact can be represented as a sum of the fields corresponding to the discrete and continuous spectrum of the wavenumber ξ (Gabrielov, 1984). The discrete spectrum at any fixed frequency is finite. The relevant displacement can be represented as a sum of normal modes

$$\sum_{k=1}^{K(\omega)} F_k \mathbf{V}_k(z) \exp(\pm i\xi_k x)$$

where the minus sign is for waves travelling toward greater x, and the plus is for those in the opposite direction.

The values of ξ^2 for the continuous spectrum exist within the range $-\infty < \xi^2 < \omega^2/b^2(Z+0)$, as has been shown in Chapter 1. We introduce a new variable defined by $\eta^2 + \xi^2 = \omega^2/b^2(Z+0)$, so that $0 < \eta < \infty$. The contribution to the displacement field due to the continuous spectrum of ξ can be written as

$$\int_0^\infty F(\eta)\mathbf{V}(\xi(\eta), z)\exp[\pm i\xi(\eta)x]\,d\eta.$$

The factor $\exp(i\omega t)$ is omitted here and below. In these formulas

$$\mathbf{V}_k(z) = V_k^{(3)}(z)\mathbf{e}_y, \qquad \mathbf{V}(\xi(\eta), z) = V^{(3)}(\xi(\eta), z)\mathbf{e}_y$$

where $V_k^{(3)}(z)$, $V^{(3)}(\xi(\eta), z)$ are the eigenfunctions of the differential operator (1.24), (1.25) at a fixed value of ω; $\xi(\eta) = \sqrt{\omega^2/b^2(Z+0) - \eta^2}$. We thus have an expression for the transmitted wave field

$$\hat{\mathbf{u}}_{\text{trans}} = \sum_{k=1}^{K^+(\omega)} B_k \mathbf{V}_k^+(z)\exp(-i\xi_k^+ x) + \\ + \int_0^\infty B(\eta)\mathbf{V}^+(\xi^+(\eta), z)\exp[-i\xi^+(\eta)x]\,d\eta \tag{3.6}$$

and

$$\hat{\mathbf{u}}_{\text{refl}} = \sum_{k=1}^{K^-(\omega)} A_k \mathbf{V}_k^{-*}(z)\exp(i\xi_k^- x) + \\ + \int_0^\infty A(\eta)\mathbf{V}^{-*}(\xi^-(\eta), z)\exp[i\xi_k^-(\eta)x]\,d\eta \tag{3.7}$$

for the reflected wave field. The functions for the structure to the right of the contact will henceforth be marked with the + sign, and those to the left of it with the − sign.

The coefficients A_k, B_k may be interpreted as reflection and transmission coefficients of the relevant normal mode, $A(\eta)$ and $B(\eta)$ determine the portion of energy that is converted from the incident normal mode to the leaking modes to the left and right of the contact, respectively, or of body waves as these will be called. The problem is to evaluate A_k, B_k. Denote the displacement and stress produced on the $x = 0$ plane by body waves as $\mathbf{V}_{\text{body}}(z)$ and $\mathbf{p}_{\text{body}}(z)$, respectively. Then the requirement of continuity of stress and displacement at $x = 0$ yields

$$\mathbf{V}_s^-(z) + \sum^{K^-} A_k \mathbf{V}_k^{-*}(z) + \mathbf{V}_{\text{body}}^-(z) = \sum_{k=1}^{K^+} B_k \mathbf{V}_k^+(z) + \mathbf{V}_{\text{body}}^+(z) \tag{3.8}$$

$$\mathbf{p}_s^-(z) + \sum^{K^-} A_k \mathbf{p}_k^{-*}(z) + \mathbf{p}_{\text{body}}^-(z) = \sum_{k=1}^{K^+} B_k \mathbf{p}_k^+(z) + \mathbf{p}_{\text{body}}^+(z). \tag{3.9}$$

Here $\mathbf{p}_k$ is the traction across the $x = 0$ plane due to the kth Love mode that travels towards increasing x. We have

$$\mathbf{p}_k = -i\xi_{kL}\mu V_k^{(3)}(z)\mathbf{e}_y. \tag{3.10}$$

The analogous traction due to waves travelling in the opposite direction is a complex conjugate of $\mathbf{p}_k^+(z)$. The coefficients A_k, B_k cannot be found directly from the equations because of the body wave terms.

To solve that problem, the essential thing is to make use of the orthogonality of eigenfunctions that correspond to the same value of ω, but to different ξ_k (formula (1.26a)). Utilizing the expression (3.10) for the traction across $x = 0$, it is readily shown that (1.26a) can be written as

$$\int_0^\infty (\mathbf{p}_k^*\mathbf{V}_l - \mathbf{p}_l\mathbf{V}_k^*)\,\mathrm{d}z = 0, \quad k \neq l \tag{3.11}$$

$$\int_0^\infty (\mathbf{p}_k\mathbf{V}_l - \mathbf{p}_l\mathbf{V}_k)\,\mathrm{d}z = 0, \quad \text{for any } k, l. \tag{3.12}$$

Similar orthogonality conditions hold for normal Love modes and for the SH body waves

$$\begin{aligned} &\int_0^\infty (\mathbf{p}_k^*\mathbf{V}_{\text{body}} - \mathbf{p}_{\text{body}}\mathbf{V}_k^*)\,\mathrm{d}z = 0 \\ &\int_0^\infty (\mathbf{p}_k\mathbf{V}_{\text{body}} - \mathbf{p}_{\text{body}}\mathbf{V}_k)\,\mathrm{d}z = 0. \end{aligned} \tag{3.13}$$

Note that Equations (3.8), (3.9) and conditions (3.11)—(3.13) are also valid for Rayleigh waves and for P-SV body waves. The traction across $x = 0$ due to the kth Rayleigh mode travelling towards increasing x is equal to

$$\mathbf{p}_k(z) = \left[-(\lambda + 2\mu)\xi_{kR} V_k^{(2)} + \lambda \frac{\mathrm{d}V_k^{(1)}}{\mathrm{d}z}\right]\mathbf{e}_x - -i\left[\mu\left(\frac{\mathrm{d}V_k^{(2)}}{\mathrm{d}z} + \xi_{kR}V_k^{(1)}\right)\right]\mathbf{e}_z. \tag{3.14}$$

3.2. A REVIEW OF AVAILABLE THEORETICAL RESULTS

During the past 20 years numerous theoretical attempts have been made based on different approaches for determining reflection and transmission coefficients of surface waves that cross vertical discontinuities. This is due to the fact that the problem of surface wave reflection and transmission across a vertical contact cannot be solved exactly, owing to the intensive body waves generated at the contact. The waves are described by means of an integral taken over the continuous spectrum of ξ (see 3.1) and are not amenable to rigorous evaluation — they can merely be found to within some approximation. For this reason either numerical or approximate methods are used for the analysis and computation of reflected and transmitted normal modes of surface waves.

The numerical methods referred to are, in the first place, the finite difference and finite element methods in wide use today (Sato, 1961; Boore, 1970, 1972; Lysmer, 1970; Drake, 1972; Lysmer and Drake, 1981; Schlue, 1981). These methods yield numerical solutions for surface wave fields in models having narrow transition zones between two laterally homogeneous structures.

Another numerical method is that based on a numerical solution of integral equations (Bukchin, 1979; Bukchin and Levshin, 1980). We shall not discuss it here, because it is dealt with in detail in Section 3.4.

Approximate computational methods are only suitable for a direct evaluation of spectral reflection and transmission coefficients. They have been developed in various directions, the main ones being those based on (1) Green function technique, (2) some approximation to the boundary conditions at the vertical contact, (3) representation of the surface wave field as a superposition of plane homogeneous and inhomogeneous body waves.

Hudson and Knopoff (1964), Knopoff and Hudson (1964), Mal and Knopoff (1965) used the Green function technique. This is a mathematical formalization of the Huygens–Fresnel principle, according to which every point in the medium reached by a wave becomes a source of new elementary waves, and the field of a travelling wave can be represented as a superposition of such elementary waves. A surface wave incident at a vertical discontinuity gives rise to a disturbance that can be regarded as a source radiating reflected and transmitted surface waves. To find the reflected and transmitted surface wave fields exactly, one needs to know displacements and stresses due to a particular normal mode at the vertical discontinuity. These are in fact unknown, hence the need for approximate evaluation. The relevant Green function is assumed to be the principal part of the wave field far from the source, i.e., the part corresponding to surface waves. Some assumptions are used to derive displacement and stress fields in a particular reflected or transmitted mode at the contact. Thus a solution will be close to the true one in so far as the assumptions are adequate.

Since the subsequent argument largely depends on the Green function technique and the representation theorem as expressed for a vertically varying quarter-space bounded by a free horizontal surface and a vertical boundary, we shall state a two-dimensional analogue of the representation theorem (1.9) for the case of no body forces. As we shall deal with the harmonic waves in what follows, we write down this theorem for displacement and stress spectra.

Note that the convolution over time in (1.9) corresponds to a product of spectral functions. The theorem thus becomes

$$\hat{u}_i(\mathbf{x}_0, \omega) = \int_L [\hat{\mathbf{G}}^{(i)}(\mathbf{x}; \mathbf{x}_0; \omega)\mathbf{T}(\hat{\mathbf{u}}(\mathbf{x}, \omega), \mathbf{n}) - \\ - \hat{\mathbf{u}}(\mathbf{x}, \omega)\mathbf{T}(\hat{\mathbf{G}}^{(i)}(\mathbf{x}; \mathbf{x}_0; \omega), \mathbf{n})]\, \mathrm{d}L, \quad i = 1, 2, 3 \tag{3.15}$$

where $\hat{\mathbf{G}}^{(i)}(\mathbf{x}; \mathbf{x}_0; \omega)$ is the time-to-frequency Fourier transform of two-dimensional

Green function, the displacement spectrum at point $\mathbf{x}$ with coordinates x, z due to a concentrated line force

$$\delta(x - x_0)\delta(z - z_0)\delta(t)\mathbf{e}_i .$$

Suppose a surface wave is incident at the discontinuity $x = 0$ from the left (Figure 3.2), producing some displacement $\mathbf{V}(z)e^{i\omega t}$ and stress field $\mathbf{p}(z)e^{i\omega t}$ giving rise to a reflected wave field in the quarter-space $x < 0$, $z > 0$. The surface wave displacement at (x_0, z_0) can be expressed through the representation theorem (3.15) in terms of displacements and stresses at the contour L consisting of the parts L_1, L_2, L_3, L_4, the L_3, L_4 being placed at infinity.

The Green function is taken to be the surface wave part of the fundamental solution corresponding to the half-space having the parameters of the left-hand quarter-space, so that the stress corresponding to the Green function vanishes at L_1. As that surface is free, the stress at the surface due to the wave under consideration is zero too. The displacements and stresses corresponding to the Green function vanish on L_3. Thus the integrals over L_1 and L_2 in (3.15) vanish. In virtue of the principle of limiting attenuation and the orthogonality conditions (3.11)–(3.13), the integral over L_4 is zero too. Finally, the surface wave displacement can be represented in the form

$$(\hat{\mathbf{u}}(\mathbf{x}_0, \omega), \mathbf{e}_i) = \int_L [\hat{\mathbf{G}}^{(i)}(0, z; x_0, z_0)\mathbf{p}(z) -$$

$$- \mathbf{V}(z)\mathbf{T}(\hat{\mathbf{G}}^{(i)}(0, z; x_0, z_0), \mathbf{e}_x)] \, dz. \tag{3.16}$$

Obviously, if $\mathbf{p}(z)$, $\mathbf{V}(z)$ form the total fields of all reflected waves at $x = 0$, then the displacement at (x_0, z_0) as given by (3.16) corresponds to the sum of all reflected waves — all normal modes and body waves.

The displacement for a reflected mode can be found from (3.16) if, in virtue of

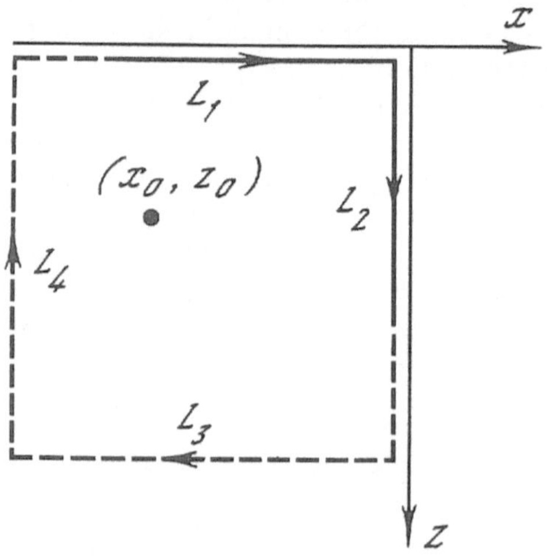

Fig. 3.2. Path of integration in the representation theorem for a reflected wave ($x < 0$).

the orthogonality conditions (3.11)—(3.13), the Green function is either replaced by the term corresponding to that mode or the contribution due to it alone is taken as the displacement and stress at the contact. Similar considerations apply to transmitted surface waves.

Knopoff and Hudson (1964) were the first to use the Green function technique to deal with reflection and transmission of Love and Rayleigh waves at the free surface of a two-dimensional wedge. However, the authors have neglected the reflected surface wave: stress and displacement corresponding to the incident wave at the discontinuity was assumed to generate a transmitted wave only. This deficiency was noticed and corrected in Mal and Knopoff (1965) where a system of equations was obtained for a joint determination of reflection and transmission coefficients.

Knopoff and Hudson (1964) have also used the Green function to deal with Love wave transmission in a layer overlying a half-space whose free surface has a step. These authors have disregarded the reflected surface wave in deriving the field of transmitted waves: the disturbance created at the discontinuity by the incident wave was considered to produce a transmitted wave alone. Lutikov (1979) makes a similar assumption in the problem of a Rayleigh wave at normal incidence on the vertical interface between two homogeneous quarter-spaces.

Levshin and Yanovskaya (1976) suggest an extension of the method based on the Huygens—Fresnel principle to find the field of both transmitted and reflected Love wave for normal incidence. The 'source' generating the transmitted wave is assumed to be the total field at the interface due to incident and reflected wave, while the 'source' generating the reflected wave is the difference of the incident and transmitted field. Its and Yanovskaya (1977, 1979a, b, 1983, 1985) have extended this treatment to Rayleigh wave reflection and transmission for oblique incidence and a curved subvertical interface. Their method will be described in more detail in Section 3.3.

Other papers (Alsop, 1966; McGarr and Alsop, 1967; Gregersen and Alsop, 1974; Malischewski, 1974, 1976) are concerned with evaluation of reflection and transmission coefficients for surface waves at a vertical contact based on the so-called method of Alsop, which assumes the intensity of body waves generated by a contact to be small compared with that of surface waves, so that the reflection and transmission coefficients can be found from the requirement that the boundary conditions – continuity of stress and displacement at the contact – should be satisfied approximately (only normal modes are to be considered).

The idea of Alsop's method can be made clear by examining the normal incidence of a harmonic Love wave at a vertical contact between two vertically varying quarter-spaces (Alsop, 1966) (see Figure 3.1).

Let the displacement in an incident Love wave be given by

$$V_1^-(z)\exp[i(\omega t-\xi_1^- x)]\mathbf{e}_y$$

(one can assume without loss of generality that the fundamental mode is incident, denoted below by the suffix 1).

Assume that $m \geqslant 1$ Love modes for a given frequency ω can exist in the medium to the left of the interface and $n \geqslant 1$ modes can exist to the right of it. The displacements in reflected and transmitted waves can then be expressed as follows:

$$\sum_{i=1}^{m} A_i(\omega) V_i^-(z) \exp[i(\omega t + \xi_i^- x)]\mathbf{e}_y,$$
$$\sum_{j=1}^{m} B_j(\omega) V_j^+(z) \exp[i(\omega t - \xi_j^+ x)]\mathbf{e}_y, \tag{3.17}$$

where $A_i(\omega)$ and $B_j(\omega)$ are reflection and transmission coefficients; V_i^-, V_j^+, ξ_i^-, ξ_j^+ are the eigenfunctions $V_i^{(3)}$, $V_j^{(3)}$ and eigenvalues ξ_{iL}, ξ_{jL} for the one-dimensional problem of 1.3 with the half-spaces having the parameters of the left (−) and right (+) quarter-space. The factor $e^{i\omega t}$ will be omitted in what follows.

We now write down the conditions for the displacement and stress to be continuous at $x = 0$, assuming that the wave field at the interface is only due to normal modes:

$$V_1^-(z) + \sum_{j=1}^{m} A_i V_i^-(z) = \sum_{i=1}^{n} B_j V_j^+(z) \tag{3.18}$$

$$-i\xi_1^- \mu^- V_1^-(z) + i\mu^- \sum_{i=1}^{m} A_i \xi_i^- V_i^-(z) = -i\mu^+ \sum_{i=1}^{n} B_j \xi_j^+ V_j^+(z). \tag{3.19}$$

The fields of two different modes in any one medium at a fixed frequency obey the orthogonality condition (1.26a):

$$\int_0^\infty \mu V_i V_j \, dz = 0, \qquad i \neq j \tag{3.20}$$

while the following normalization is conveniently imposed on the eigenfunctions:

$$\xi_i \int_0^\infty \mu V_i^2 \, dz = 1. \tag{3.21}$$

Because the interface will generate body waves, in addition to normal modes, the boundary conditions written in the form (3.18), (3.19) will hold only approximately. We suggest using any of the following procedures to construct equations that will satisfy the conditions approximately.

One consists in multiplying (3.19) by V_j^+ and integrating from 0 to ∞. The transmission coefficients B_j can then be expressed through the A_j using (3.20), (3.21) as follows:

$$B_j = P_{1j} - \sum_{i=1}^{m} A_i \xi_i P_{ij} \tag{3.22}$$

where

$$P_{ij} = \xi_i^- \int_0^\infty \mu^- V_i^- V_j^+ \, dz = \frac{1}{i} \int_0^\infty p_i^- V_j^+ \, dz \tag{3.23}$$

relates the stress $\mathbf{p}_i^- = p_i^- \mathbf{e}_y$ at the interface $x = 0$ due to the ith mode in medium 1 to the displacement due to the jth mode in medium 2.

The A_i are found from the requirement that (3.18) be satisfied approximately in such a manner as to minimize the mean (over the depth) square difference of displacement to the right and left of the interface, i.e., the integral

$$\Phi = \int_0^\infty [\delta V]^2 \, dz = \int_0^\infty \left(V_1^- + \sum_{j=1}^{m} A_i V_i^- - \sum_{i=1}^{n} B_j V_j^+ \right)^2 dz. \qquad (3.24)$$

The B_j as given by (3.22) are substituted in (3.24) and the A_i are found from the equations which emerge on setting partial derivatives of Φ with respect to the A_i equal to zero.

There is another procedure commonly considered to be preferable for the case in which some area of the contact surface is free from stress (that is, when the free surface of the half-space involves a step). It determines the coefficients by requiring (3.19), continuity of stress across the contact, to be approximately true. Stress then vanishes over the area corresponding to the step, while the displacement does not, so it seems more adequate to minimize the mean squared stress difference across the contact. The condition equivalent to (3.22) is obtained by multiplying (3.18) by $\xi_j^+ \mu^+ V_j^+$ and integrating from 0 to ∞. This yields the following expression for transmission coefficients in terms of reflection coefficients

$$B_j = S_{j1} + \sum_{i=1}^{n} A_i S_{ji} \quad \text{where}$$
$$S_{ji} = \xi_j^+ \int_0^\infty \mu^+ V_i^- V_j^+ \, dz. \qquad (3.25)$$

The transmission coefficients thus obtained are expressed through the reflection coefficients A_i and substituted into the functional to be minimized:

$$\Psi = \int_0^\infty [\delta p]^2 \, dz = \int_0^\infty \left(\xi_1^- \mu^- V_1^- - \mu^- \sum_{i=1}^{m} A_i \xi_i^- V_i^- - \right.$$
$$\left. - \mu^+ \sum_{j=1}^{n} B_j \xi_j^+ V_j^+ \right)^2 dz. \qquad (3.26)$$

Differentiating Ψ with respect to the A_i, we get a set of equations in unknown reflection coefficients.

These two procedures produce different results, so that the approach developed by Alsop does not determine the desired coefficients uniquely. It must also be noted that (3.18), (3.19) can be used to derive other, internally inconsistent relations connecting the A_i and the B_j. One can add to (3.22) and (3.24) relations which can be obtained either by multiplying (3.18) by $\xi_i^- \mu^- V_i^-$ with subsequent integration, thus expressing the A_i through a linear combination of the B_j:

$$A_i = \sum_j B_j P_{ij} - \delta_{i1} \qquad (3.27)$$

or by multiplying (3.19) by V_i^- with a similar integration, yielding the following relation:

$$A_i = \delta_{i1} - \sum_j B_j S_{ji}. \tag{3.28}$$

We see that the method is fairly arbitrary; it remains uncertain in which situation which procedure is to be preferred.

Malischewski (1976) used the Alsop method to deal with Love and Rayleigh waves at oblique incidence to the interface between the media, but the method has since not been developed. This is due on the one hand to the above mentioned arbitrariness (hence nonuniqueness) and, on the other, to the fact that the method is a poor approximation for the case where the contrast between the media in contact is not small. Fairly large body waves are then generated which this method altogether ignores. For this reason even its inventors themselves have suggested evaluating the resulting difference in surface wave energy flux to the left and right of the interface, which must be equal to the energy of disregarded body waves, and using estimates given by the method only when the body wave energy is relatively small.

We now discuss some of the work in which the field of surface waves is represented as superposed plane homogeneous and inhomogeneous waves and the reflection and transmission coefficients for these body waves are determined over individual areas of the contact, enabling the fields of reflected and transmitted waves to be found over the entire contact (Alsop, Goodman and Gregersen, 1974; Gregersen, 1978; Gregersen and Alsop, 1976). The reflection and transmission coefficients of surface waves can then be found using the representation theorem.

The idea of the method will as before be illustrated for normal incidence of a plane Love wave at a vertical interface. The quarter-spaces in contact will be assumed to be plane-stratified. The Love wave in each layer is represented as a superposition of homogeneous and inhomogeneous plane waves that travel toward positive and negative z (called downgoing and upgoing waves, respectively). Love waves in the half-space are represented by a downgoing inhomogeneous wave alone. The set of such waves is shown schematically in Figure 3.3.

The waves that make up the surface wave are reflected at the top and bottom of the kth layer at an angle θ_k which is determined by phase velocity C:

$$\sin \theta_k = b_k^-/C. \tag{3.29}$$

The angle of incidence at the contact $x = 0$, φ_k, for the waves that make up the surface wave in the kth layer, is given by

$$\sin \varphi_k = \cos \theta_k = \sqrt{1 - (b_k^-/C)^2}. \tag{3.30}$$

The refraction angle ψ_q for the area where that layer is in contact with the qth layer of medium 2 having the shear wave velocity b_q^+ is found from Snell's law:

$$\sin \psi_q = b_q^+ \sqrt{(1/b_k^-)^2 - 1/C^2}. \tag{3.31}$$

In a similar fashion, a wave can be assumed to be incident on the contact at an angle φ_N given by

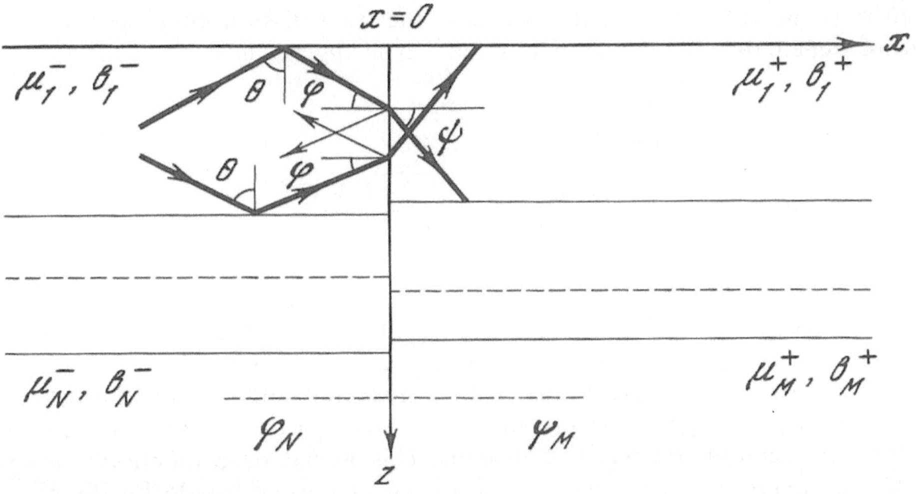

Fig. 3.3. Transmission across the $x = 0$ discontinuity of homogeneous (in the top layer) and inhomogeneous (in the half-space) plane waves that make up a surface wave.

$$\sin \varphi_N = \sqrt{1 - (b_N^-/C)^2} \tag{3.32}$$

but then the angle of incidence would be imaginary. Accordingly, the angle of refraction must be imaginary too:

$$\sin \psi_M = b_M^+ \sqrt{(1/b_N^-)^2 - 1/C^2}. \tag{3.33}$$

The wave in a layer is represented by a superposition of upgoing and downgoing waves. This is also true for the case in which the wave in a layer is represented by a superposition of inhomogeneous waves. This can be demonstrated by expressing the displacement of an incident Love wave in a layer in the form $V^-(z)\exp[i(\omega t - \xi^- x)]$,

$$V^-(z) = \begin{cases} A_1 e^{\beta_1 z} + B_1 e^{-\beta_1 z}, & 0 < z < z_1 \\ A_2 e^{\beta_2 z} + B_2 e^{-\beta_2 z}, & z_1 < z < z_2 \\ \cdots\cdots\cdots\cdots\cdots\cdots\cdots\cdots & \\ B_N e^{-\beta_N z}, & z > z_N \end{cases} \tag{3.34}$$

where $\beta_k = \sqrt{\xi^2 - \omega^2/b_k^2}$ (the first term on a right-hand side is for the upgoing wave, the second for the downgoing). For the sake of definiteness, we assume the A_k, B_k to obey the normalization (3.21). Similarly, the Rayleigh wave displacement in a layer is represented as a superposition of upgoing and downgoing P- and SV-waves. Each of these is reflected and transmitted at the vertical interface. In this way, the displacement of a reflected SH wave, say, at the contact between two layers of media 1 and 2 can be written as

$$V^{\text{refl}}(z) = k(\xi)V(z), \qquad V^{\text{trans}}(z) = \kappa(\xi)V(z) \tag{3.35}$$

where k is the reflection coefficient; κ is the transmission coefficient of an SH wave incident at the contact of media 1 and 2 at an angle of φ:

$$k = \frac{\mu^-/b^- \cos\varphi - \mu^+/b^+ \cos\psi}{\mu^-/b^- \cos\varphi + \mu^+/b^+ \cos\psi}$$

$$= \frac{\mu^-/C - \mu^+ \sqrt{1/(b^+)^2 - 1/(b^-)^2 + 1/C^2}}{\mu^-/C + \mu^+ \sqrt{1/(b^+)^2 - 1/(b^-)^2 + 1/C^2}}, \tag{3.36}$$

$$\kappa = 1 + k. \tag{3.37}$$

For a Rayleigh wave, the reflected wave field at the contact consists of reflected P and SV waves due to the incidence of upgoing and downgoing P and SV waves.

It is easy to see that the fields of both reflected and transmitted waves are discontinuous at the places where the contact cuts through horizontal interfaces, which is unacceptable on physical grounds. This, as has been mentioned above, is due to our ignoring waves that are diffracted at corners made by intersections between interfaces. If these waves were added to the total field of reflected (transmitted) waves, that would satisfy the boundary conditions and make the fields at the vertical contact continuous. Thus, the principal restriction in the method discussed consists in disregarding the field of diffracted waves. The displacements and stresses in reflected (transmitted) waves at the contact can be viewed as the source of reflected (transmitted) surface wave modes. Proceeding on the basis of the representation theorem (3.16) and the orthogonality conditions (3.11)–(3.13), one can get the following expressions for reflection (A_i) and transmission (B_j) coefficients at the contact:

$$A_i = \frac{1}{2i}\int_0^\infty [i\xi_1^-\mu^-(z)k(z)V_1^-(z)V_i^-(z) +$$

$$+ i\xi_i^-\mu^-(z)k(z)V_1^-(z)V_i^-(z)]\,dz$$

$$= \frac{\xi_1^- + \xi_i^-}{2}\int_0^\infty \mu^-(z)k(z)V_1^-(z)V_i^-(z)\,dz \tag{3.38}$$

$$B_i = \frac{1}{2}\left[\xi_j^+\int_0^\infty \mu^+(z)V_j^+(z)V_1^-(z)\kappa(z)\,dz + \right.$$

$$\left. + \xi_1^-\int_0^\infty \mu^-(z)V_j^+(z)V_1^-(z)(1 - k(z))\,dz\right]. \tag{3.39}$$

3.3. APPROXIMATE CALCULATION OF SURFACE WAVE REFLECTION AND TRANSMISSION COEFFICIENTS AT A CONTACT

The idea of the approximate method to be discussed below for calculating surface wave reflection and transmission coefficients is due to Levshin and Yanovskaya

(1976), who examine the simplest case of surface wave reflection and transmission: normal incidence of a Love wave on a vertical contact between two vertically varying structures. Its and Yanovskaya (1977) used the method to deal with a Rayleigh wave normally incident on a vertical interface; Its and Yanovskaya (1979a, b) considered the general case of any type of wave, whether Rayleigh or Love, at oblique incidence. In a later paper, Its and Yanovskaya (1983) extended this treatment to the case of a subvertical interface with an arbitrary curved profile. The method is based on Green function.

We begin by considering a plane Love or Rayleigh surface wave at normal incidence on the vertical interface $x = 0$. For simplicity we assume the incident wave to contain a single sth surface wave mode, while the field of reflected and transmitted surface waves is represented as a superposition of all modes for a fixed frequency ω. If the displacement in the incident wave is $\hat{\mathbf{u}}_s^-(x, z, \omega)\exp(i\omega t)$, then that in the reflected waves is given by

$$\sum_i A_i \hat{\mathbf{u}}_i^{-*}(x, z, \omega)\exp(i\omega t)$$

and for the transmitted waves

$$\sum_j B_j \hat{\mathbf{u}}_j^{+}(x, z, \omega)\exp(i\omega t).$$

Here, the A_i and B_j are the reflection and transmission coefficients to be determined. The factor $\exp(i\omega t)$ will be omitted below.

The total wave field to the left and right of the interface is a superposition of normal modes and body waves which are attenuated away from the contact.

The displacement to the left of the interface can then be represented as the sum

$$\hat{\mathbf{u}}_s^-(x, z, \omega) + \sum_i A_i \hat{\mathbf{u}}_i^{-*}(x, z, \omega) + \hat{\mathbf{u}}_{\text{body}}^-(x, z, \omega) \tag{3.40}$$

and that to the right as

$$\sum_j B_j \hat{\mathbf{u}}_j^{+}(x, z, \omega) + \hat{\mathbf{u}}_{\text{body}}^+(x, z, \omega) \tag{3.41}$$

where $\hat{\mathbf{u}}_{\text{body}}^+$ and $\hat{\mathbf{u}}_{\text{body}}^-$ are body wave displacements. The continuity of displacement and stress at the interface $x = 0$ can be written as

$$\mathbf{V}_s^-(z) + \sum_i A_i \mathbf{V}_i^{-*}(z) - \sum_j B_j \mathbf{V}_j^{+}(z) + \mathbf{V}_{\text{body}}^- - \mathbf{V}_{\text{body}}^+ \tag{3.42}$$

$$\mathbf{p}_s^-(z) + \sum_i A_i \mathbf{p}_i^{-*}(z) - \sum_j B_j \mathbf{p}_j^{+}(z) + \mathbf{p}_{\text{body}}^- - \mathbf{p}_{\text{body}}^+. \tag{3.43}$$

Here,

$$\begin{aligned} \hat{\mathbf{u}}_i(x, z) &= \mathbf{V}_i(z)\exp(-i\xi x) \\ \mathbf{T}(\hat{\mathbf{u}}_i(x, z), \mathbf{e}_x) &= \mathbf{p}_i(z)\exp(-i\xi x) \\ \hat{\mathbf{u}}_{\text{body}}(0, z) &= \mathbf{V}_{\text{body}}(z). \end{aligned} \tag{3.44}$$

The representation theorem for the qth component of displacement in the ith reflected mode is, in accordance with (3.16),

$$A_i(\hat{\mathbf{u}}_i^{-*}(x_0, z_0), \mathbf{e}_q) = \int_0^\infty [\hat{\mathbf{G}}^{(q)-}(0, z; x_0, z_0)\mathbf{p}_i^{\text{refl}}(z) - \\ - \mathbf{V}_i^{\text{refl}}(z)\mathbf{T}(\hat{\mathbf{G}}^{(q)-}(0, z; x_0, z_0), \mathbf{e}_x)]\,dz \tag{3.45}$$

where $\hat{\mathbf{G}}^{(q)}$ is the Fourier transform of the Green function, i.e., the displacement at (x, z) due to a point source at (x_0, z_0), which is a unit force acting along the q-axis $(q = x, y, z)$. $\mathbf{V}_i^{\text{refl}}(z)$ and $\mathbf{p}_i^{\text{refl}}(z)$ that enter the integrand are the amplitudes of displacement and stress in the ith reflected mode at the surface $x = 0$.

The surface wave part of the Green function for a vertically inhomogeneous half-space (two-dimensional case) can be found in Herrera (1964). It has the form

$$\hat{\mathbf{G}}^{(q)}(x, z; x_0, z_0) = \begin{cases} \sum_k \dfrac{(\hat{\mathbf{u}}_k^*(x_0, z_0), \mathbf{e}_q)\hat{\mathbf{u}}_k(x, z)}{J_k}, & x > x_0 \\ \sum_k \dfrac{(\hat{\mathbf{u}}_k(x_0, z_0), \mathbf{e}_q)\hat{\mathbf{u}}_k^*(x, z)}{J_k}, & x < x_0 \end{cases} \tag{3.46}$$

where

$$J_k = \int_0^\infty [(\mathbf{T}^*(\hat{\mathbf{u}}_k, \mathbf{e}_x), \hat{\mathbf{u}}_k) - (\mathbf{T}(\hat{\mathbf{u}}_k, \mathbf{e}_x), \hat{\mathbf{u}}_k^*)]\,dz. \tag{3.46a}$$

We have

$$\hat{\mathbf{u}}_k(x, z) = \mathbf{V}_k(z)\exp(-i\xi_{kD}x)$$

in (3.46); when $q = x, z$, D = R (Rayleigh wave), we have $\mathbf{V}_k(z) = V_k^{(1)}(z)\mathbf{e}_z - iV_k^{(2)}(z)\mathbf{e}_x$, and when $q = y$, D = L (Love wave), we have $\mathbf{V}_k(z) = V_k^{(3)}(z)\mathbf{e}_y$. Accordingly, the $\mathbf{T}(\hat{\mathbf{G}}^{(q)}(0, z; x_0, z_0), \mathbf{e}_x)$ in (3.45) is determined from (3.46) using (3.44), $\mathbf{p}(z)$ being expressed in terms of $\mathbf{V}_i(z)$ either through (3.10) (Love wave case) or through (3.14) (Rayleigh wave). In virtue of the orthogonality condition (3.11), one can replace in the representation theorem (3.45) the complete expression for the Green function as given by (3.46) by the term corresponding to the ith mode alone. Dividing both sides by $(\hat{\mathbf{u}}_i^{-*}(x_0, z_0), \mathbf{e}_q)$, one gets the reflection coefficient A_i:

$$A_i = \frac{1}{J_i^-}\int_0^\infty [\mathbf{p}_i^{\text{refl}}(z)\mathbf{V}_i^-(z) - \mathbf{V}_i^{\text{refl}}(z)\mathbf{p}_i^-(z)]\,dz. \tag{3.47}$$

Let us now express $\mathbf{p}_i^{\text{refl}}$ and $\mathbf{V}_i^{\text{refl}}$ that enter the integrand of (3.45) through the boundary conditions (3.42), (3.43). The relation (3.47) then becomes

$$A_i = \frac{1}{J_i^-}\int_0^\infty \left[\mathbf{V}_i^-\left(\sum_j B_j\mathbf{p}_j^+ - \mathbf{p}_s^- - \sum_{k \neq i} A_k\mathbf{p}_k^{-*} + \mathbf{p}_{\text{body}}^+ - \mathbf{p}_{\text{body}}^-\right) - \\ - \mathbf{p}_i^-\left(\sum_j B_j\mathbf{V}_j^+ - \mathbf{V}_s^- - \sum_{k \neq i} A_k\mathbf{V}_k^{-*} + \mathbf{V}_{\text{body}}^+ - \mathbf{V}_{\text{body}}^-\right)\right]dz. \tag{3.48}$$

Different modes in any one medium obey the orthogonality condition (3.11), (3.12). Using these, and introducing a notation for the coefficient of coupling between the ith reflected mode in medium 1 and the jth transmitted mode in medium 2

$$P_{ij} = \frac{1}{J_i^-} \int_0^\infty (\mathbf{p}_i^- \mathbf{V}_j^+ - \mathbf{V}_i^- \mathbf{p}_j^+) \, dz \tag{3.49}$$

one can write (3.48) in the form

$$A_i = -\sum_j B_j P_{ij} - \\ - \frac{1}{J_i^-} \int_0^\infty [\mathbf{p}_i^-(\mathbf{V}_{\text{body}}^+ - \mathbf{V}_{\text{body}}^-) - \mathbf{V}_i^-(\mathbf{p}_{\text{body}}^+ - \mathbf{p}_{\text{body}}^-)] \, dz. \tag{3.50}$$

The second term on right-hand side of (3.50) expresses the coupling between the ith reflected mode and the body waves at the contact of the two media. Since the normal mode field is orthogonal to the body wave field in any one medium (formula (3.13)), the term in question takes the form

$$\frac{1}{J_i^-} \int_0^\infty [\mathbf{p}_i^- \mathbf{V}_{\text{body}}^+ - \mathbf{V}_i^- \mathbf{p}_{\text{body}}^+] \, dz$$

giving the coupling between the ith reflected mode in medium 1 and the body wave field in medium 2. Assuming the coupling to be weak, so that the above integral can be neglected, one gets

$$A_i = -\sum_j B_j P_{ij}. \tag{3.51}$$

The same procedure can be carried out for the transmission coefficient B_j. We express the field of a transmitted surface wave at a point far enough from the interface using the representation theorem (3.45), and then write the stress and displacement in the jth transmitted mode at the contact in terms of all the other waves on the basis of the relevant boundary conditions, assuming $\int_0^\infty [\mathbf{p}_j^+ \mathbf{V}_{\text{body}}^- - \mathbf{V}_j^+ \mathbf{p}_{\text{body}}^-] \, dz$ to be small (this integral expresses the coupling between the body wave field in medium 1 and the jth transmitted mode in medium 2); we then get a relation similar to (3.51)

$$B_j = S_{sj}^* + \sum_i A_i P_{ij}^* \tag{3.52}$$

where

$$S_{sj} = \frac{1}{J_j^+} \int_0^\infty (\mathbf{p}_s^{-*} \mathbf{V}_j^+ - \mathbf{V}_s^{-*} \mathbf{p}_j^+) \, dz \tag{3.53}$$

is the coefficient of coupling between the sth incident mode in medium 1 and the jth transmitted mode in medium 2.

Equations (3.51), (3.52) constitute a set of linear equations to determine reflection and transmission coefficients.

We now write down explicit expressions for coupling coefficients between modes separately for Love and Rayleigh waves.

Love wave. Making use of the expression (3.46a) for J_k, as well as of the fact that $\mathbf{V}_k = V_k^{(3)}(z)\mathbf{e}_y$ and $\mathbf{p}_k(z) = -i\xi_{k\mathrm{L}}\mu V_k^{(3)}(z)\mathbf{e}_y$, we get

$$J_k = 2i\xi_{k\mathrm{L}} \int_0^\infty \mu(z)\,(V_k^{(3)})^2\,\mathrm{d}z.$$

If the eigenfunctions $V_k^{(3)}(z)$ are normalized in accordance with (3.21), we shall have $J_k = 2i$; therefore

$$P_{ij} = \frac{1}{2}\left\{\xi_{j\mathrm{L}}^+ \int_0^\infty \mu^+ V_i^{(3)-} V_j^{(3)+}\,\mathrm{d}z - \xi_{i\mathrm{L}}^- \int_0^\infty \mu^- V_i^{(3)-} V_j^{(3)+}\,\mathrm{d}z\right\}$$

$$S_{ij} = \frac{1}{2}\left\{\xi_{i\mathrm{L}}^- \int_0^\infty \mu^- V_i^{(3)-} V_j^{(3)+}\,\mathrm{d}z + \xi_{j\mathrm{L}}^+ \int_0^\infty \mu^+ V_i^{(3)-} V_j^{(3)+}\,\mathrm{d}z\right\}. \quad (3.54)$$

Rayleigh wave. In this case, the displacement and stress in a mode (with the suffix R dropped in (3.55) through (3.57)) that travels towards increasing x (i.e., both in the incident and transmitted waves) have the form

$$\mathbf{V} = -iV^{(2)}\mathbf{e}_x + V^{(1)}\mathbf{e}_z, \qquad \mathbf{p} = \bar{\sigma}_{11}\mathbf{e}_x - i\bar{\sigma}_{13}\mathbf{e}_z \quad (3.55)$$

where

$$\bar{\sigma}_{11} = -(\lambda + 2\mu)\xi_R V^{(2)} + \lambda\frac{\partial V^{(1)}}{\partial z},$$

$$\bar{\sigma}_{13} = \mu\left(\frac{\partial V^{(2)}}{\partial z} + \xi_R V^{(1)}\right). \quad (3.56)$$

Substituting these expressions in (3.46a), we get

$$J = 2i\left[\xi_R \int_0^\infty (\lambda + 2\mu)\,(V^{(2)})^2\,\mathrm{d}z + \int_0^\infty \mu\frac{\partial V^{(2)}}{\partial z}V^{(1)}\,\mathrm{d}z + \right.$$

$$\left. + \xi_R \int_0^\infty \mu (V^{(1)})^2\,\mathrm{d}z + \int_0^\infty \lambda\frac{\partial V^{(1)}}{\partial z}V^{(2)}\,\mathrm{d}z\right]$$

$$= 2i\,[\xi_R(I_R^{(1)} + I_R^{(2)}) + I_R^{(3)} + I_R^{(4)}]. \quad (3.57)$$

Similarly to the previous treatment, we impose a normalization on the eigenfunctions to make $J = 2i$, i.e., the expression within the square brackets to equal unity. The coupling coefficients P_{ij} and S_{ij} will under this normalization take the form

$$P_{ij} = \{Q^-_{ij} - Q^+_{ij} - M^-_{ij} + M^+_{ij} - L^-_{ij} + L^+_{ij} - R^-_{ij} + R^+_{ij}\},$$
$$S_{ij} = \{Q^-_{ij} + Q^+_{ij} + M^-_{ij} + M^+_{ij} + L^-_{ij} + L^+_{ij} - R^-_{ij} - R^+_{ij}\}, \tag{3.58}$$

where

$$Q^-_{ij} = \xi^-_{iR} \int_0^\infty (\lambda^- + 2\mu^-) V_i^{(2)-} V_j^{(2)+} \, dz;$$

$$Q^+_{ij} = \xi^+_{jR} \int_0^\infty (\lambda^+ + 2\mu^+) V_i^{(2)-} V_j^{(2)+} \, dz;$$

$$M^-_{ij} = \int_0^\infty \mu^- \frac{\partial V_i^{(2)-}}{\partial z} V_j^{(1)+} \, dz; \qquad M^+_{ij} = \int_0^\infty \mu^+ \frac{\partial V_j^{(2)+}}{\partial z} V_i^{(1)-} \, dz;$$

$$L^-_{ij} = \xi^-_{iR} \int_0^\infty \mu^- V_i^{(1)-} V_j^{(1)+} \, dz; \qquad L^+_{ij} = \xi^+_{iR} \int_0^\infty \mu^+ V_i^{(1)-} V_j^{(1)+} \, dz; \tag{3.59}$$

$$R^-_{ij} = \int_0^\infty \lambda^- \frac{\partial V_j^{(1)+}}{\partial z} V_i^{(2)-} \, dz; \qquad R^+_{ij} = \int_0^\infty \lambda^+ \frac{\partial V_i^{(1)-}}{\partial z} V_j^{(2)+} \, dz.$$

Since all coupling coefficients are real, the reflection and transmission coefficients are real too.

Now consider an sth Rayleigh (or Love) surface wave mode at oblique incidence at an angle of θ_s^- to a vertical interface. The difference from the case of normal incidence is that, firstly, different modes are reflected and transmitted at different angles ($\theta_{k\mathrm{D}}^-$ and $\theta_{k\mathrm{D}}^+$, respectively) that are determined by the phase velocity of the relevant mode in accordance with Snell's law

$$\xi^{\pm}_{k\mathrm{D}} \sin \theta^{\pm}_{k\mathrm{D}} = \xi^-_s \sin \theta^-_s, \tag{3.60}$$

and, secondly, a Rayleigh wave obliquely incident on the interface gives rise to Love waves (D = L), as well as to Rayleigh waves, (D = R) and vice versa.

The case of oblique incidence cannot be treated within the framework of two-dimensional wave propagation. The Green function in the representation theorem should then be that for a three-dimensional problem (see 1.2), and the integration should be performed over a closed surface rather than along a contour. The surface in question is taken to be that bounding the relevant quarter-space, i.e., the $x = 0$ and $z = 0$ planes, and closing the surface at infinity. Similarly to the two-dimensional case, it follows from the boundary conditions and the principle of limiting attenuation that it is only the integral over the $x = 0$ plane which is not zero. The integration over y is done approximately, using the method of stationary phase (Copson, 1965), which is justified when the observation site is at a great distance from the interface. The condition for the phase to be stationary leads to (3.60). Subsequent treatment proceeds on the lines of the normal incidence case, but the field of reflected and transmitted waves is represented as a sum of both Rayleigh and Love modes. In contrast to the case of normal incidence, the coupling coefficients between different modes in any one medium are no longer zero. For convenience of notation Rayleigh and Love modes will be numbered consecutively, the mode index ranging from 1 to m_R for Rayleigh waves, and from $1 + m_R$ to $m_R + m_L$ for Love waves in the same medium. The result is a set of equations for reflection and transmission coefficients that is similar to (3.51), (3.52):

$$\begin{aligned} A_i \cos\theta_i^- &= -\sum_j B_j P_{ij} + \sum_{k \neq i} A_k Q_{ik} + T_{is} \\ B_j \cos\theta_j^+ &= S_{sj}^* + \sum_i A_i P_{ij}^* - \sum_{l \neq j} B_l Y_{lj}^* \end{aligned} \tag{3.61}$$

where the P_{ij}, S_{ij} are given by formulas like (3.49), (3.53) in which

$$V_k(z) = V_k^{(3)}(\mathbf{e}_y \cos\theta_k - \mathbf{e}_x \sin\theta_k) \tag{3.62}$$

for Love waves and

$$\mathbf{V}_k(z) = V_k^{(1)}\mathbf{e}_z + V_k^{(2)}(\mathbf{e}_x \cos\theta_k + \mathbf{e}_y \sin\theta_k) \tag{3.63}$$

for Rayleigh waves;

$$\mathbf{p}_k(z) = \mathbf{T}(\mathbf{V}_k(z) \exp[-i\xi_{k\mathrm{D}}(x\cos\theta_k + y\sin\theta_k)], \mathbf{e}_x)$$

at $x = 0$, $y = y_{st}$ (y_{st} is the point of stationary phase). Here, Q_{ik} is the coupling coefficient between the ith and kth reflected mode in medium 2:

$$Q_{ik} = \frac{1}{J_i^-} \int_0^\infty (\mathbf{p}_i^- \mathbf{V}_k^{-*} - \mathbf{V}_i^- \mathbf{p}_k^{-*})\, \mathrm{d}z \tag{3.64}$$

T_{is} being the coupling coefficient between the ith reflected mode and the sth incident mode

$$T_{is} = \frac{1}{J_i^-} \int_0^\infty (\mathbf{p}_i^- \mathbf{V}_s^- - \mathbf{V}_i^- \mathbf{p}_s^-)\, \mathrm{d}z \tag{3.65}$$

and Y_{lj} is that between the lth and jth transmitted mode

$$Y_{lj} = \frac{1}{J_j^+} \int_0^\infty (\mathbf{p}_l^{+*} \mathbf{V}_j^{+*} - \mathbf{V}_l^{+*} \mathbf{p}_j^{+*})\, \mathrm{d}z. \tag{3.66}$$

For normal incidence all these coefficients vanish because of the orthogonality condition. We can arrive at an understanding of what form a coupling coefficient between two modes takes by examining the simplest case of a Rayleigh wave incident at an angle of θ^- on a vertical interface between two homogeneous quarter-spaces. This situation involves a single reflected and a single transmitted Rayleigh wave. The angle of refraction is denoted θ^+. The Rayleigh wave displacement vector in the system of coordinates rotated relative to the original system so as to make the x-axis coincide with the direction of wave propagation has the components $(-iV^{(2)}, 0, V^{(1)})$. The traction acting on the plane perpendicular to the propagation direction has the components $(\bar{\sigma}_{11}, 0, -i\bar{\sigma}_{13})$, where

$$\bar{\sigma}_{11} = -(\lambda + 2\mu)\xi V^{(2)} + \lambda \frac{\partial V^{(1)}}{\partial z}$$
$$\bar{\sigma}_{13} = \mu \left(\frac{\partial V^{(2)}}{\partial z} + \xi V^{(1)} \right) \tag{3.67}$$

while the traction components across the plane of incidence are $(0, \bar{\sigma}_{11} + 2\mu\xi V^{(2)}, 0)$, respectively. The traction across the $x = 0$ plane thus has the form

$$\bar{\sigma}_{11}\mathbf{e}_x - i\bar{\sigma}_{13}\mathbf{e}_z \cos\theta - 2\mu\xi V^{(2)} \sin\theta(\mathbf{e}_y \cos\theta - \mathbf{e}_x \sin\theta).$$

Substituting this and the displacement vector into (3.49), (3.53), we obtain the coupling coefficients:

$$P = [I_1 \cos\theta^- - I_2 \cos\theta^+ + 2\xi^- \sin\theta^- \sin(\theta^- + \theta^+)I_3]/2$$
$$S = [-I_1 \cos\theta^- - I_2 \cos\theta^+ + 2\xi^- \sin\theta^- \sin(\theta^- - \theta^+)I_3]/2 \tag{3.68}$$

where

$$I_1 = \int_0^\infty (\bar{\sigma}_{11}^+ V^{(2)-} + \bar{\sigma}_{13}^- V^{(1)+})\, dz$$
$$I_2 = \int_0^\infty (\bar{\sigma}_{11}^- V^{(2)+} + \bar{\sigma}_{13}^+ V^{(1)-})\, dz \tag{3.69}$$
$$I_3 = \int_0^\infty (\mu^+ - \mu^-) V^{(2)+} V^{(2)-}\, dz.$$

Energy relations. The energy of a surface wave incident on an interface between two media is partitioned: some part is converted into transmitted surface wave energy, part into reflected energy and, lastly, some part is converted to the body waves unaccounted for. The boundary conditions of continuity of displacement and stress at the interface cannot be satisfied, unless the field of body waves has displacement and stress discontinuities at $x = 0$ that are equal respectively to

$$[\mathbf{V}] = \mathbf{V}^+_{\text{body}} - \mathbf{V}^-_{\text{body}} = \mathbf{V}^-_s + \sum_i A_i \mathbf{V}^-_i - \sum_j B_j \mathbf{V}^+_j$$
$$[\mathbf{p}] = \mathbf{p}^+_{\text{body}} - \mathbf{p}^-_{\text{body}} = \mathbf{p}^-_s + \sum_i A_i \mathbf{p}^-_i - \sum_j B_j \mathbf{p}^+_j \tag{3.70}$$

as can be deduced from the boundary conditions (3.42)—(3.43). These boundary conditions are for normal incidence, but hold for oblique incidence too, except that $\mathbf{V}_i(z)$ and $\mathbf{p}_i(z)$ are to be determined in accordance with (3.62), (3.63). The work that should be done during a cycle of oscillation to generate these displacements and stresses at the portion of the interface formed by a semi-infinite strip of unit width is given by the integral (Its and Yanovskaya, 1977)

$$W = \frac{\pi}{2i} \int_0^\infty ([\mathbf{p}]^*[\mathbf{V}] - [\mathbf{V}]^*[\mathbf{p}])\, dz. \tag{3.71}$$

That work can be shown to be equal to the difference in surface wave energy flux (during a cycle of oscillation) to the left and right of the interface. One can see this by writing down the energy carried by a surface wave across a vertical semi-infinite strip of unit width:

$$\mathfrak{P} = \frac{\pi}{2i} \int_0^\infty (\mathbf{p}^*\mathbf{V} - \mathbf{V}^*\mathbf{p})\, dz \tag{3.72}$$

where $\mathbf{V}(z)$ and $\mathbf{p}(z)$ are the vertical distributions of displacement and traction, respectively, on a vertical plane of x = constant. Substituting into (3.72) the expressions for displacement and stress to the left (incident and reflected wave) and to the right (transmitted wave) of the interface and subtracting the one from the other, we get the difference of the energy fluxes to the left and to the right

$$\mathfrak{P}^- - \mathfrak{P}^+ = \frac{\pi}{2i} \int_0^\infty ([\mathbf{p}]^*[\mathbf{V}] - [\mathbf{V}]^*[\mathbf{p}])\, dz \tag{3.73}$$

which is exactly (3.71).

Thus, the energy loss in surface waves equals the work done by the stress discontinuity $[\mathbf{p}]$ at the $x = 0$ interface to produce the displacement discontinuity $[\mathbf{V}]$. It is just this work which becomes the energy of body waves unaccounted for.

It can further be shown that energy satisfies the condition of stationarity with respect to the reflection and transmission coefficients of surface waves, i.e., the coefficients A_i and B_j determined by the above method satisfy the equations

$$\partial W/\partial A_i = 0, \qquad \partial W/\partial B_j = 0. \tag{3.74}$$

That this is so can readily be verified by substituting (3.70) into (3.71).

3.4. A NUMERICAL METHOD FOR DETERMINING THE DISPLACEMENT FIELD

Consider the problem of a plane harmonic Love wave of frequency ω at normal incidence on a vertical welded contact $x = 0$ (see Figure 3.1) between two horizontally stratified quarter-spaces. The incident wave propagates from $x = -\infty$ along the x-axis as given by (3.1). The displacement field due to it then has the form

$$\mathbf{u}(x, z, t) = u(x, z) \exp(i\omega t)\mathbf{e}_y.$$

The function $u(x, z)$ is the solution of the boundary-value problem described in 3.1. Gabrielov (1984) has carefully examined the solution for existence and uniqueness; Bukchin (1979), Bukchin and Levshin (1980) have put forward a numerical method for computing the displacement function $u(x, z)$ to be described below.

Suppose we know the displacement at the contact, $u(0, z) = f(z)$. In that case (see (3.6), (3.7)), $u(x, z)$ can be represented in the form

$$u(x,z) = \begin{cases} V_s^-(z)\exp(-i\xi_s^- x) + \sum_i A_i V_i^-(z)\exp(i\xi_i^- x) + \\ + \int_0^\infty A(\eta)V_\eta^-(z)\exp[i\xi^-(\eta)x]\,\mathrm{d}\eta, \quad x \leqslant 0, \\ \sum_j B_j V_j^+(z)\exp(-i\xi_j^+ x) + \\ + \int_0^\infty B(\eta)V_\eta^+(z)\exp[-i\xi^+(\eta)x]\,\mathrm{d}\eta, \quad x \geqslant 0, \end{cases} \tag{3.75}$$

where

$$A(\eta) = \int_0^\infty \mu^-(z)f(z)V_\eta^{-*}(z)\,\mathrm{d}z, \quad A_i = \int_0^\infty \mu^-(z)f(z)V_i^-(z)\,\mathrm{d}z - \delta_{is}$$
$$B(\eta) = \int_0^\infty \mu^+(z)f(z)V_\eta^{+*}(z)\,\mathrm{d}z, \quad B_j = \int_0^\infty \mu^+(z)f(z)V_j^+(z)\,\mathrm{d}z. \tag{3.76}$$

The sets of eigenfunctions $V_\eta^-(z)$, $V_i^-(z)$ and $V_\eta^+(z)$, $V_j^+(z)$ form an orthogonal basis in the space of square integrable functions (see Kazi, 1976; Gabrielov, 1984). They are normalized as follows:

$$\int_0^\infty \mu^-(z)V_\eta^-(z)V_\lambda^{-*}(z)\,\mathrm{d}z = \delta(\eta - \lambda),$$

$$\int_0^\infty \mu^-(z)V_i^-(z)V_k^-(z)\,\mathrm{d}z = \delta_{ik},$$

$$\int_0^\infty \mu^+(z)V_\eta^+(z)V_\lambda^{+*}(z)\,\mathrm{d}z = \delta(\eta - \lambda),$$

$$\int_0^\infty \mu^+(z)V_j^+(z)V_q^+(z)\,\mathrm{d}z = \delta_{jq}.$$

Formulas (3.75) and (3.76) at $x = 0$ provide expansions of $f(x)$ on this basis ($f(z)$ is assumed to be a square integrable function). For any such function the displacement $u(x, z)$ of the form (3.75) satisfies the differential equation (1.4), the conditions of continuity of displacement and stress at layer boundaries, the absence of stress at the horizontal portion of the free surface, the limiting attenuation at infinity, and the continuity of displacement at the vertical contact $x = 0$. It is only the continuity of stress at the contact which will not hold for an arbitrary function $f(z)$:

$$\mu^-(z)\frac{\partial u(0-0, z)}{\partial x} = \mu^+(z)\frac{\partial u(0+0, z)}{\partial x}. \tag{3.77}$$

Substituting into (3.77) the derivatives $\partial u(x, z)/\partial x$ at the left and right of $x = 0$ as found from (3.75), (3.76), we obtain an equation in $f(z)$. This however is an integral equation of the first kind, so cannot be solved numerically because it is incorrect, unless it is first stabilized (Tikhonov and Arsenin, 1974). We replace the condition of continuity of stress (3.77) by its finite-difference analog

$$\mu^-(z)\,[u(0, z) - u(-\varepsilon, z)]/\varepsilon = \mu^+(z)\,[u(\varepsilon, z) - u(0, z)]/\varepsilon + O(\varepsilon),$$

where ε is a small positive quantity. Discarding the quantities $O(\varepsilon^2)$, we get

$$f(z) = u(0, z) = \frac{\mu^-(z)u(-\varepsilon, z) + \mu^+(z)u(\varepsilon, z)}{\mu^-(z) + \mu^+(z)}. \tag{3.78}$$

Substituting into (3.78) $u(-\varepsilon, z)$ and $u(\varepsilon, z)$ expressed by means of (3.75) and (3.76), we obtain an integral equation in $f(z)$:

$$\begin{aligned} f(z) = {} & [\mu^-(z)/(\mu^-(z) + \mu^+(z))]\Bigg\{ V_s^-(z)\,[\exp(i\xi_s^-\varepsilon) - \exp(-i\xi_s^-)] + \\ & + \sum_i V_i^-(z)\exp(-i\xi_i^-\varepsilon)\int_0^\infty \mu^-(\tau)f(\tau)V_i^-(\tau)\,\mathrm{d}\tau + \\ & + \int_0^\infty V_\eta^-(z)\exp[-i\xi^-(\eta)\varepsilon]\int_0^\infty \mu^-(\tau)f(\tau)V_\eta^{-*}(\tau)\,\mathrm{d}\tau\,\mathrm{d}\eta\Bigg\} + \\ & + [\mu^+(z)/(\mu^-(z) + \mu^+(z))]\Bigg\{\sum_j V_j^+(z)\exp(-i\xi_j^+\varepsilon) \times \\ & \times \int_0^\infty \mu^+(\tau)f(\tau)V_j^+(\tau)\,\mathrm{d}\tau + \int_0^\infty V_\eta^+(z)\exp[-i\xi^+(\eta)\varepsilon] \times \\ & \times \int_0^\infty \mu^+(\tau)f(\tau)V_\eta^{+*}(\tau)\,\mathrm{d}\tau\,\mathrm{d}\eta\Bigg\}. \end{aligned} \tag{3.79}$$

It can be shown that, if $f(z)$ satisfies (3.79) to within $O(\varepsilon^2)$, it differs from the true displacement function $u(0, z)$ at the contact by a quantity of order ε. This is also true when ε is a small complex quantity, $\varepsilon_0(1 + i)$, where ε_0 is real. Bukchin (1979) showed that under this choice of ε ($\varepsilon_0 > 0$), the method of successive approximations applied to (3.79) converges.

If $f_{n-1}(z)$ and $f_n(z)$ are two iterates of $f(z)$, and their difference is of order

$O(\varepsilon_0^2)$, then $f_n(z)$ differs from $u(0, z)$ by $O(\varepsilon_0)$. To sum up, when a solution $f(z)$ of (3.79) has been found to a given accuracy using the method of successive approximations, formulas (3.75) and (3.76) can be used to evaluate the displacement $u(x, z)$ at any point, as well as the reflection and transmission coefficients, A_i and B_j.

3.5. OBLIQUE CONTACT AND OTHER POSSIBLE REFINEMENTS

Oblique contact. The models discussed in the preceding sections involve a vertical contact between vertically varying quarter-spaces, and obviously are very crude approximations to actual situations at junctions of blocks of different structure. As a rule, fault zones, to say nothing of the transition from oceanic to continental crust, are obliquely dipping. Hence a more adequate model for such transitional zones is an obliquely dipping interface between blocks. For this reason papers have appeared dealing with transmission of a surface wave across a nonvertical (obliquely dipping and curved) interface. Lutikov (1979) investigates the effect of an oblique interface between two media on Rayleigh wave propagation and shows, in particular, that the reflection and transmission coefficients are in this case complex. Malischewski (1974) solves the problem of reflection and transmission of a Rayleigh wave across a curved interface using Alsop's method, and obtains complex reflection and transmission coefficients. Its and Yanovskaya (1983) have shown that the Green function technique function as described in 3.3 can be readily extended to deal with a nonvertical interface. Before we consider it, we would like to discuss the restrictions to be imposed on the configuration of the interface. This method, like other studies in surface wave propagation in media involving curved or obliquely dipping interfaces, assumes all surface waves (incident, reflected, and transmitted) to propagate to the left and right of the contact in the same fashion as in a half-space having the relevant parameter values. This approximation can, however, be considered valid for the case of a vertical contact only. When the interface is curvilinear within the interval that contains the abscissas of the interface points, the structure corresponds neither to medium 1 nor to medium 2 (Figure 3.4 shows this interval bounded by two vertical dashed lines). For the case of an infinitely long oblique interface this discrepancy occurs for the entire medium on the side of the acute angle. It seems natural, therefore, to impose certain restrictions on interface configuration, assuming the interface to be vertical down from some depth H (Figure 3.4). An approximation of this sort will not affect the properties of surface waves whose depth of penetration is shallower that H. Besides, this restriction is not very serious for modelling transition zones in a real earth, since the structures on both sides of the contact usually become less markedly different at greater depths, so that one seems entitled to extend the contact downward along the vertical.

Another obvious restriction is that the interval between the interface points farthest from one another should not anyway exceed the wavelength. Otherwise, this interval will give rise to waves that are different from those which propagate to the left and right of the interval, and the field of these waves will affect the wave field at greater distances.

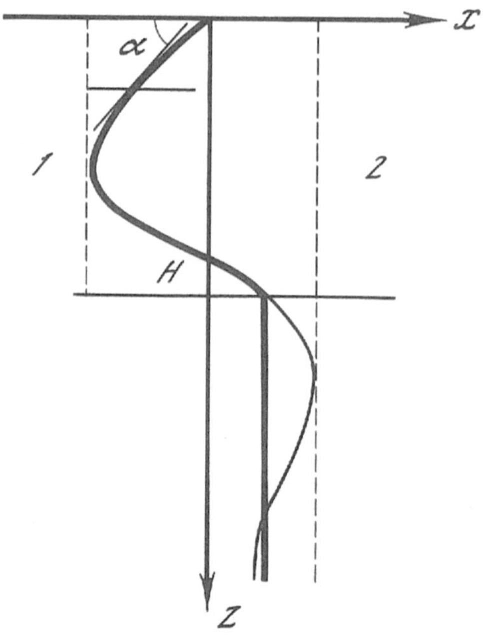

Fig. 3.4. Model of a curved discontinuity. The heavy line shows the adopted discontinuity approximation for waves with depths of penetration less than H.

Interfaces that satisfy those two conditions will be called subvertical, and all subsequent discussion will be confined to subvertical interfaces.

The equation that defines an interface between two media is taken in the form $x = x(z)$, i.e., any interface is assumed to be a cylindrical surface along the y-axis. The angle of dip at a point of the interface is given by $\alpha = \arctan(dx/dz)$ (Figure 3.4). For convenience we place the origin at the point of intersection between the interface and the free surface $z = 0$, so that $x(0) = 0$. Consider the general case of the sth surface wave mode incident at some angle θ_s to the interface. It can readily be shown, using the method described in 3.3, that the equations for determining reflection and transmission coefficients are in this case exactly those derived for a vertical interface (Equations (3.61)), the coupling coefficients between modes being given by formulas like (3.49), (3.53), (3.62)–(3.64). The difference is that the integration is performed along the interface contour L rather than along z. Thus, for example, the coupling coefficient between the ith reflected mode and the jth transmitted mode is

$$P_{ij} = \frac{1}{2i} \int_L [\mathbf{T}^*(\hat{\mathbf{u}}_i^{\text{refl}}, \mathbf{n})\hat{\mathbf{u}}_j^{\text{trans}} - \hat{\mathbf{u}}_i^{\text{refl}}\mathbf{T}(\hat{\mathbf{u}}_j^{\text{trans}}, \mathbf{n})]\, dl \tag{3.80}$$

where $\mathbf{T}(\hat{\mathbf{u}}, \mathbf{n})$ is the traction at a given point of the interface. If the normal $\mathbf{n}$ to the interface is directed so as to make an acute angle with the x-axis, then

$$\mathbf{T}(\hat{\mathbf{u}}, \mathbf{n}) = \mathbf{T}(\hat{\mathbf{u}}, \mathbf{e}_x)\cos\alpha + \mathbf{T}(\hat{\mathbf{u}}, \mathbf{e}_z)\sin\alpha. \tag{3.81}$$

The integration in (3.80) is performed along the interface, so that $dl = dz/\cos\alpha(z)$.

The other coupling coefficients between modes are also given by formulas like (3.80).

In calculating a coupling coefficient one should remember that the values of $\mathbf{T}(\hat{\mathbf{u}}, \mathbf{n})$ and $\hat{\mathbf{u}}$ in the integrand are to be determined at the points of $x(z)$. The dependence of $\mathbf{T}(\hat{\mathbf{u}}, \mathbf{n})$ and $\hat{\mathbf{u}}$ on x is controlled by an exponential function of the form $\exp[\pm\xi_i^{\pm} x \cos\theta_i^{\pm}]$. From this it can be seen that the coupling coefficients must be complex, accordingly the reflection and transmission coefficients are complex too.

Unwelded contact. Complex coefficients also arise in cases where elastic parameters vary rapidly within a transition zone compared with the host media, so that a zone of this kind can hardly be modelled by a welded contact. Investigations show that faulting, cracking, and moistening lead to a dramatic drop in the shear modulus within fault zones (Nikolaev, 1978). Pod'yapolsky (1963) has studied elastic wave propagation across a thin layer having a lower shear modulus. He shows that this problem can be solved approximately by replacing the layer with a plane unwelded contact at which the following boundary condition are fulfilled:

$$\begin{aligned}
&\hat{\mathbf{u}}_{n\,\mathrm{inc}}(0, z, \omega) + \hat{\mathbf{u}}_{n\,\mathrm{refl}}(0, z, \omega) = \hat{\mathbf{u}}_{n\,\mathrm{trans}}(0, z, \omega)\\
&\hat{\mathbf{u}}_{\tau\,\mathrm{inc}}(0, z, \omega) + \hat{\mathbf{u}}_{\tau\,\mathrm{refl}}(0, z, \omega) + m\mathbf{T}((\hat{\mathbf{u}}_{\tau\,\mathrm{inc}} + \hat{\mathbf{u}}_{\tau\,\mathrm{refl}}), \mathbf{e}_x)|_{x=0}\\
&\quad = \hat{\mathbf{u}}_{\tau\,\mathrm{trans}}(0, z, \omega)\\
&\mathbf{T}((\hat{\mathbf{u}}_{\mathrm{inc}} + \hat{\mathbf{u}}_{\mathrm{refl}}), \mathbf{e}_x)|_{x=0} = \mathbf{T}(\mathbf{u}_{\mathrm{trans}}, \mathbf{e}_x)|_{x=0}.
\end{aligned} \tag{3.82}$$

The normal and tangential (with respect to the interface) displacement components are denoted by $\mathbf{n}$ and τ, respectively. A 'non-weldedness' parameter is defined as the ratio of the transition zone thickness h_0 to the shear modulus μ_0 within it:

$$m = \frac{h_0}{\mu_0}. \tag{3.83}$$

Its and Malischewski (1987a, b) show that the Green function technique can be extended to deal with unwelded contacts. The solution is constructed in a similar fashion to the case of welded contact. Here is a brief account of the derivation of a set of equations to evaluate reflection and transmission coefficients for normal incidence of the sth Love mode at an unwelded vertical contact. In this case a Love wave has displaement and stress components alone that are tangent to the $x = 0$ plane, and conditions (3.82) can be written as

$$\begin{aligned}
&V_s^-(z) + \sum_i A_i V_i^{-*}(z) + V_{\mathrm{body}}^- + m\left(p_s^-(z) + \sum_i A_i p_i^{-*}(z) + p_{\mathrm{body}}^-\right)\\
&\quad = \sum B_j V_j^+(z) + V_{\mathrm{body}}^+;\\
&p_s^-(z) + \sum_i A_i p_i^{-*}(z) + p_{\mathrm{body}}^- = \sum B_j p_j^+(z) + p_{\mathrm{body}}^+.
\end{aligned} \tag{3.84}$$

We now use the expression (3.47) for the reflection coefficient A_i derived from the representation theorem, solve (3.84) for p_i^{refl} and V_i^{refl} and, disregarding as before the relation between normal modes and body waves, obtain equations that can again be represented in the form (3.51), but with different complex coefficients P_{ij}:

$$P_{ij} = \tilde{P}_{ij} + imZ_{ij} \tag{3.85}$$

where

$$Z_{ij} = \frac{i}{J_i^-}\int_0^\infty p_i^- p_j^+ \,\mathrm{d}z. \tag{3.86}$$

Similarly, the expressions for B_j have the form (3.52), but involve the coefficients S_{sj} given by

$$S_{sj} = \tilde{S}_{sj} + imX_{sj} \tag{3.87}$$

where

$$X_{sj} = \frac{i}{J_j^+}\int_0^\infty p_s^{-*} p_j^+ \,\mathrm{d}z. \tag{3.88}$$

The $\tilde{P}_{ij}$ and $\tilde{S}_{sj}$ which enter (3.85) and (3.87) are defined by (3.49), (3.53).

The possibilities offered by the introduction of unwelded contact in modelling a transition zone are made clearer by defining a dimensionless parameter of unweldedness, for instance, in the form

$$\tilde{m} = \frac{h_0}{\mu_0}\,\xi_{sL}^- \mu^-(0) \tag{3.89}$$

where $\mu^-(0)$ is the shear modulus near the free surface. The expressions for P_{ij} and S_{sj} can then be rewritten as

$$P_{ij} = \tilde{P}_{ij} + i\tilde{m}\tilde{Z}_{ij}$$

$$S_{sj} = \tilde{S}_{sj} + i\tilde{m}\tilde{X}_{sj}$$

where

$$\tilde{Z}_{ij} = -\frac{\xi_i^- \xi_j^+}{2\xi_{sL}^-}\int_0^\infty \frac{\mu^- \mu^+}{\mu^-(0)} V_i^{(3)-} V_j^{(3)+} \,\mathrm{d}z$$

$$\tilde{X}_{sj} = \frac{\xi_j^+}{2}\int_0^\infty \frac{\mu^- \mu^+}{\mu^-(0)} V_s^{(3)-} V_j^{(3)+} \,\mathrm{d}z \tag{3.90}$$

$\tilde{X}_{sj}$ and $\tilde{Z}_{ij}$ thus defined are dimensionless coefficients like $\tilde{P}_{ij}$ and $\tilde{S}_{sj}$.

The conditions necessary for modelling a transition zone by a discontinuity are described by the inequality $h\xi \ll 1$. The parameter of unweldedness is therefore bounded only when $\mu_0 \ll \mu_1(\bar{0})$. Otherwise $\tilde{m} \Rightarrow 0$, and P_{ij}, S_{sj} become the coupling coefficients (3.85), (3.87) for welded contact.

In considering the normal incidence of Rayleigh waves, as well as the general case of a surface wave at oblique incidence on an unwelded contact, we should remember that displacements at the contact have both a tangential and a normal component and use boundary conditions in the general form to derive the same set of equations (3.61) in which T_{is}, Q_{ik}, Y_{lj} are given by (3.64) through (3.66), while P_{ij} and S_{sj} can be represented in the form (3.85), (3.88). The coefficients Z_{ij} and X_{sj} which these expressions involve have the form

$$Z_{ij} = \frac{i}{J_i^-} \int_0^\infty \mathbf{p}_{\tau i}^- \mathbf{p}_{\tau j}^+ \, dz$$
$$X_{sj} = \frac{i}{J_j^+} \int_0^\infty \mathbf{p}_{\tau s}^{-*} \mathbf{p}_{\tau j}^+ \, dz. \tag{3.91}$$

However, the notion of a fault as an interface that is unwelded throughout its depth is also simplistic, since we are aware of the fact that pronounced inhomogeneities occur in the Earth's upper layers. In dealing with wavelengths longer than the fault depth, this approximation could cause bias in the estimates of reflection and transmission coefficients. For this reason a more adequate model would be one describing a narrow low-velocity zone of finite depth. A model of this kind can be constructed by combining a welded and an unwelded contact. As both boundary conditions can be mathematically represented in a unified form as

$$\hat{\mathbf{u}}_{n\,\mathrm{inc}}(0, z, \omega) + \hat{\mathbf{u}}_{n\,\mathrm{refl}}(0, z, \omega) = \hat{\mathbf{u}}_{n\,\mathrm{trans}}(0, z, \omega)$$
$$\hat{\mathbf{u}}_{\tau\,\mathrm{inc}}(0, z, \omega) + \hat{\mathbf{u}}_{\tau\,\mathrm{refl}}(0, z, \omega) + m(z)\mathbf{T}((\hat{\mathbf{u}}_{\tau\,\mathrm{inc}} + \hat{\mathbf{u}}_{\tau\,\mathrm{refl}}), \mathbf{e}_x)|_{x=0}$$
$$= \hat{\mathbf{u}}_{\tau\,\mathrm{trans}}(0, z, \omega) \tag{3.92}$$
$$\mathbf{T}((\hat{\mathbf{u}}_{\mathrm{inc}} + \hat{\mathbf{u}}_{\mathrm{refl}}), \mathbf{e}_x)|_{x=0} = \mathbf{T}(\hat{\mathbf{u}}_{\mathrm{trans}}, \mathbf{e}_x)|_{x=0}$$

where the parameter of unweldedness is a function of depth and vanishes beginning at some value of depth, it is obvious that the set of equations for estimating reflection and transmission coefficients at a contact described by (3.92) will have the form (3.61) with the coefficients P_{ij}, S_{sj}, the only difference being that Z_{ij} and X_{sj} now involve the parameter of unweldedness as a function of depth:

$$Z_{ij} = \frac{i}{J_i^-} \int_0^\infty m(z) \mathbf{p}_{\tau i}^- \mathbf{p}_{\tau j}^+ \, dz$$
$$X_{sj} = \frac{i}{J_j^+} \int_0^\infty m(z) \mathbf{p}_{\tau s}^{-*} \mathbf{p}_{\tau j}^+ \, dz. \tag{3.93}$$

Another possible refinement to the model that makes it more realistic is the assumption that the medium to the right and left of the interface has slight lateral inhomogeneity and that the interface is not a straight line in plan (Figure 3.4). In that case both the incident and the reflected and transmitted waves travel along curved paths as given by the ray equation (2.52). An incident wave will be reflected and transmitted, and the intensity of derived surface waves can only be evaluated far enough from the interface, according to what we have learned in this chapter. Wave intensity is controlled both by the value of the relevant coefficient (either reflection or transmission one) and by the change in geometrical spreading on crossing the interface. The geometrical spreading of space-time rays can be found by the method of Section 2.5, an approximate evaluation procedure for the coefficients being described in this chapter. The variation of geometrical spreading along a ray is given by the sets of differential equations (2.85), (2.92).

Note that the space-time ray method can give, not the spectral amplitude, but wave amplitude directly on a seismogram, thus enabling one to construct a seismogram without having recourse to spectral transformations. At the same time, reflection and transmission coefficients as evaluated by this method correspond to harmonic waves, i.e., they are spectral coefficients. This is, however, not critical for constructing a theoretical seismogram by the above procedure. That this is so can be readily seen when the surface wave field is expressed as a Fourier integral, the integral being evaluated by the stationary phase method. The spectral reflection or transmission coefficient which appears as a factor in the integrand is a slowly varying function, so that its value at the frequency corresponding to the stationary point enters as a factor in signal amplitude. Thus, although wave propagation in a model that involves both slight lateral inhomogeneities and interfaces between blocks is described within the framework of the space-time ray method, while the energy loss due to reflection and transmission at interfaces is evaluated by the spectral method, it is still possible to combine and in this way to calculate all the necessary features that are required for constructing seismograms in models approximating real structures. This would of course involve considerable computational difficulties, but seems to be feasible with fifth-generation computers. Algorithms for some steps of the calculation will be described in Chapter 4.

4. COMPUTATION TECHNIQUES FOR SURFACE WAVES

This chapter describes some techniques for computing surface wave fields with examples of their use. Since the theory of surface waves in laterally varying media set forth in Chapters 2 and 3 is essentially based on the properties of eigenfunctions and eigenvalues in one-dimensional boundary value problems for Love and Rayleigh waves discussed in Chapter 1, we begin by discussing solutions to those problems (Sections 4.1 and 4.2).

Considering that there is a large amount of literature devoted to this problem (see, for example, Thomson (1950), Haskell (1953), Molotkov (1961, 1972, 1984), Knopoff (1964), Vilkovich *et al.* (1966), Neigauz and Shkadinskaya (1966), Schwab and Knopoff (1970), Shkadinskaya (1971), Levshin (1973), Wiggins (1976), Abo-Zena (1979), Kennett and Clarke (1983)) we shall only discuss some aspects that contain technically novel ideas. The two sections that follow are concerned with computational techniques in use for laterally varying media, namely, for dynamic ray tracing (4.3) and for calculating reflection and transmission coefficients for a vertical contact (4.4).

4.1. ON THE MATRIX METHOD FOR ONE-DIMENSIONAL BOUNDARY VALUE PROBLEMS

Eigenvalues and eigenfunctions for one-dimensional boundary value problems corresponding to Love and Rayleigh waves are most frequently determined by matrix methods first developed by Thomson (1950) and Haskell (1953). In these methods, a vertically varying medium is modelled by a set of n homogeneous layers overlying a homogeneous half-space.

The eigenfunction in a homogeneous layer is represented by a linear combination of exponential functions, or trigonometric functions in case the exponents are imaginary. The coefficients of exponential functions in adjacent layers are connected by means of linear relations that express continuity of stress and displacement at interfaces. The coefficients for the uppermost layer can therefore be expressed in terms of those for the half-space using the machinery of matrix algebra. Taking into account the boundary condition that stresses should vanish at the free surface, one can construct a function that is the left-hand side of a dispersion relation

$$\Phi(\omega, \xi) = 0. \tag{4.1}$$

The usual form of $\Phi(\omega, \xi)$ is a product of two 2×2 matrices for Love waves or 4×4 ones for Rayleigh waves. There are detailed accounts of how these products should be formed, so we skip these technicalities. A few remarks will be made on computational difficulties encountered in solving the dispersion relation.

In problems associated with the determination of reflection and transmission coefficients for subvertical discontinuities, all roots of the dispersion relation ξ_k should be found within the finite interval $[\xi_{\min}, \xi_{\max}]$ corresponding to the existence region of normal modes. On the other hand, when a given mode propagates in a medium with weak lateral inhomogeneities, it is required to calculate a value of ξ for this mode in models corresponding to a vertical distribution of elastic parameters at consecutive points of a wave path. In both cases difficulties arise in numerical calculation of the roots at high frequencies ω where many modes exist whose velocities asymptotically approach one and the same value. The roots of the dispersion relation get closer to one another, so that there may occur missing roots in practical computations and jumps from one branch of a dispersion curve to another.

Another disadvantage in solving the Rayleigh wave dispersion relation (4.1) whose left-hand side is written as a product of 4×4 matrices is loss of accuracy in numerical computations for $\xi H_i \operatorname{Re} \sqrt{\xi^2 - \omega^2/a_i^2} \gg 1$ and $\operatorname{Re} \sqrt{\xi^2 - \omega^2/a_i^2} \gg \operatorname{Re} \sqrt{\xi^2 - \omega^2/b_i^2}$ (i is the number of the layer and H_i its thickness). This disadvantage has been removed by L. A. Molotkov (1961, 1972, 1984). He represents the left-hand side of the dispersion relation as a product of 5×5 matrices (for Rayleigh waves) instead of 4×4 ones. We shall not discuss this method, as it is described in detail in Molotkov (1984). There are also other modifications of the Thomson—Haskell method that are free from the above disadvantage (Knopoff, 1964; Dunkin, 1965; Schwab and Knopoff, 1970; Abo-Zena, 1979; Kennet and Clarke, 1983).

4.2. A NUMERICAL METHOD FOR SOLVING ONE-DIMENSIONAL BOUNDARY VALUE PROBLEMS

The necessity of modelling media with continuous variation of velocity with depth by plane-stratified analogs, the difficulties in accommodating sphericity effects and identifying roots of the dispersion relation at high frequencies, all these factors make it preferable to use techniques of numerical integration which would be free from the above defects. One is a modification of the fast recurrence technique put forward by Lidsky and Neigauz (1962) and applied by Neigauz and Shkadinskaya (1966), Shkadinskaya (1971) to solve the one-dimensional problem for Rayleigh waves.

Below we describe a procedure using this scheme for Love wave calculations in a spherically symmetrical sphere, a number of advantages over previous techniques (Vilkovich *et al.*, 1966; Levshin, 1973). The fact that an elastic sphere instead of a spherical shell is considered is not essential in our case, because surface waves do not penetrate into the region around the core-mantle boundary.

The differential operator for displacement $V = V^{(3)}$ in the range $0 \leqslant R \leqslant R_0$ is

$$\frac{d}{dR}\left[\mu\left(\frac{dV}{dR} - \frac{V}{R}\right)\right] + \frac{3\mu}{R}\frac{dV}{dR} +$$

$$+\left[\omega^2\rho - \mu\frac{N^2+1}{R^2}\right]V = 0, \tag{4.2}$$

the boundary conditions being

$$\bar{\sigma}_{\varphi R} = \mu(R)\left(\frac{dV}{dR} - \frac{V}{R}\right) = 0|_{R=R_0} \tag{4.3}$$

$$V = 0|_{R=0}. \tag{4.4}$$

Here $N^2 = \nu(\nu + 1)$, where ν (the index of the spherical harmonic) takes on integer values. Solutions for noninteger ν are found by interpolation. It is required to determine the eigenvalues $\omega_{k\mathrm{L}}(\nu)$ and eigenfunctions $V_k(R, \nu)$ for fixed k and ν.

The solution is found by the fast recurrence method (Lidsky and Neigauz, 1962). We define a function F as follows:

$$\mu\left(\frac{dV}{dR} - \frac{V}{R}\right) = -lFV, \tag{4.5}$$

where l is a normalizing constant, and define a complex function $W(R)$ as

$$W = (i - F)(i + F)^{-1} \tag{4.6}$$

W can obviously be represented as

$$W = \exp[i\psi(R)].$$

Differentiating (4.5) with respect to R, substituting into (4.2), and eliminating dV/dR everywhere, we get

$$\frac{dF}{dR} = \frac{1}{\mu}F^2 - \frac{4F}{R} + \frac{\mathfrak{R}}{l}, \tag{4.7}$$

where

$$\mathfrak{R} = \frac{\mu(2 - N^2)}{R^2} + \omega^2\rho.$$

We convert (4.5) from F to W, thereby obtaining a recurrence equation for W:

$$dW/dR = iS(W)W, \tag{4.8}$$

where

$$S(W) = \left(\frac{l}{\mu} + \frac{\mathfrak{R}}{l}\right) + \left(\frac{\mathfrak{R}}{l} - \frac{l}{\mu}\right)\frac{W + W^*}{2} + \frac{2i(W - W^*)}{R}.$$

The boundary conditions for W are

$$W(R_0) = 1 \tag{4.9}$$

$$W(0) = -1. \tag{4.10}$$

Substituting the expression for $W = \exp[i\psi(R)]$ into (4.6), we get a recurrence equation for ψ:

$$\frac{d\psi}{dR} = S(e^{i\psi}) \tag{4.11}$$

with the boundary conditions

$$\psi(R_0) = 2m\pi, \qquad m = 0, 1, 2, 3, \ldots, \tag{4.12}$$

$$\psi(0) = -\pi. \tag{4.13}$$

It can be shown that the eigenvalue number k and the integer m in (4.12) are identically equal, whence comes a numerical technique to determine $\omega^2_{kL}(\nu)$ for fixed k and ν. Take a trial value of ω^2 and integrate (4.8) numerically from $R = 0$ to $R = R_0$, calculating $\psi(R) = \arg W$ at each step. Thereby we determine the quantity

$$\Phi(\omega^2, v) = \psi(\omega^2, R_0) - 2k\pi.$$

Because $\varphi(\omega^2)$ steadily increases with ω^2, the value of ω^2 for which $|\varphi(\omega^2)| < \varepsilon$, where ε is a small constant, is easily found (by division into halves, Newton's method etc.). (4.8) is integrated using the difference scheme

$$W_{n+1} = \left(1 - \frac{ih_n}{2} S(W_{n+1/2})\right)^{-1} \left(1 + \frac{ih_n}{2} S(W_{n+1/2})\right) W_n,$$

$$W_{n+1/2} = \left(1 - \frac{ih_n}{2} S(W_n)\right)^{-1} \left(1 + \frac{ih_n}{2} S(W_n)\right) W_n$$

with the variable step $h_n = \min(\delta, H_0)$, where $\delta = \pi/(nS(W_n))$. The value of n is chosen so that ψ may be followed without jumps of 2π in ψ. The choice of H_0 is governed by considerations of the accuracy of integration.

When ω^2_{kL} has been found, the eigenfunction $V(R, \nu)$ is sought. Define the function

$$\eta = \frac{2l}{1 + W^*} V_k. \tag{4.14}$$

Using (4.2), (4.5) and (4.6), we find an equation for η:

$$d\eta/dR = \Psi(W)\eta, \tag{4.15}$$

where

$$\Psi(W) = \frac{2W^* - 1}{R} + \frac{i}{2}\left[\frac{l}{\mu} + \frac{\Re}{l} + \left(\frac{\Re}{l} - \frac{l}{\mu}\right) W^*\right];$$

$$\eta(0) = 1. \tag{4.16}$$

Equation (4.15) is solved numerically in the interval between $R = 0$ and $R = R_0$ using the difference scheme

$$\eta_{n+1} = \left(1 - \frac{h_n}{2}\Psi(W_{n+1})\right)^{-1}\left(1 + \frac{h_n}{2}\Psi(W_n)\right)\eta_n$$

together with the equation for W_n with the same step h_n. When $\eta(R)$ has been found, it is easy to determine $V_k(\nu, R)$ and its derivative with respect to R:

$$V_k = \frac{1 + W^*}{2l}\eta,$$
$$\frac{dV_k}{dR} = \frac{1}{2}\left(\frac{1 + W^*}{Rl} + i\frac{1 - W^*}{\mu}\right)\eta. \tag{4.17}$$

Large values of ν are typical of surface waves, $\nu > 10$. Equation (4.2) for such ν and the distributions of μ and ρ typical of the Earth's mantle have a turning point R_t below which the solution falls off exponentially as $R \Rightarrow 0$. An interval $[R_{in}, 0]$ therefore exists in which the actual behavior of the solution does not affect that for $R > R_{in}$. Naturally, exclusion of this interval from the numerical scheme, as was done in the matrix method, makes the computation significantly faster.

With this end in view, we first find the turning point R_t: the smallest R such that

$$\gamma(R) = \frac{\mu(R)(N^2 - 2)}{R^2} - \omega^2\rho > 0$$

for all $R < R_t$. From here on, we construct an upper-bound solution for our equation towards decreasing R

$$Y(R) = \exp\left[\int_{R_t}^{R}\sqrt{\gamma(R)}\,dR\right],$$

until this quantity becomes less than $\exp(-M)$ (usually $M = 10$ to 12). The resulting value of R is taken as R_{in}, the starting point for integration.

Although one can in principle apply the boundary condition (4.10) at R_{in}, better results are achieved by 'freezing' μ and ρ below R_{in}:

$$\mu(R) = \mu(R_{in} - 0)$$
$$\rho(R) = \rho(R_{in} - 0) \qquad R_{in} > R \geqslant 0.$$

The solution to (4.2) with the boundary condition (4.4) within this interval is a spherical Bessel function (Kuznetsov, 1965):

$$V(R) = \frac{1}{\sqrt{R}}J_{\nu+1/2}\left(\omega\sqrt{\frac{\rho}{\mu}}R\right),$$

where $J_{\nu+1/2}$ is a Bessel function of the first kind. Knowing $V(R_{in})$ and $dV/dR(R_{in})$, one can determine the boundary condition at R_{in} for W in (4.5) and (4.6). In addition to the eigenfunction V_k and its derivative with respect to R, the values of $\partial C_{kL}/\partial b$, $\partial C_{kL}/\partial\rho$ are determined at the grid points. The integration also yields $I^{(0)}_{kL}$, $I^{(1)}_{kL}$, $I^{(2)}_{kL}$ (formulas (1.43), (1.49)), and phase and group velocity C_{kL}, U_{kL} found from these by using (1.47), (1.48). Good agreement between the 'integral' phase velocity and that found from the eigenvalue $\omega^2_{kL}(\nu)$ as $C_{kL} = R_0\omega_{kL}(\nu)/(\nu + 1/2)$ is a criterion of reliable calculation.

The above technique for solving the one-dimensional boundary value problem for Love waves on a sphere does not require trigonometric functions to be computed and for this reason is more efficient than recurrence techniques previously proposed for Love waves (Vilkovich, Levshin and Neigauz, 1966). A similar technique of matrix recurrence for Rayleigh waves on a sphere is described by Shkadinskaya (1971).

4.3. DYNAMIC RAY TRACING

Calculation of space-time rays and wave packet amplitudes for surface waves by numerical integration of (2.52), (2.85), (2.92) is difficult, because one must be able to find, at each point of the ray, phase and group velocity, as well as the first and second derivative of phase velocity with respect to x, y and the first derivative of group velocity with respect to both frequency and coordinates. To do this in the general case, one has to solve the dispersion relation at each point of the ray, to construct the eigenfunction of the one-dimensional boundary value problem, and to calculate derivatives of phase and group velocity; all these tasks require numerical differentiation. However, when we take the simplest case of a homogeneous layer on a homogeneous half-space with a slightly curved interface between these, the right-hand sides of (2.52), (2.85), (2.92) can be easily calculated, permitting numerical modelling to be performed. This model is also of practical interest, because the variation of surface wave velocities and amplitudes in a real Earth within the period range 15—50 sec is largely controlled by changes in crustal thickness rather than by the variation of elastic moduli in the crust and upper mantle.

We proceed to examine the problem for this model in some more detail.

We assume the top of the layer to be horizontal: $z = 0$ and the bottom to be given by $z = Z(x, y)$. Then

$$\nabla C = \frac{\partial C}{\partial Z}\nabla Z, \qquad \nabla U = \frac{\partial U}{\partial Z}\nabla Z.$$

The second derivatives of phase velocity with respect to coordinates are expressed in terms of the derivatives with respect to Z:

$$\frac{\partial^2 C}{\partial x^2} = \frac{\partial^2 C}{\partial Z^2}\left(\frac{\partial Z}{\partial x}\right)^2 + \frac{\partial C}{\partial Z}\,\frac{\partial^2 Z}{\partial x^2}$$

$\partial C/\partial Z$, $\partial^2 C/\partial Z^2$, $\partial U/\partial Z$, and $\partial U/\partial \omega$ can be easily calculated for this model, because C and U are functions of the product $\omega Z = \zeta$:

$$C(\omega, Z) = \tilde{C}(\zeta), \qquad U(\omega, Z) = \tilde{U}(\zeta).$$

In such a case, if these functions have been calculated as functions of ζ, then the velocities can be found at any point for a fixed wave packet frequency ω from

$$C(x, y, \omega) = \tilde{C}(\omega Z(x, y)); \qquad U(x, y, \omega) = \tilde{U}(\omega Z(x, y)).$$

Hence the derivatives that are needed to calculate the right-hand sides are determined in terms of the derivatives of these functions with respect to ζ:

$$\frac{\partial C}{\partial Z} = \omega \tilde{C}', \quad \frac{\partial U}{\partial Z} = \omega \tilde{U}', \quad \frac{\partial^2 C}{\partial Z^2} = \omega^2 \tilde{C}'', \quad \frac{\partial U}{\partial \omega} = Z(x, y)\tilde{U}',$$

where the prime denotes differentiation with respect to ζ.

In order to be able to calculate $\tilde{C}'$, $\tilde{U}'$, $\tilde{C}''$ from closed formulas, rather than by numerical integration, which gives poorer accuracy, group velocity should be expressed in terms of phase velocity using (1.45). To do this, one should express the eigenfunction of the one-dimensional boundary value problem $V_k^{(3)}$ for Love waves or $(V_k^{(1)}, V_k^{(2)})$ for Rayleigh waves in terms of phase velocity, and then find expressions for the integrals entering (1.45). The resulting formula for Love wave group velocity is quoted in Section 1.4. A similar formula can be derived for Rayleigh waves but, being much more cumbersome, as has been mentioned in Section 1.4, it is not given here.

From (1.36) one can see that the formula for group velocity has the structure

$$\tilde{U}(\zeta) = F(\zeta, \tilde{C}(\zeta))$$

so that

$$\tilde{U}' = \frac{\partial F}{\partial \zeta} + \frac{\partial F}{\partial C} \tilde{C}'. \tag{4.18}$$

The derivative $\tilde{C}'$ is derived by using the relation connecting phase and group velocity:

$$\frac{1}{U} = \frac{1}{C} - \frac{\omega}{C^2} \frac{\partial C}{\partial \omega}.$$

Remembering that $\partial C/\partial \omega = \check{Z}C'$, we get

$$\tilde{C}' = \tilde{C}^2(\tilde{C}^{-1} - \tilde{U}^{-1})/\zeta. \tag{4.19}$$

Hence (4.18) can be written as

$$\tilde{U}' = F_\zeta + F\tilde{C}^2(\tilde{C}^{-1} - \tilde{U}^{-1})/\zeta. \tag{4.20}$$

Thus, finding an analytical expression for $\tilde{U}(\zeta) = F(\zeta, \tilde{C})$, one can obtain an analytical expression for $\tilde{U}'$. To find an expression for $\tilde{C}''$, we differentiate (4.19) and use the expression for $\tilde{C}'$

$$\tilde{C}'' = -2\tilde{C}^3(\tilde{C}^{-1} - \tilde{U}^{-1})/(\zeta^2\tilde{U}) + \tilde{C}^2\tilde{U}'/(\zeta\tilde{U}^2). \tag{4.21}$$

Thus, having calculated phase velocity $\tilde{C}(\zeta)$ by solving the dispersion relation, one can calculate group velocity and all the necessary derivatives ($\tilde{C}'$, $\tilde{U}'$ and $\tilde{C}''$) from closed formulas.

We now perform some manipulations with the right-hand sides of (2.52), (2.85), (2.92).

First of all, instead of a slowness vector **k** we shall use the angle φ made by **k** with the x-axis. Then (2.52) can be written as a set of three equations for the coordinates x, y of the vector **r** and the angle φ:

$$\frac{dx}{dt} = \tilde{U}(\omega Z(x, y)) \cos \varphi$$

$$\frac{dy}{dt} = U(\omega Z(x, y)) \sin \varphi \tag{4.22}$$

$$\frac{d\varphi}{dt} = \frac{\omega U(\omega Z(x, y))}{\tilde{C}(\omega Z(x, y))} \tilde{C}'(\omega Z(x, y)) \left[-\frac{\partial Z}{\partial x} \sin \varphi + \frac{\partial Z}{\partial y} \cos \varphi \right].$$

The initial value for φ is the ray azimuth at the source, φ, which was defined in 2.2 as a 'ray' coordinate.

Below, for the sake of brevity, we shall write $\tilde{C}$, $\tilde{U}$ and derivatives of the velocities without appending an argument, with the understanding that these functions are to be calculated for the argument $\omega Z(x, y)$.

We denote:

$$F_1(x, y, \varphi) = \frac{\partial Z}{\partial y} \cos \varphi - \frac{\partial Z}{\partial x} \sin \varphi$$

$$F_2(x, y, \varphi) = \frac{\partial Z}{\partial x} \cos \varphi + \frac{\partial Z}{\partial y} \sin \varphi$$

$$f_{11}(x, y, \varphi) = -\frac{\partial^2 Z}{\partial x\, \partial y} \sin 2\varphi + \frac{\partial^2 Z}{\partial x^2} \sin^2 \varphi + \frac{\partial^2 Z}{\partial y^2} \cos^2 \varphi$$

$$f_{12}(x, y, \varphi) = \frac{\partial^2 Z}{\partial x\, \partial y} \cos 2\varphi + \sin \varphi \cos \varphi \left(\frac{\partial^2 Z}{\partial y^2} - \frac{\partial^2 Z}{\partial x^2} \right)$$

$$R_H(x, y, \omega) = \frac{\omega}{\tilde{C}} (\tilde{U}'\tilde{C} - \tilde{C}'\tilde{U})$$

$$R_1(x, y, \varphi, \omega) = \frac{\omega}{\tilde{C}^2} [f_{11} \tilde{U}\tilde{C}' + F_1(F_1 R_H \tilde{C}' + \omega \tilde{U} F_1 \tilde{C}'')]$$

$$R_2(x, y, \varphi, \omega) = \frac{\omega}{\tilde{C}^2} [f_{12} \tilde{U}\tilde{C}' + F_2(F_1 R_H \tilde{C}' + \omega \tilde{U} F_1 \tilde{C}'')].$$

Then Equations (4.22) together with (2.85), (2.92) form the system

$$\frac{dx}{dt} = \tilde{U} \cos \varphi$$

$$\frac{dy}{dt} = \tilde{U} \sin \varphi$$

$$\frac{d\varphi}{dt} = -\frac{\omega \tilde{U} \tilde{C}'}{\tilde{C}} F_1$$

$$\frac{dQ_1^\omega}{dt} = \tilde{U}\tilde{C}P_1^\omega + \frac{\omega \tilde{U}\tilde{C}'}{\tilde{C}} F_1 Q_2^\omega$$

$$\frac{dQ_2^\omega}{dt} = F_1 R_H Q_1^\omega + \omega \tilde{U}' F_2 Q_2^\omega + Z\tilde{U}' \tag{4.23}$$

$$\frac{dP_1^\omega}{dt} = -R_1 Q_1^\omega - R_2 Q_2^\omega - \frac{\omega Z \tilde{U}' \tilde{C}'}{\tilde{C}^2} F_1 -$$

$$- \frac{\tilde{U}}{\tilde{C}^2} \left(\tilde{C}' + \omega Z \tilde{C}' - \frac{\omega Z}{\tilde{C}} (\tilde{C}')^2 \right) F_1$$

$$\frac{dQ_1^\varphi}{dt} = \tilde{U}\tilde{C}P_1^\varphi + \frac{\omega \tilde{U}\tilde{C}'}{\tilde{C}} F_1 Q_2^\varphi$$

$$\frac{dQ_2^\varphi}{dt} = F_1 R_H Q_1^\varphi + \omega \tilde{U}' F_2 Q_2^\varphi$$

$$\frac{dP_1^\varphi}{dt} = -R_1 Q_1^\varphi - R_2 Q_2^\varphi.$$

Thus, the procedure to determine Love wave amplitude at some point $t = T$ of a space-time ray with the parameters ω, φ is as follows:

(1) integrate the system (4.23) in the interval $[0, T]$;
(2) find the geometrical spreading from (2.63);
(3) calculate $V^{(3)}(0)/\sqrt{\int_0^\infty \rho [V^{(3)}]^2 \, dz}$ which enters (2.66) and determines wave energy redistribution along the vertical; according to 1.4, this is $[(\rho_1/2\omega)(\omega Z + \mathfrak{L})]^{-1/2}$;
(4) fix the source radiation pattern $K_L(\varphi, \omega)$ from some considerations.

EXAMPLE. We shall describe the results of dynamic ray tracing for Love waves in a layer of variable thickness overlying a half-space. The depth of the layer bottom is given by the following function of the horizontal coordinates x, y:

$$Z(x, y) = H_0 + h \exp(-Ax^2 - By^2)$$

with the parameter values $H_0 = 20$ km, $h = 50$ km, $A = 5.8 \times 10^{-7}$ km^{-2}, $A = 3.8 \times 10^{-7}$ km^{-2}. Shear velocities in the layer and the half-space were $b_1 = 3.0$ km/sec, $b_2 = 4.5$ km/sec, the densities $\rho_1 = 2.0$ g/cm^3, $\rho_2 = 3.0$ g/cm^3. Figure 4.1 shows isolines of the layer thickness. The model represents crustal structure in

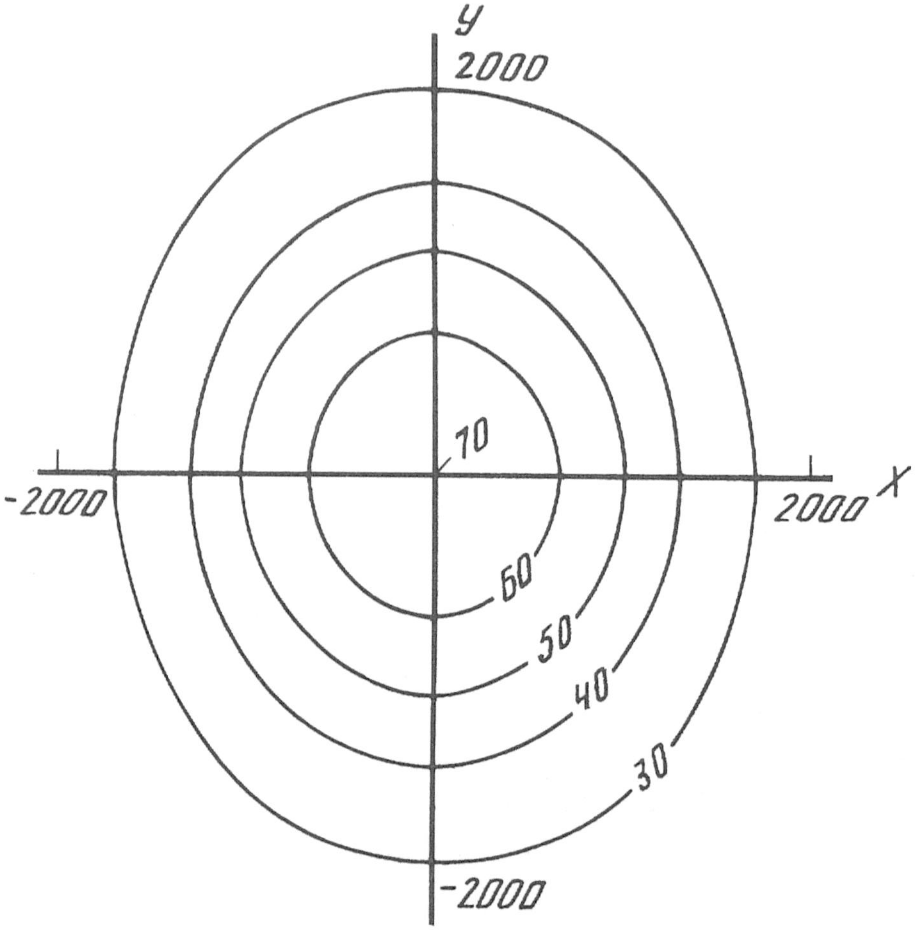

Fig. 4.1. A contour map of layer thickness in the model for calculations.

Central and East Asia, where the crustal thickness varies from 20 km near the Pacific coast to 70 km in Tibet. The function $K_L(\varphi, \omega)$ was taken to be equal to 1.

Space-time rays were calculated for fixed frequencies travelling from the source towards the region of greater layer thickness. Lower phase velocity corresponds to this region, the lowering being particularly pronounced at the frequencies 0.1 to 0.2 rad/sec. It is to be expected therefore that rays that traverse this region would focus or at least get closer.

Figure 4.2 shows space-time rays projected onto the $z = 0$ plane for three frequencies and isolines of arrival times for the wave packet or, which is the same thing, cross-sections of a set of space-time rays for the same frequency by planes of $\tau =$ constant. Adjacent isolines are at intervals of $\Delta t = 160$ sec.

Rays on the $z = 0$ plane are more strongly curved for the frequency $\omega = 0.1$ rad/sec. This is due to the fact that phase velocity has the greatest lateral variation at this frequency. This can be seen from Figure 4.3 presenting the variation of phase velocity along x-axis for the three frequencies.

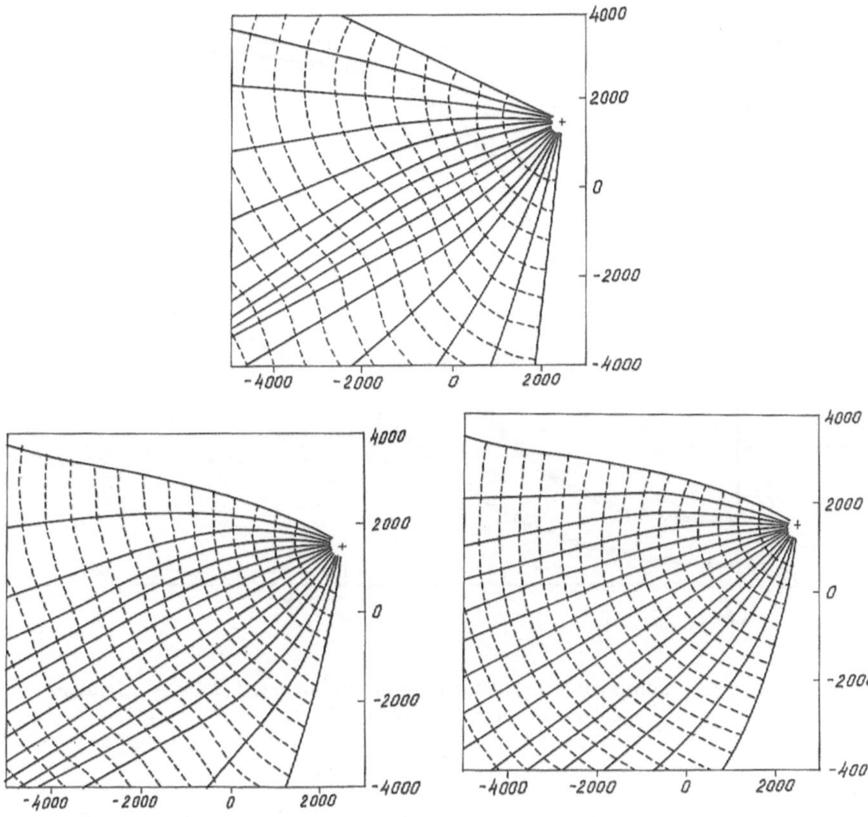

Fig. 4.2. Projections of space-time Love wave rays for the model in Figure 4.1. (a) $\omega = 0.1$ rad/sec ($T = 62.8$ sec); (b) $\omega = 0.2$ rad/sec ($T = 31.4$ sec); (c) $\omega = 0.3$ rad/sec ($T = 20.9$ sec).

Wave packet amplitude was calculated along the space-time rays; it is shown in Figure 4.4 for the frequencies 0.1 to 0.4 rad/sec as a function of epicentral distance along a ray that passes approximately through the center of the study area. Also given for comparison is an amplitude-distance curve in a laterally homogeneous medium (the amplitude falls off as $1/r$). The first thing one notices is the presence of two spikes in the curve for $\omega = 0.3$ rad/sec. Similar spikes (at different points) exist in the curves for frequencies around 0.3 rad/sec.

It is at these points that rays are tangent to a space-time caustic. This is degenerate in a laterally homogeneous medium, being a conical surface formed by rays corresponding to ω_0 for which group velocity has a minimum (Figure 2.3). In our case of a laterally varying medium, a caustic is the envelope surface of a set of space-time rays. Points where rays are tangent to the caustic correspond to the Airy phase on seismograms. Obviously, the Airy phase periods are different at different points of the x, y-surface.

A comparison of the amplitude curves presented here to the curve corresponding to a laterally homogeneous medium shows that the amplitude decays more slowly than $1/r$, also for frequencies away from the Airy phase frequency.

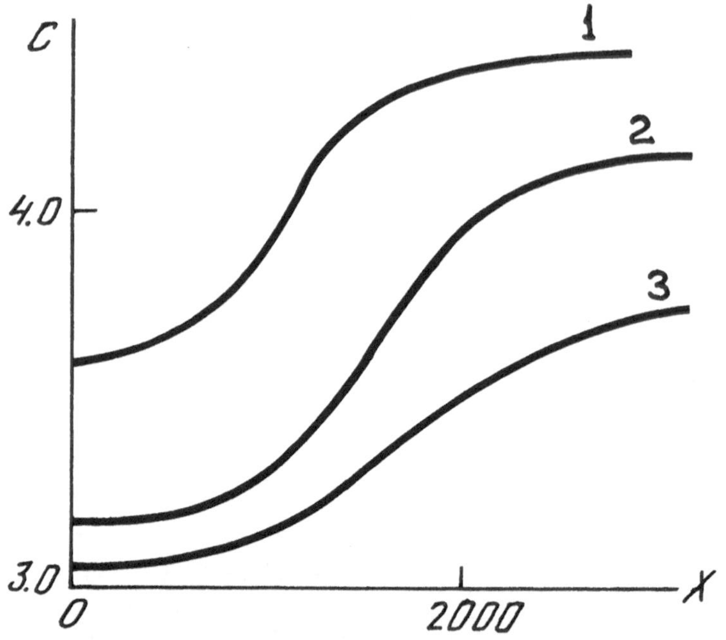

Fig. 4.3. Phase velocity variation along x-axis for different frequencies: (1) $\omega = 0.1$ rad/sec, (2) $\omega = 0.2$ rad/sec, (3) $\omega = 0.3$ rad/sec.

This is partly accounted for by focusing of rays on the plane (see Figure 4.2). However, the bending of rays in space is not sufficient to account for the observed attenuation: the variation of amplitude as a function of distance being also strongly affected by 'focusing' of space-time rays in the time domain, an effect that is particularly pronounced at frequencies around the Airy phase frequency.

When the distribution of surface wave phase velocity $C = C(x, y)$ or $C = C(\vartheta, \varphi)$ can be regarded as fixed for some period, ray tracing presents no difficulty; it is done by integrating numerically the set of differential ray equations with group velocity U eliminated by the change of variable $\mathrm{d}s = U\,\mathrm{d}t$. In the plane case the set of equations becomes

$$\frac{\partial x}{\partial s} = \cos\varphi, \qquad \frac{\partial y}{\partial s} = \sin\varphi$$

$$\frac{\partial \varphi}{\partial s} = C^{-1}\left(\frac{\partial C}{\partial x}\sin\varphi - \frac{\partial C}{\partial y}\cos\varphi\right)$$

and in the spherical case it is

$$\frac{\mathrm{d}\theta}{\mathrm{d}s} = \cos\alpha/R, \qquad \frac{\mathrm{d}\varphi}{\mathrm{d}s} = \sin\alpha/R$$

$$\frac{\mathrm{d}\alpha}{\mathrm{d}s} = \frac{1}{C}\left(\frac{\sin\alpha}{R}\frac{\partial C}{\partial \theta} - \frac{\cos\alpha}{R\sin\theta}\frac{\partial C}{\partial \varphi}\right) - \frac{\sin\alpha\,\mathrm{cotan}\,\theta}{R},$$

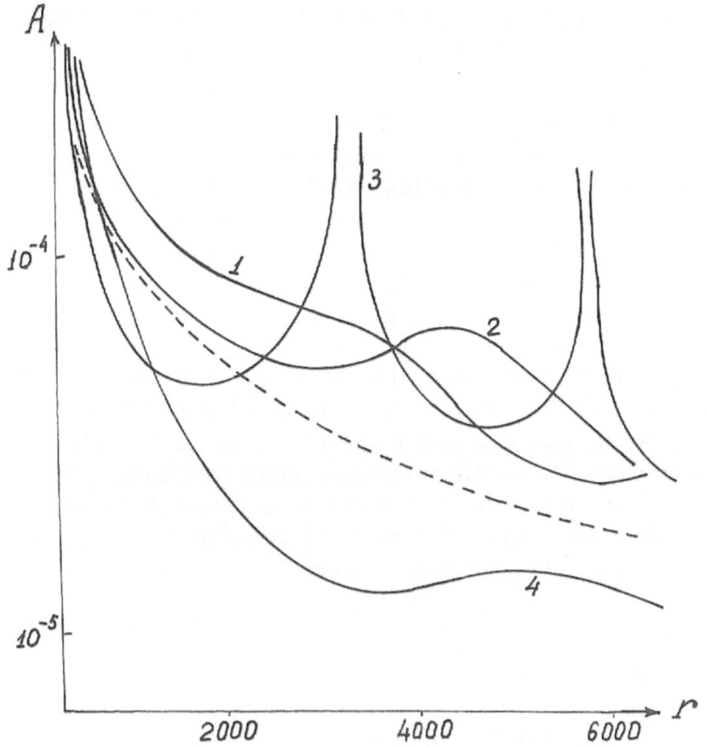

Fig. 4.4. Wave packet amplitude along a ray passing through the origin: (1) ω 0.1 rad/sec, (2) ω = 0.2 rad/sec (3) ω = 0.3 rad/sec, (4) ω = 0.4 rad/sec. The dashed line shows amplitude decay with distance in a laterally homogeneous medium.

which is readily derived from (2.110) by the substitution

$$k_\theta = \cos \alpha / C, \qquad k_\varphi = \sin \alpha / C.$$

Jobert and Jobert (1983), Lay and Kanamori (1985), Sobel and von Seggern (1978) have performed ray tracing on a spherical surface for fixed velocity distributions corresponding to different periods. Inhomogeneities can produce large deviations of surface wave paths from great circles, hence multipathing.

The usual procedure with such calculated ray schemes is to use them for qualitative conclusions about at the distribution of surface wave amplitude anomalies. Places where rays concentrate are likely to have increased amplitude and those where they are rarefied must have a lower amplitude. However, this treatment does not take into account the dispersive spreading of a wave packet, this being different along different paths, and the change in the wave energy profile along the vertical. Lazareva and Yanovskaya (1975) have calculated surface wave amplitude anomalies due to spatial geometrical spreading alone the theoretical anomalies being noticeably below the observed ones. The authors hypothesized that the discrepancy was due to the fact that the evaluation of the amplitude did not incorporate the change in wave packet amplitude caused by its spreading in time. The above model calculations do demonstrate that one must accommodate all factors controlling surface wave amplitude.

4.4. CALCULATION OF REFLECTION AND TRANSMISSION COEFFICIENTS

Reflection and transmission coefficients of surface waves were calculated using the technique described in 3.3. The principal computational difficulty with this technique is the calculation of coupling coefficients P_{ij}, S_{ij}, Q_{ij}, T_{ij}, Y_{ij}; these are expressible as integrals involving the eigenfunctions of the relevant modes. To simplify calculations, therefore, it would be natural to use a model whose eigenfunctions are expressible explicitly, so that the coupling coefficients too can be obtained in explicit form. The simplest model in question is a plane-stratified medium, eigenfunctions in layers being expressible as combinations of exponential functions as mentioned above. Thus, the medium is assumed to be plane-stratified to the right and left of the contact, the right and left horizontal interfaces being possibly at different depths, except for the free surface $z = 0$ which is common to both media. In other words, no step is assumed at the free surface.

The eigenfunctions are to be in some way normalized, before reflection and transmission coefficients can be calculated. All calculations below are based on the normalization introduced in Section 3.2, namely,

$$\xi_{kL} I_{kL}^{(1)} = 1 \tag{4.24}$$

for Love waves and

$$\xi_{kR}(I_{kR}^{(1)} + I_{kR}^{(2)}) + (I_{kR}^{(3)} + I_{kR}^{(4)}) = 1 \tag{4.25}$$

for Rayleigh waves. On the other hand, in order to be able to analyse the results and compare them to experimental data, as well as to the results of calculation by other methods, one needs values of reflection and transmission coefficients as would be independent of the normalization. With this aim in view, one usually considers the ratio of displacements due to the relevant modes at the free surface too (the surface ratio). The surface ratios will be denoted A_{si}^{surf} and B_{sj}^{surf}. Obviously, $A_{ii}^{\text{surf}} = A_{ii}$ is independent of the normalization. For this reason we shall omit the superscript in subsequent discussions of the reflection coefficient for the same mode.

The reflection and transmission coefficients are calculated as follows. Eigenvalues ξ_k are found for all modes at a fixed frequency ω in each of the two media in accordance with the algorithm described in 4.1.

Further, coupling coefficients between modes are determined by summing expressions corresponding to integrals along individual portions of the vertical discontinuity. These expressions are readily found by integrating the products of exponential functions. Further, according to the particular problem in hand, one uses the Gauss method to solve the set of linear equations (3.51) or (3.61), yielding reflection and transmission coefficients for eigenfunctions under the normalization (4.24), (4.25). Lastly, surface ratios are found from

$$A_{si}^{\text{surf}} = A_{si} V_i^{(3)-}(0)/V_s^{(3)-}(0)$$

$$B_{sj}^{\text{surf}} = B_{sj} V_j^{(3)+}(0)/V_s^{(3)-}(0)$$

for Love waves and

$$A_{si}^{\text{surf}} = A_{si} V_i^{(1)-}(0)/V_s^{(1)-}(0)$$

$$B_{sj}^{\text{surf}} = B_{sj} V_j^{(1)+}(0)/V_s^{(1)-}(0)$$

for Rayleigh waves.

The efficiency of this method was evaluated by comparing the results of numerical modelling by it to the results of other methods described in Chapter 3, as well as to the results of ultrasonic modelling (Gegechkori and Yanovskaya 1979; Sikharulidze and Mandzhgaladze, 1977).

The most reliable test of the approximate estimates is their agreement with the results obtained by numerical methods. We have compared the approximate estimates of reflection and transmission coefficients to the results obtained by the numerical procedure described in 3.4 (Bukchin, 1980) using a simple model of the contact between continental and oceanic crustal blocks. The model consisted of a uniform layer overlying a uniform half-space with a jump in thickness at $x = 0$ from 5 km (ocean) to 40 km (continent). The layer parameters were b = 3.5 km/sec, ρ = 2.7 g/cm^3, the half-space parameters were b = 4.5 km/sec, ρ = 3.3 g/cm^3. The calculation was performed for a fundamental mode of Love wave at normal incidence on the $x = 0$ discontinuity from the ocean block. Figure 4.5 presents the surface ratios B_{11}^{surf} and reflection coefficients A_{11} for different periods. The results obtained by the two different methods are consistent, leading one to conclude that the approximate method for evaluating the amplitudes of Love waves excited at a contact of two significantly different quarter-spaces such as the ocean–continent transition is efficient. Gegechkori and Yanovskaya (1979) have compared reflection coefficients of Rayleigh waves as calculated by the method

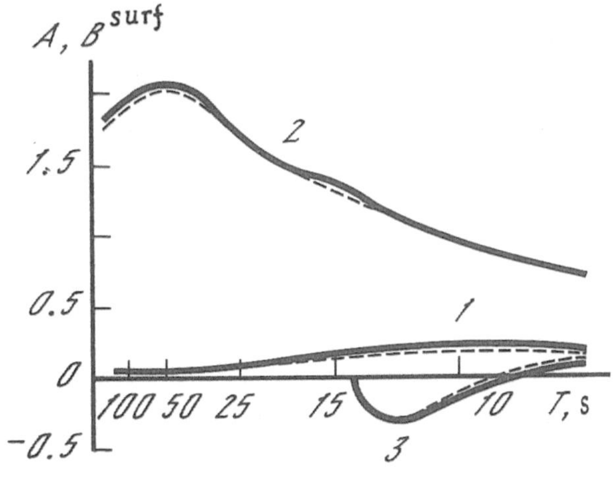

Fig. 4.5. Calculations of reflected and transmitted Love waves for the fundamental mode at normal incidence on the ocean-continent boundary from the ocean. (1) reflection coefficient for fundamental mode; (2, 3) surface transmission coefficients corresponding to the fundamental and first higher transmitted mode. The solid line represents the results based on the Green function tecnique, the dashed line is based on the method described in Section 3.4.

and those determined experimentally using three-dimensional ultrasonic modelling for two models of the contact between two uniform quarter-spaces.

One model is a contact between the media with a weak velocity contrast, the materials in contact being gypsum and sealing wax whose the Rayleigh wave velocities C_R = 1.3 and 1.21 km/sec, respectively. The results of calculations and experimental studies shown in Figure 4.6 demonstrate a satisfactory agreement between theory and experiment throughout the entire range of angles of incidence.

The other model represents a strong velocity contrast, the materials in contact being sealing wax (C_R = 1.21 km/sec) and duralumin (C_R = 2.61 km/sec). The results of theory and experiment are presented in Figure 4.7; also shown are reflection coefficients calculated by the Alsop method. A comparison of these three curves shows that the Alsop coefficients are too high while those obtained by the Green function technique are in good agreement with the experiment in the range of subcritical angles. The considerable discrepancy between theoretical and experimental curves in the overcritical range seems to be due to the fact that, as the intensity of the reflected wave increases, the coupling coefficient between the reflected wave and body waves in medium 2, which has not been incorporated in the calculation, increases too.

The results obtained by the two approximate methods (Gregersen—Alsop method and that described in 3.3) were compared for two models of the ocean–continent contact F and L (Figure 4.8) used in Gregersen and Alsop (1976). Gregersen (1978). Models L and F represent the structure of the crustal and upper mantle blocks in contact with varying degrees of detail. The principal difference between the models is that model L involves a low-velocity layer in the depth range 50 to 215 km.

The surface ratios B_{11}^{surf} for Love waves at normal incidence as functions of

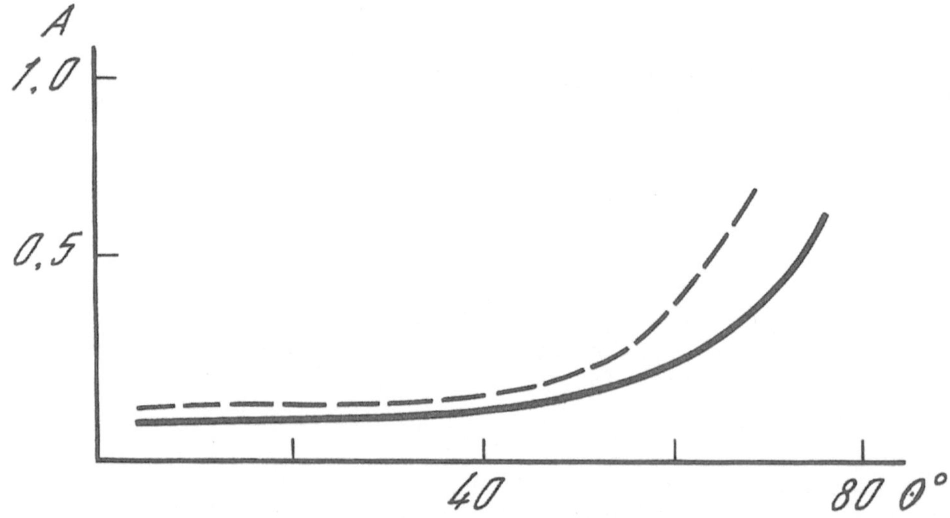

Fig. 4.6. Rayleigh wave reflection coefficient for the sealing wax/gypsum contact as a function of the angle of incidence. Solid line shows the results based on the Green function technique, dashed line represents the experiment.

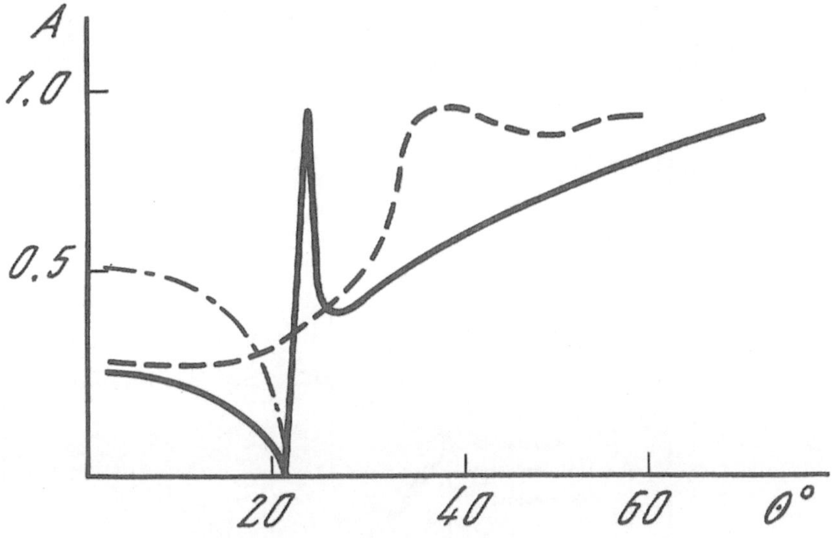

Fig. 4.7. Rayleigh wave reflection coefficient for the sealing wax/duralumin contact. Legend as for Figure 4.5. The dot-and-dash line shows the results by Alsop's method.

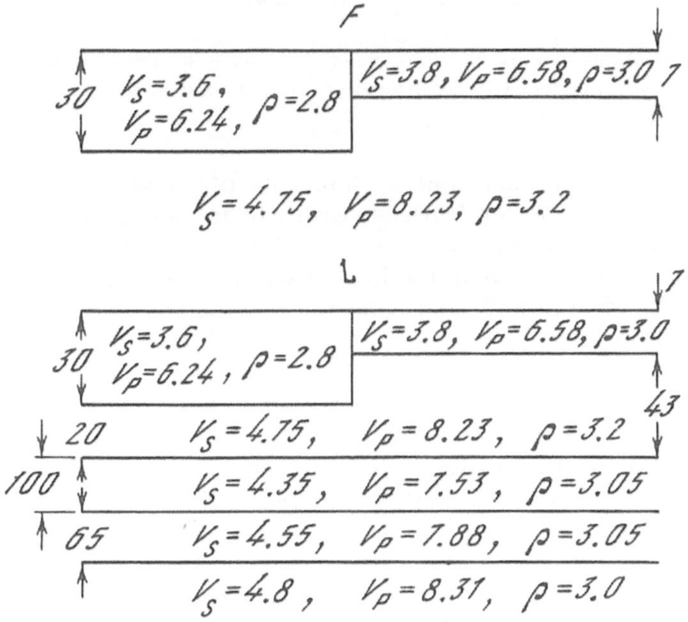

Fig. 4.8. Ocean-continent contact models. V_s is shear velocity, km/sec; V_p is compressional velocity, km/sec; ρ is density, g/cm^3; layer thicknesses are in km.

frequency are presented in Figure 4.9. The curves obtained by the two different methods are nearly identical within the entire frequency range considered. It is interesting to note a sharp difference between the two models in the values of the

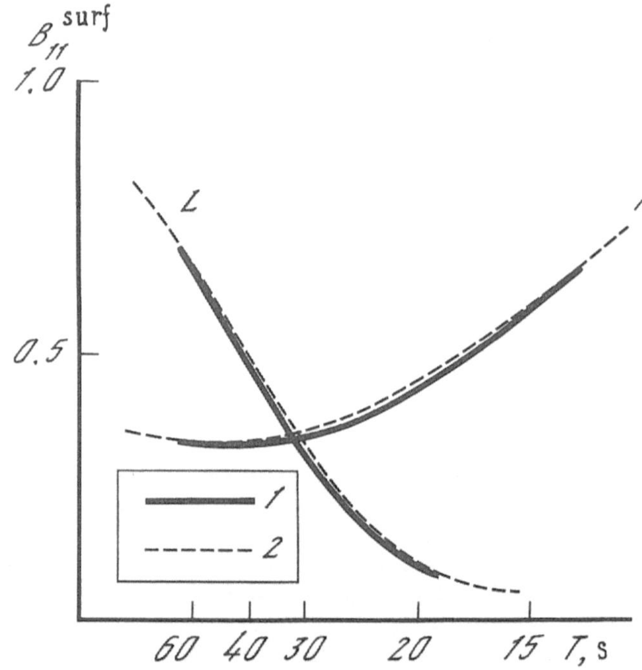

Fig. 4.9. Surface ratios of fundamental Love mode for models F and L. Seaward incidence. The calculations were done by Gregerson-Alsop (1) and Green function (2) method.

coefficients and in their behaviour as functions of frequency, all the more so because the surface ratios for Rayleigh waves in models F and L presented in Figure 4.10 are practically identical. This seems to be due to a stronger influence of the thick low-velocity layer on the Love wave field as compared with Rayleigh waves. The surface ratio B_{11}^{surf} and reflection coefficient A_{11} of Love waves at period $T = 30$ sec as a function of the angle of incidence for model L are presented in Figure 4.11. One can see that, as the angle of incidence increases, the surface ratios $B_{11}^{\text{surf}}(\vartheta)$ calculated by the two methods do not vary and are nearly identical, while the $A_{11}(\vartheta)$ diverge for large angles.

It would be of interest to investigate how the energy of an incident surface wave is redistributed among the transmitted modes. The portion of incident energy carried by the transmitted (or reflected) wave can be characterized by the ratio of energy fluxes transported by the respective waves across a vertical semi-infinite strip of unit width: $0 < z < \infty$, $0 < y < 1$. When a Love wave is at normal incidence on the $x = 0$ discontinuity, the energy flux in any of the waves is

$$\frac{U\omega^2}{2} \int_0^\infty \rho[V(z)]^2 \, dz$$

where $V(z)$ is displacement amplitude in the relevant wave ($V_s^{(3)-}(z)$ in the sth incident mode, $A_{si}V_i^{(3)-}(z)$ in the ith reflected mode, $B_{sj}V_j^{(3)+}(z)$ in the jth transmitted mode, $V_k^{(3)\pm}(z)$ being an eigenfunction that is normalized in accord-

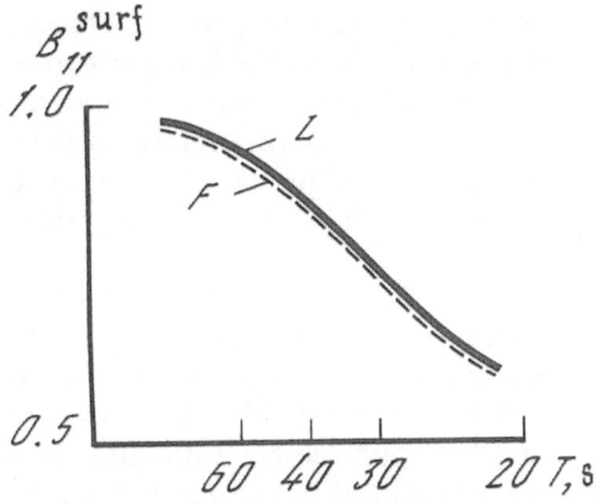

Fig. 4.10. Surface ratios of fundamental Rayleigh mode for models F and L based on Green function technique. Seaward incidence.

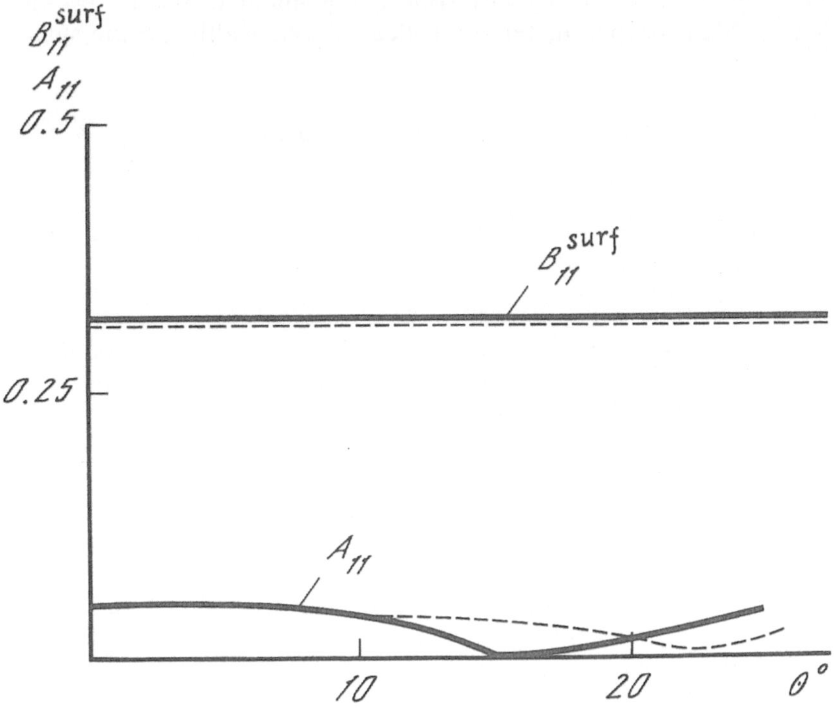

Fig. 4.11. B_{11}^{surf} and A_{11} coefficients of Love waves calculated for model L using Green function technique (dashed line) and Gregersen–Alsop method (solid line). Seaward incidence.

ance with (4.24)). Using the expression for Love wave group velocity, one can readily show that the ratio of transmitted and incident energy flux equals B_{sj}^2. This also holds for normal incidence of Rayleigh waves.

Figure 4.12 illustrates increasing fundamental mode Love wave energy to be transferred to higher modes as the frequency increases. The curves were calculated for normal incidence of a Love wave in model L using two approximate methods: the Gregersen—Alsop method and the Green function technique. The results indicate an increasing portion of energy E transferred to the first higher mode and decreasing energy of the fundamental mode, the energy found by the Green function technique for periods less than 20 sec being smaller than that by the Gregersen—Alsop method.

All the above estimates of efficiencies for the approximate methods are for welded contacts. The reliability of the approach described in 3.5 for unwelded contacts was tested by using the finite element method to compute reflected and transmitted Love wave fields in a model with a low-velocity transition zone and comparing the results with the approximate estimates of reflection and transmission coefficients. The model we used in the finite-element calculations is shown in Figure 4.13. The calculations were performed for the frequency range 0.25–0.35 rad/sec as allowed by the model parameters and the capabilities of our computer. The approximate model to be compared with the above involved a vertical discontinuity between two homogeneous quarter-spaces with parameters of zones I_1 and I_2. The low-velocity zone I_0 was imitated by an unwelded contact with the unweldedness parameter controlled by zone width and the shear modulus in it:

$$m(z) = \begin{cases} \dfrac{h}{\mu_0}, & 0 < z < H_2 \\[2ex] \dfrac{h}{\mu_0} \dfrac{z - H_1}{H_2 - H_1}, & H_2 < z < H_1 \\[2ex] 0, & z > H_1 \end{cases}$$

The results (Figure 4.14) demonstrate fairly satisfactory agreement between reflection and transmission coefficients obtained by approximate and finite-element techniques (although one cannot help remarking that the approximate reflection coefficients are slightly overestimated compared with those determined by finite elements). It supports the hypothesis that a low-velocity zone and an unwelded contact affect the amplitudes of reflected and transmitted surface waves in a similar way.

To sum up, the method described in 3.3 for evaluating reflected and transmitted surface waves yields results that are in good agreement with those obtained by alternative approximate and numerical procedures, hence it can be used for numerical modelling of observed surface waves.

Sikharulidze (1978) provides evidence of intensive surface waves reflected from deep faults. In this connection it would be of interest to investigate numerically how deep faults affect surface wave propagation. To a first approximation (without a low-velocity zone), a fault can be modelled by an interface where the layers are displaced relative to one another and their thicknesses change discontinuously. A similar model shown in Figure 4.15 was used for calculations assuming a fundamental mode Love wave at normal incidence from the left to the interface, the top layer thickness H to the right of the contact being varied in the range 1.5 to 10 km.

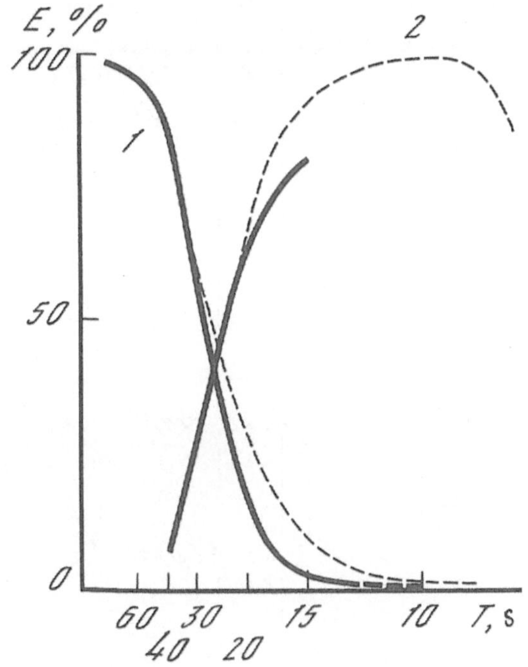

Fig. 4.12. Energy flux ratio of transmitted fundamental (1) and first higher mode (2) to incident Love wave in model L, for the landward incidence of fundamental mode.

The results are given in Figure 4.16a, b. For this model, as *H* and correspondingly the fault depth increase, the portion of energy that goes to the fundamental mode of the transmitted wave which is, as has been shown, determined by the square of the transmission coefficient B_{11}, decreases towards higher frequencies, whereas the coefficient B_{12} which controls the portion of energy supplied to the second mode becomes much greater (Figure 4.16a). This is related to the fact that phase velocities of incident fundamental mode and transmitted second mode become close in value for certain frequencies (Figure 4.17). The reflection coefficients A_{1i} (Figure 4.16b) increase with increasing frequency more steeply for deeply extending faults, but always remain below the transmission coefficients in value. Thus modelling of faults by contacts of the type described fails to produce intensive reflected waves.

This result was confirmed by calculating reflection and transmission coefficients of Love waves for normal incidence in a model of the Caspian deep-seated fault (Figure 4.18). The transmission coefficient B_{11} within the period range 12.5 to 5.5 sec varied in the limits 0.97—1.0 (Figure 4.19), the limits for the reflection coefficient were 0.01—0.05 (Figure 4.20). These values enable us to conclude that our model of welded contact to be inadequate to represent the true fault, since the Caucasian faults typically have intensive reflected waves and strongly decreased amplitudes of transmitted waves with a characteristic dependence on frequency, namely, a rapid fall-off in the transmission coefficient with increasing frequency in the longer period range and a slow one for high frequencies (Bagramyan *et al.*, 1978).

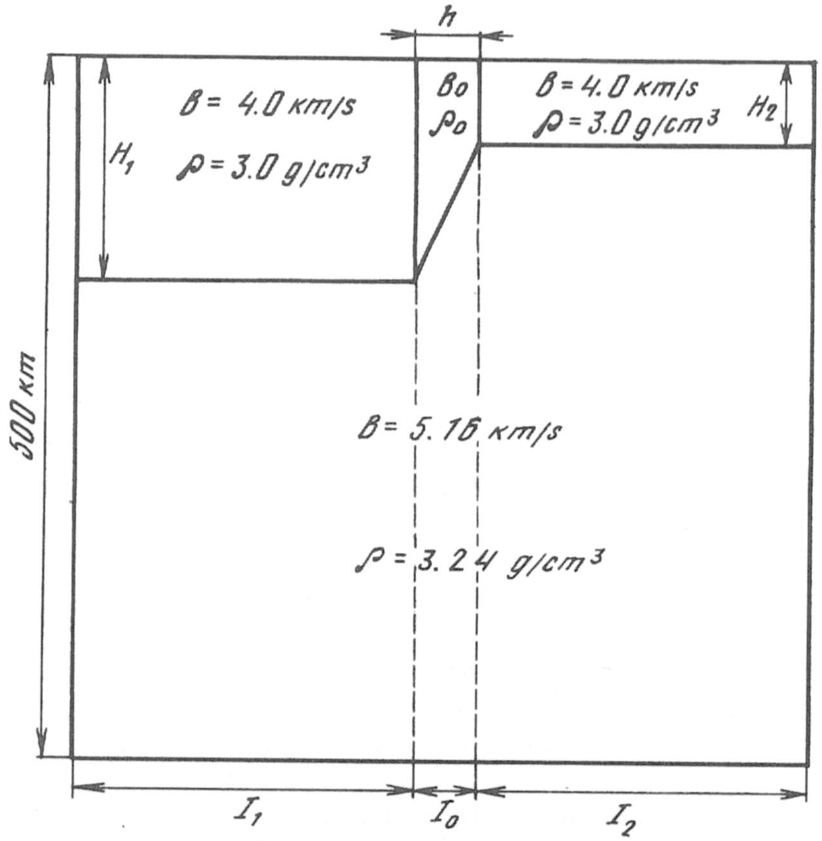

Fig. 4.13 Model of a low-velocity fault used in finite-element calculations. I_1 and I_2 are laterally homogeneous zones of 50 km width with the elastic parameters given in the figure; $H_1 = 40$ km, $H_2 = 16$ km. I_0 is a laterally varying zone of 10 km width involving a wedging-out low-velocity inclusion with the elastic parameters $b_0 = 2$ km/sec, $\rho_0 = 2.5$ g/cm^3, $\mu_0 = 10^{10} = 10^{10}$ dyne/cm^2.

The theoretical and observed coefficients can be made consistent within an order of magnitude by describing the fault as an unwelded contact with an unweldedness parameter whose dependence on depth is given by (4.26). Reflection and transmission coefficients of Love waves were calculated for six models of unwelded contact which can be interpreted as a low-velocity inclusion with the parameters

(1) $h = 2$ km $\quad H_1 = 22$ km $\quad H_2 = 15$ km

(2) $h = 2$ km $\quad H_1 = 22$ km $\quad H_2 = 31$ km

(3) $h = 2$ km $\quad H_1 = 31$ km $\quad H_2 = 31$ km

(4) $h = 4$ km $\quad H_1 = 22$ km $\quad H_2 = 15$ km

(5) $h = 4$ km $\quad H_1 = 22$ km $\quad H_2 = 31$ km

(6) $h = 4$ km $\quad H_1 = 38$ km $\quad H_2 = 31$ km.

Shear velocity was $b_0 = 1.6$ km/sec and density $\rho_0 = 1.6$ g/cm^3 for all six models.

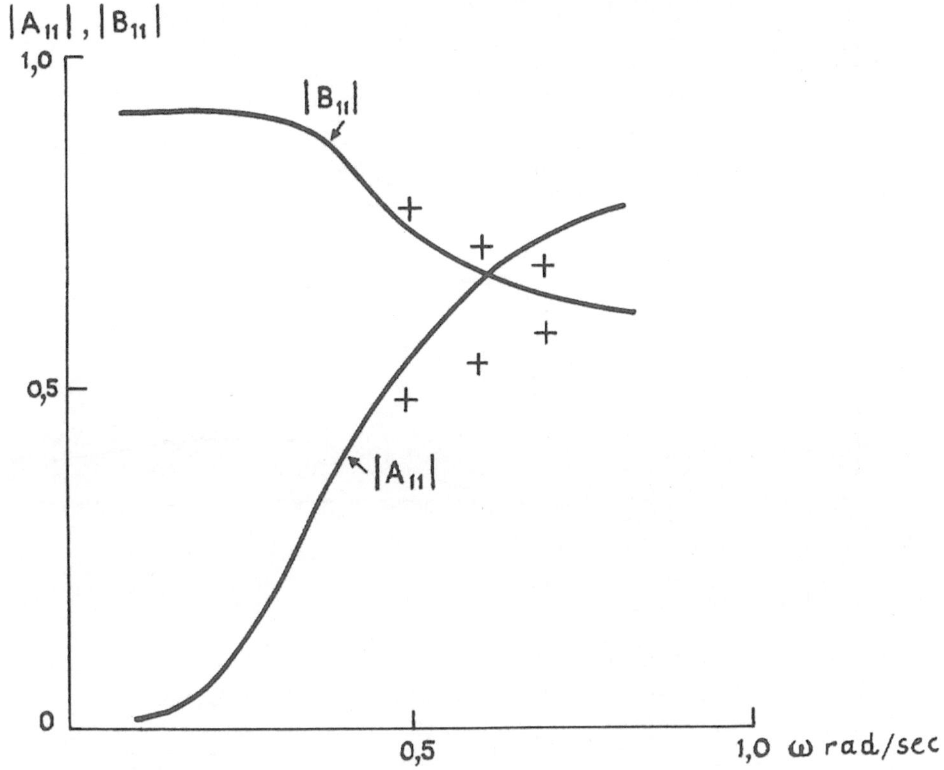

Fig. 4.14. Reflection and transmission coefficients of fundamental Love mode for a low-velocity fault. The calculations were done by finite elements (crosses) and Green function technique (solid line).

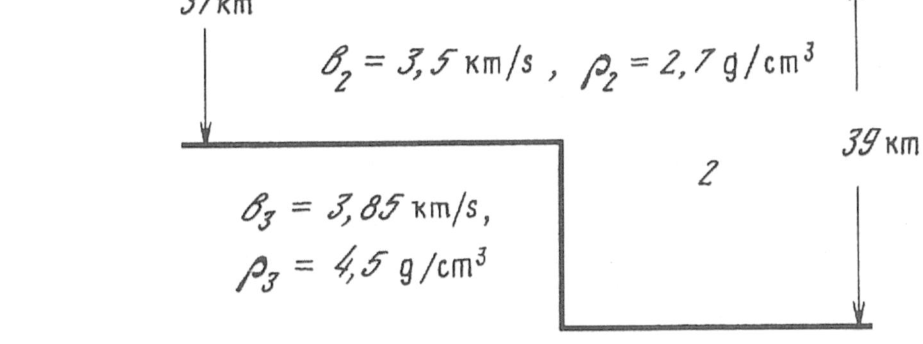

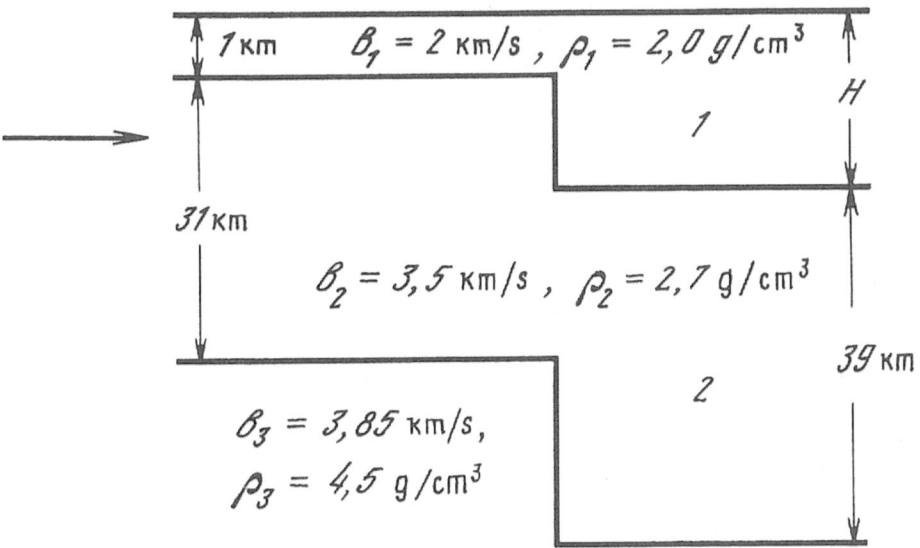

Fig. 4.15. Deep fault model.

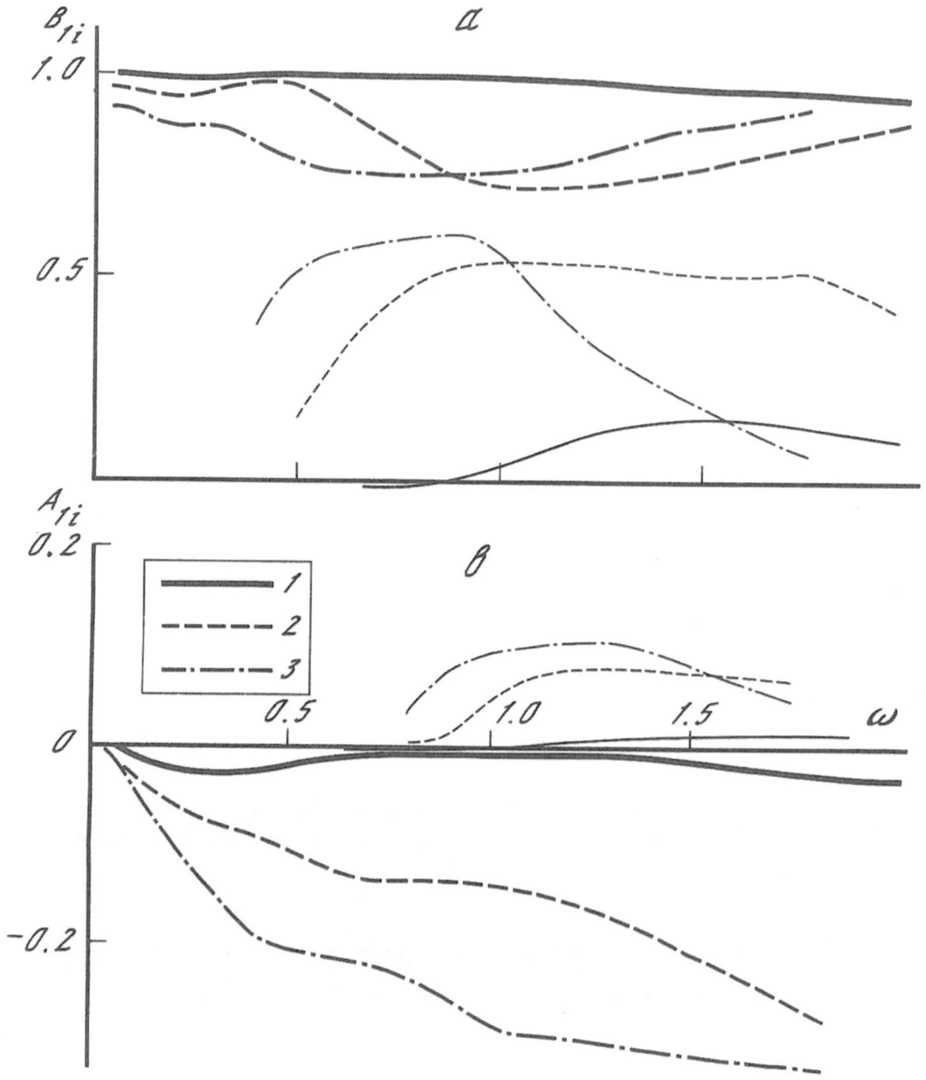

Fig. 4.16. (a) Transmission B_{1i} and (b) reflection A_{1i} coefficients for Love waves. The curves for the fundamental mode of transmitted or reflected waves ($i = 1$) are shown by heavy lines, those for the first higher mode ($i = 2$) by light lines. (1) $H = 1.5$ km, (2) $H = 5$ km, (3) $H = 10$ km.

Transmission coefficients were calculated as functions of frequency (period) for all the variants. The results are presented in Figure 4.19. The low-velocity layer $h = 2$ km thick strongly affects the transmission coefficient as a function of frequency (curves 1 to 3). However, the difference between welded and unwelded contact for the value $h = 2$ km is not large enough to make the curves of $B_{11}(\omega)$ calculated for low-velocity transition zones at different depths to be significantly different and to have well-pronounced inflexion regions from which to infer the depth of the fault. When the width of the low-velocity zone becomes as large as

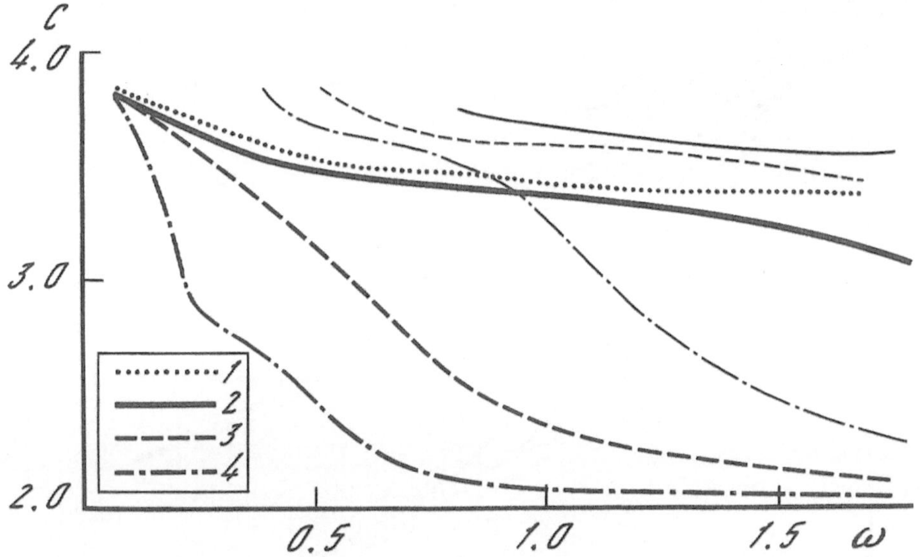

Fig. 4.17. The phase velocities of incident (1) and transmitted (2—4) Love waves. The curves for the fundamental mode are shown by heavy lines, those for the first higher mode by light lines. (2) H = 1.5 km, (3) $H = 5$ km, (4) $H = 10$ km.

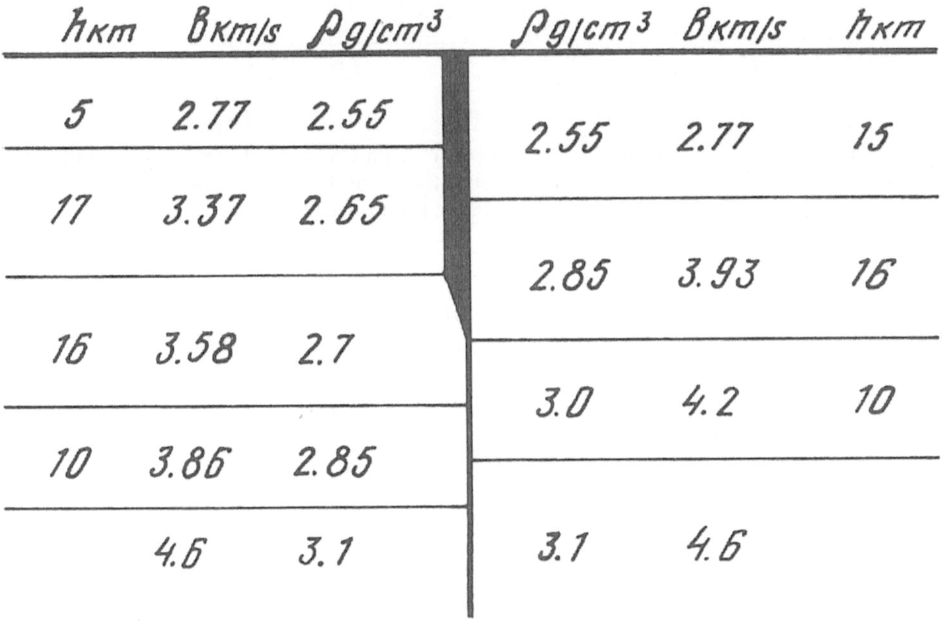

Fig. 4.18. Model of the Caspian fault. A low-velocity transition zone modelled by an unwelded contact, shown as a black wedging-out inclusion.

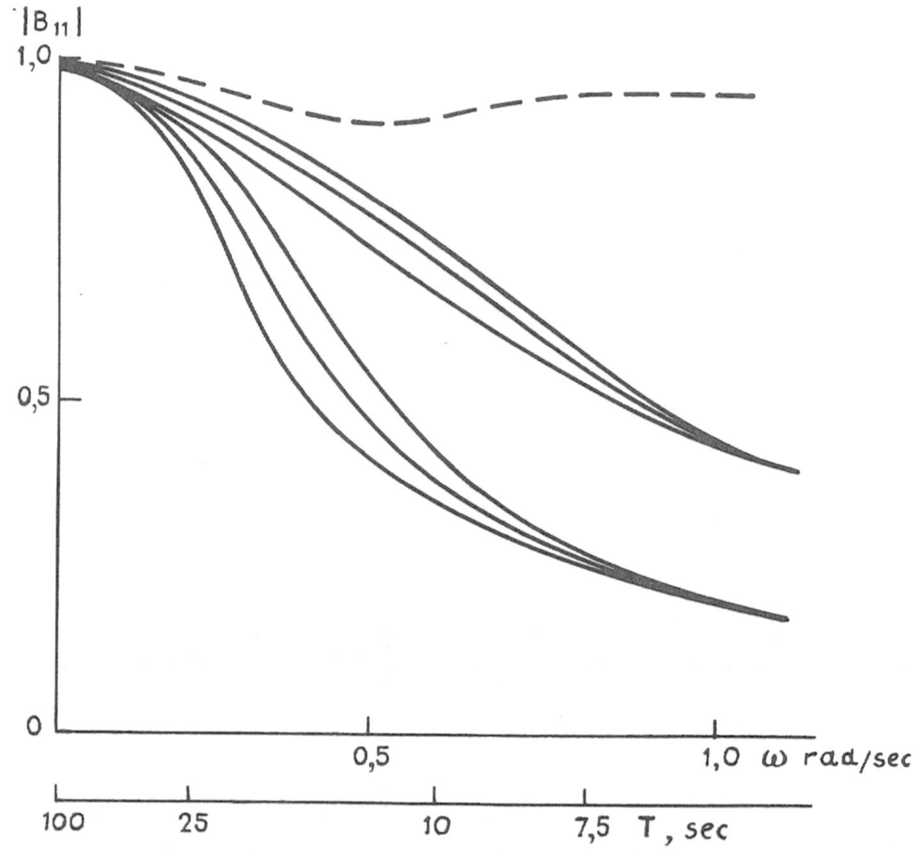

Fig. 4.19. Transmission coefficients of Love waves for various models of the Caspian fault. Welded contact is shown by dashes. Unwelded contact is shown by solid lines, the curve number corresponding to the number of the model.

4 km, the dependence of transmission coefficient on frequency changes (curves 4 to 6), the curves for faults of different depths are well separated in the period range 10 sec $< T <$ 50 sec, three ranges being noticeable in each curve. From the physical point of view these can be interpreted as follows: the fault has practically no effect on surface wave propagation for long periods (large wavelengths), the transmission coefficient being nearly unity for all fault models. As the period decreases, the low-velocity zone described by the unwelded contact becomes progressively more significant, the transmission coefficients decreasing practically linearly with increasing frequency. For long periods, when most of the wave energy concentrates in the layers where the low-velocity fault zone is located, its effect on the propagation of transmitted waves stabilizes, the transmission coefficient is nearly independent of fault depth, the curves coalesce and flatten out. Thus, all the curves have ranges of slope variation that separate zones of strong and weak fault influence on the amplitude of transmitted waves. As has been mentioned, a similar dependence is noted for observed transmission coefficients of

surface waves across deep faults by Bagramyan *et al.* (1978) who hypothesize that the fault depths can be inferred from wavelengths in the ranges of slope change. These ranges in Figures 4, 5, 6 can be estimated by the following frequency intervals and the mean wavelengths for the intervals

4: $0.5 \text{ rad/sec} < \omega < 0.75 \text{ rad/sec}, \quad \bar{\lambda}_{inc} = 28.9 \text{ km}$

5: $0.4 \text{ rad/sec} < \omega < 0.65 \text{ rad/sec}, \quad \bar{\lambda}_{inc} = 34.5 \text{ km}$

6: $0.3 \text{ rad/sec} < \omega < 0.6 \text{ rad/sec}, \quad \bar{\lambda}_{inc} = 41.4 \text{ km}.$

A comparison between the above wavelength estimates and fixed fault depths suggests that fault depth can be estimated by about 2/3 of the wavelength in the range of slope change for the $B_{11}(\omega)$ curve.

Reflection coefficients as functions of frequency for the six models are presented in Figure 4.20. The introduction of low-velocity inclusions has produced a dramatic rise in the reflection coefficients with increasing frequency, providing a clue to interpretation of observed reflected waves. It is interesting to note that the difference between the curves for models 1—3 and 4—6 is not as great as for

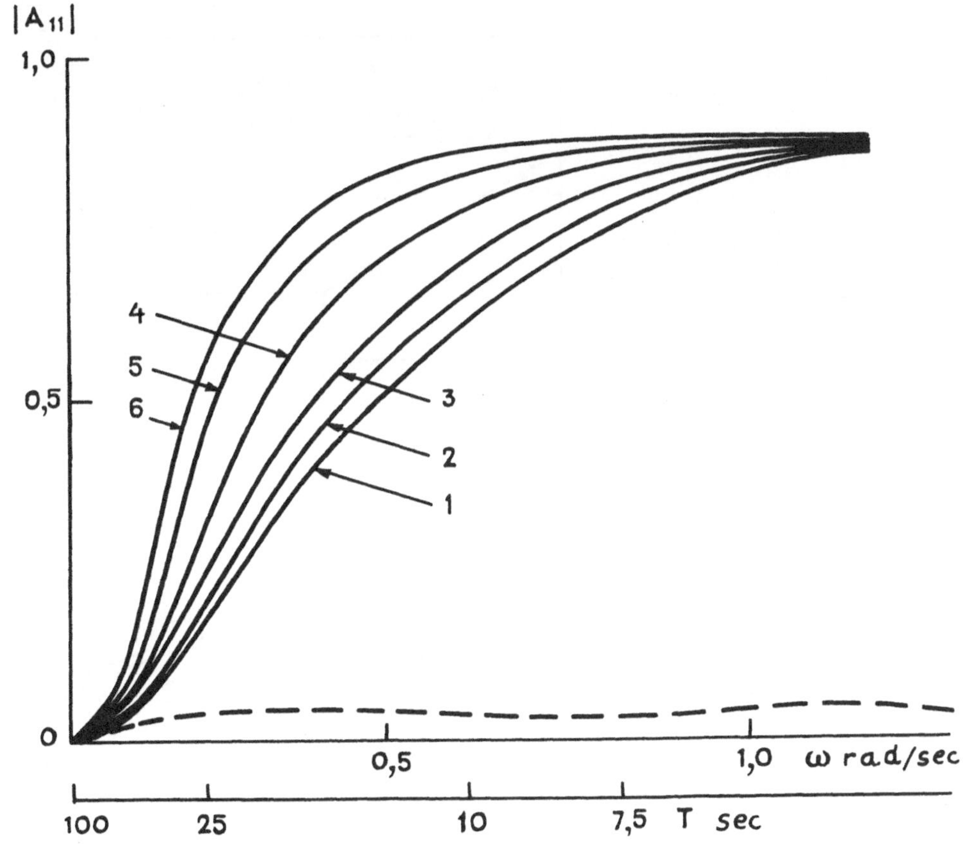

Fig. 4.20. Reflection coefficients of Love waves for models of the Caspian fault. Legend as for Figure 4.16.

transmission coefficients. This seems to be due to the fact that the approximate technique somewhat overestimates reflection coefficients, as shown by a comparison with the calculations by finite elements.

The effect of inclined discontinuities on transmission and reflection coefficients of surface waves was studied by using formulas (3.80), (3.81) derived in 3.5 to calculate these coefficients for normal incidence of Love waves at an inclined step formed by the interface between the layer and the half-space. The model is shown in Figure 4.21. The parameters for the layer and the half-space are the same on both sides of the step, so that the difference between the two media consists in the different layer thicknesses. The calculations were done for waves incident on the step both from the right and from the left for various angles of inclination of the discontinuity.

According to the restrictions indicated in 3.5, the coefficients were calculated up to frequencies for which the wavelengths would be equal to the lateral extent of the step. Figure 4.22 shows the results for the reflection coefficient alone, since the

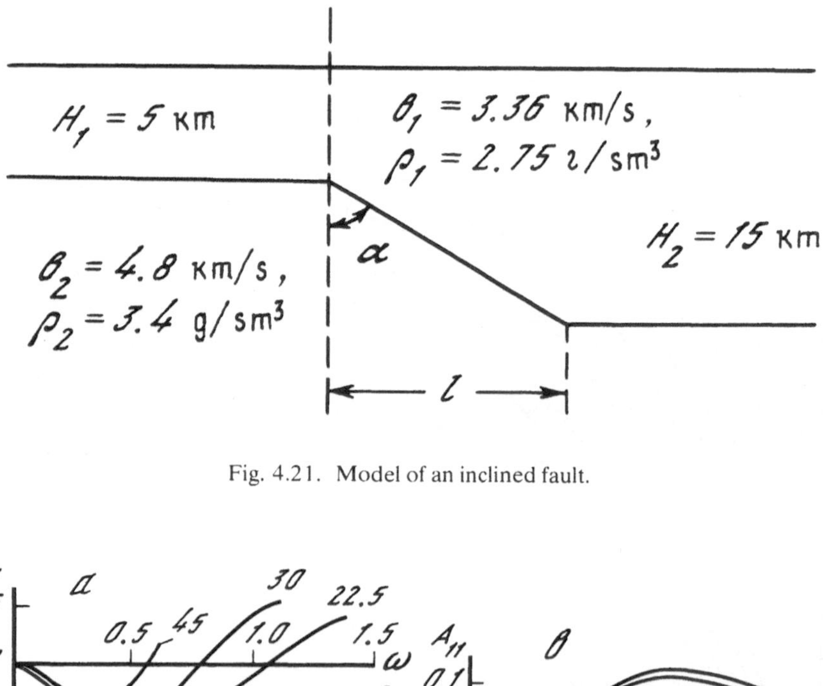

Fig. 4.21. Model of an inclined fault.

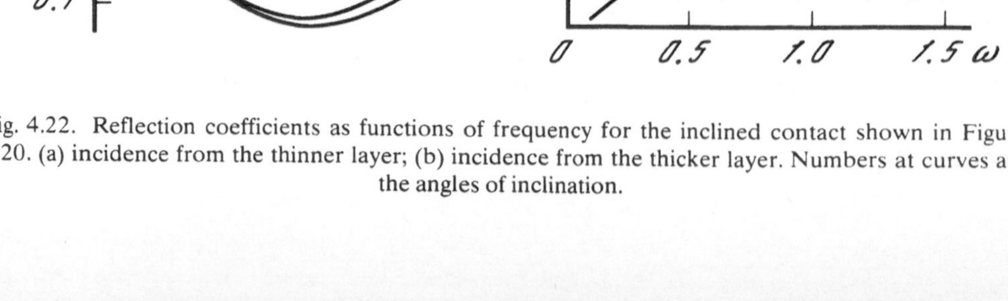

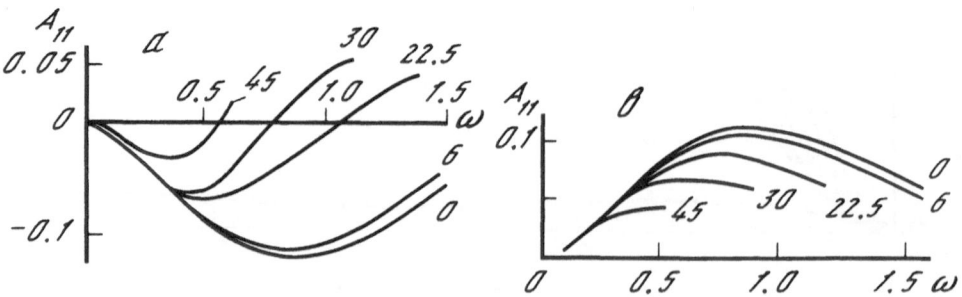

Fig. 4.22. Reflection coefficients as functions of frequency for the inclined contact shown in Figure 4.20. (a) incidence from the thinner layer; (b) incidence from the thicker layer. Numbers at curves are the angles of inclination.

transmission coefficient is practically unaffected by the step inclination in the entire frequency range. The reflection coefficient is practically unaffected by the step inclination at low frequencies. When $l < \lambda/10$, the $A_{11}(\omega)$ curves deviate from the relevant curve for a vertical discontinuity by 15 percent at most for all the angles considered here. The discrepancy becomes more serious with increasing frequency.

The above examples demonstrate the adequacy of the approximate method described in 3.3, 3.5 for evaluating reflection and transmission coefficients of surface waves produced at a welded and unwelded contact between two media that are not excessively different in structure. However, in numerical modelling of reflected and transmitted surface waves one should remember that this model is still far from being an adequate representation of real deep faults and junctions between differently structured lithospheric blocks. Reflected and transmitted waves seem to be strongly affected by factors, not incorporated in the present model, such as increased attenuation and scattering at the contact, and conversion of most of surface wave energy into body waves scattered at local inhomogeneities.

PART 2

INTERPRETATION OF SURFACE WAVE OBSERVATIONS

5. RECORDING, IDENTIFICATION, AND MEASUREMENT OF SURFACE WAVE PARAMETERS

The interpretation of surface wave data to study the Earth's interiors in different regions and to determine the structure of seismic sources must involve, as an initial step, routine measurement of surface-wave spectral characteristics along a great number of epicenter-station paths (Levshin, 1980). Such measurements call for the collection and standardized analysis of extensive experimental data, viz., seismograms that are recorded at seismic stations. In this process one transforms the data to a unified format, corrects them for instrument response, identifies signals of interest, measures spectral signal parameters. The present chapter is concerned with every step of this analysis of raw data, the emphasis being on methods in use for identifying and measuring signal parameters.

5.1. RECORDING TECHNIQUES AND EQUIPMENT IN SURFACE WAVE OBSERVATIONS

Seismic observations show that, depending on excitation and recording conditions, seismic surface waves that propagate in the crust and upper mantle involve periods between 5—10 sec and 300—500 sec (sometimes as long as 700 or 800 sec (Dziewonski, 1971)). According to the terminology generally adopted in seismology, such oscillations are called long period ones, as compared with the 0.5- to 5-sec oscillations represented by body waves, even though the same usage not infrequently applies to periods of 50 sec or longer.

The recording conditions for such waves are controlled by the intensity of the seismic sources that generate them, by attenuation within the Earth, and by the intensity of site-related seismic noise, either natural or man-made, which is not connected with the signals.

We shall not discuss source properties in any great detail, apart from noting the enormous dynamic range of seismic sources (at least ten decimal orders in radiated elastic energy). Wave attenuation for periods longer than 10 sec is comparatively small (less than a decimal order over distances of a few thousand kilometers) and does not affect the recording process in a comparable degree. Lastly, seismic noise in the frequency range of interest has a complicated structure (Figure 5.1) (see Feofilaktov (1977), Aki and Richards (1980), Kolesnikov and Toksöz (1984)). Ocean microseisms (microseisms of the first kind) produce a high level of noise in the 5 to 15 sec period range. The noise is low in the 20 to 40 sec

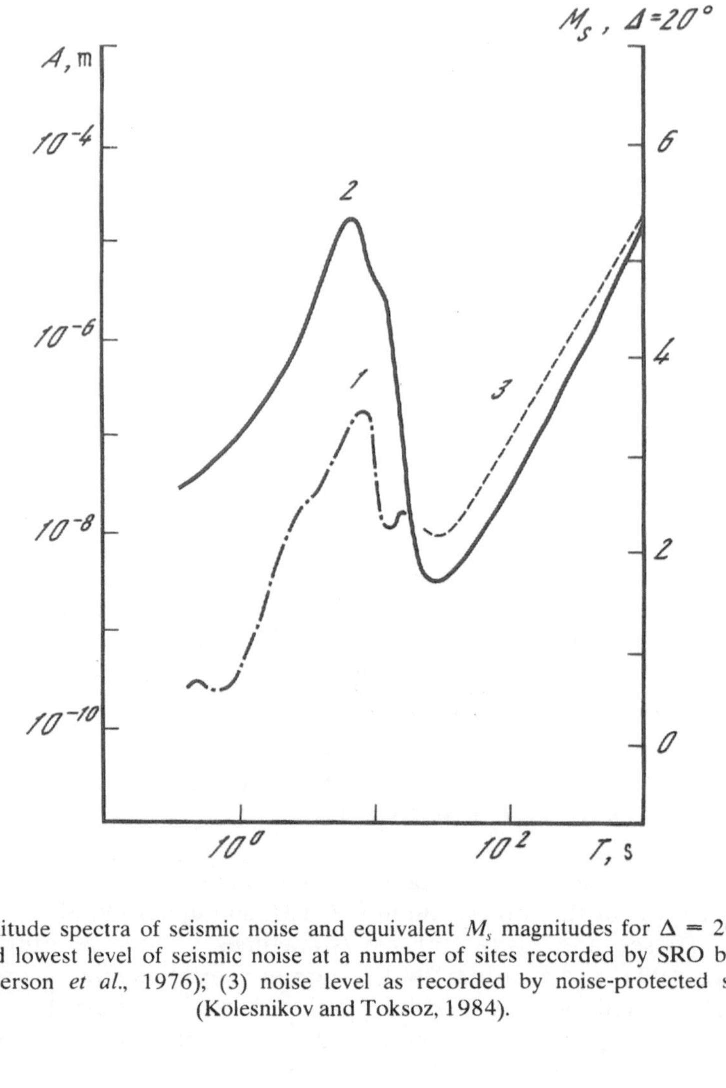

Fig. 5.1. Amplitude spectra of seismic noise and equivalent M_s magnitudes for $\Delta = 20°$. (1, 2) are the highest and lowest level of seismic noise at a number of sites recorded by SRO borehole seismographs (Peterson *et al.*, 1976); (3) noise level as recorded by noise-protected seismographs (Kolesnikov and Toksoz, 1984).

region, hence that 'window' is being widely used to record surface waves from weak sources. With longer periods the noise level progressively grows owing to pressure (wind) fluctuations that are acting both directly on the recording instruments and on the ground at recording site or over adjacent area (microseisms of the second kind). Noise level on seismograph records may vary by one or two decimal orders in response to weather conditions. Various methods that are used for suppression of pressure-induced seismometer oscillations (hermetic sealing, vacuum, thermal insulation and compensation) counteract the nonseismic component of the noise. The noise that acts through the ground is suppressed by installing instruments in boreholes a few hundred meters deep; another method for seismic noise suppression consists in correlation filtering using coherent "pressure" oscillations (recordings of pressure or equivalent temperature fluctuations). Experiments testify to great efficiency achieved with this method of pressure noise suppression.

Recording procedures. The instruments employed in seismology to record long period seismic oscillations are mostly seismometers, that is, pendulum sensors coupled with electrodynamic or capacitance transducers transforming displacement into electric current (Matsievsky, 1980; Plešinger and Horalek, 1976; Wielandt and Streckeisen, 1982), less frequently gravimeters (Agnew *et al.*, 1976), quartz and laser strainmeters (Aki and Richards, 1980; Starovoit *et al.*, 1971), and other devices (Aki and Richards, 1980; Lin'kov, Tipisev and Butsenko, 1980; Blum and Gaulon, 1971). Until recently seismic oscillations had been mainly recorded by analog methods (on photographic paper), but today digital recording is increasingly employed. Figure 5.2 shows typical instrument response curves for seismographs in use in the USSR and abroad.

In connection with the need for joint standardized analyses of data coming from several different stations, networks of standard long-period or broadband seismographs have been created: world gravimeter network IDA (Agnew *et al.*, 1976), world seismographic networks GDSN (based on modified WWSSN stations and on the observatories SRO and ASRO) (Peterson *et al.*, 1976), GEOSCOPE (Roult and Romanovicz, 1983), regional western European networks ORPHEUS (Dost *et al.*, 1984), Chinese digital network (Peterson J., Kexin Qu *et al.*, 1987) etc. Long-period waves are usually recorded on a continuous basis, data collection being handed through magnetic tapes sent by mail; satellite data acquisition systems are being planned.

A number of countries created station arrays with large numbers of sensors in the late '60s and early '70s, including long-period ones where data are telemetered to a processing center (LASA, NORSAR, ALPA). The long-period sensors in such arrays are three-component seismometers whose resonant response curves have a peak around 20 sec (Figure 5.2). In this monograph we are using data from an array of this kind, NORSAR (Bungum, Husebye and Ringdal, 1971), located in southern Norway (until 1977 that array had 22 long-period stations aligned in three concentric circles, the greatest diameter being 110 km). Readings are sent through telephone lines at intervals of 1 second to a processing center. Because of large costs involved in maintaining such arrays, they have by now either completely or partially discontinued operation.

Some Soviet stations employ a system of digital magnetic-tape recording (Grudeva *et al.*, 1973) that writes data in digital code on a magnetic tape of 35 mm width with subsequent rewriting on computer magnetic tapes. A team working at the Institute of Physics of the Earth of the Academy of Sciences of the USSR has developed a tentative microprocessor system recording digitally on magnetic tape in computer format. Semiautomatic digitizing of paper records is also widely used, but the narrow dynamic range (about 30 dB) and low digitizing accuracy due to drum speed irregularities limit the usefulness of these digital data compared with those available through direct recording on magnetic tape.

5.2. BASICS OF SURFACE WAVE PROCESSING

The next four sections are devoted to the principles underlying the digital processing of surface wave records. We wish to emphasize the specific features of problems that arise and the appropriate techniques. Completeness is not claimed

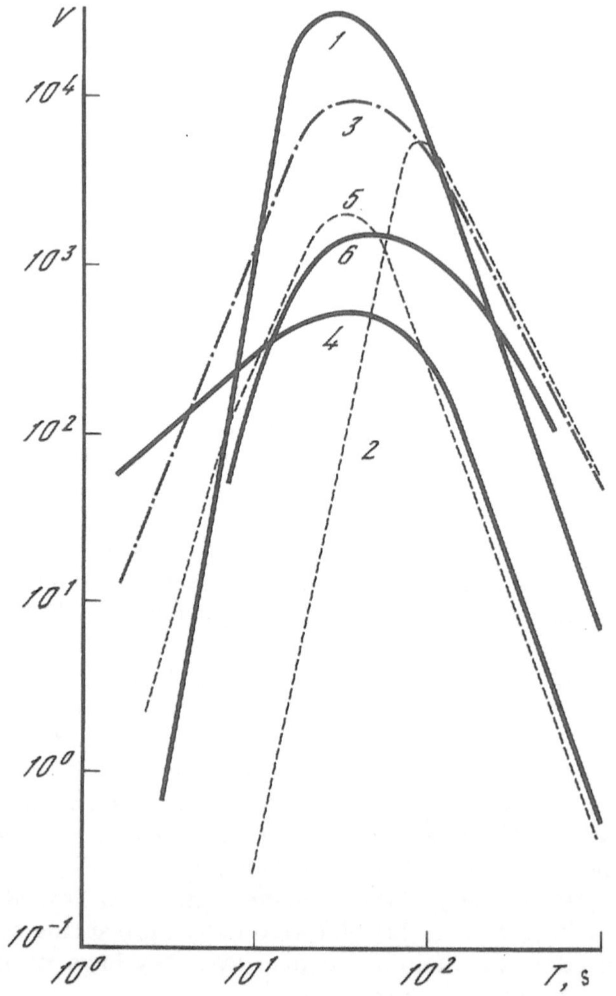

Fig. 5.2. Amplitude response curves of modern long-period seismographs. (1) SRO borehole seismograph (Peterson *et al.*, 1976); (2) IDA gravity meter (Agnew *et al.*, 1976); (3) noise-protected seismograph (Kolesnikov and Toksoz, 1982, 1984); (4) WWSSN seismograph (Oliver and Murphy, 1971); (5) seismic channel at NORSAR (Bungum, Husebye and Ringdal, 1971); (6) Wielandt-Streckeisen seismograph (Wielandt and Streckeisen, 1982).

for this exposition, in the first place as regards technicalities. The only two exceptions which are treated at more length than the rest are the frequency-time analysis, this being the most graphic method of surface wave representation, and some polarization problems as being less thoroughly discussed in the seismological literature.

Processing of a single record is dealt with in Sections 5.2 and 5.3. Measurement of the polarization and the potential of a spatial surface wave analysis are considered in 5.4 and 5.5.

The notation employed in these sections is not quite consistent with that of the preceding chapters, it is only for a few concepts that are central to the whole of the book that notation is treated uniformly. The term 'measurement' implies digital calculation of signal characteristics rather than the physical act of measurement.

If the information contained in a seismogram is to be used in geophysical interpretation, it should be translated into the language of concepts and quantities that form a theoretical model; this is the principal aim of data processing techniques.

It is easy to see that the problem has two aspects; the one is exploratory, model building analysis, while the other is more in the nature of routine calculations within the framework of a chosen model. The concept of a model is here understood in a broad sense, involving many different problems. These may include the question of laterally homogeneous approximation to a structure or something specific to a record, like the problem of ascribing some particular wave train to the signal or regarding it as the noise. It goes without saying that many 'model statements' with respect to surface waves in the Earth have been repeatedly corroborated; yet, nearly every seismogram presents some unexpected feature that demands informal (frequently personally biased) decision making. Such situations are best handled by trying a different representation of data, rather that by applying a different data processing theory. Mathematical techniques mainly belong to the latter aspect, i.e., routine processing.

It is with the direct processing of a one-dimensional surface wave record that the present section is concerned. We shall list the assumptions relating to surface waves when these are regarded as 'signals', discuss the parameters that describe their general properties, and see how the principal processing techniques follow from these.

Measurable characteristics. Definitions. We first state explicitly what the 'model language' is into which we wish to 'translate' a trace recorded by a seismometer.

In theory the phrase 'surface wave' denotes a certain term in the expansion of the total wave motion. Corresponding to a 'wave' is the term 'signal' in processing theory. The signal is understood to be that portion of a record which carries information on the wave of interest, all the rest being considered as noise with respect to that signal. We first identify a signal (that is, recognize the typical wave features), separate it from out the background noise and, lastly, identify its characteristics with the corresponding characteristics of the wave. The ultimate end of data processing is thus to measure characteristics of a signal. If the medium is assumed to be 'linear' and 'stationary' (that is, its properties do not change with time), then the most convenient representation of a wave is in terms of its spectrum. Such a medium affects a signal as a linear filter does whose response equals Green's function of the relevant differential equations. A filter is conveniently fixed by giving its frequency response, which in the simplest case of a laterally homogeneous medium has the form

$$M(\omega, r)\exp[-i\xi(\omega)r] \tag{5.1}$$

where $\xi(\omega)$ is the wavenumber, r the distance the wave has travelled. $M(\omega, r)$ and $\xi(\omega)$ are controlled by the medium alone and do not depend on wave shape; they contain the information on the medium provided by travelling waves. We have seen above that many problems in seismic source studies are conveniently stated in spectral form.

The most easily interpretable characteristic is the phase response of the filtering medium, which controls the time features of a propagating wave. It is usually given by fixing phase velocity

$$C(\omega) = \omega/\xi(\omega). \tag{5.2}$$

We shall see below that surface wave processing requires measuring another phase parameter, namely group velocity

$$U(\omega) = 1 \Big/ \frac{d\xi}{d\omega}(\omega). \tag{5.3}$$

The functions $U(\omega)$ and $C(\omega)$ are called dispersion curves. They are related by

$$\frac{1}{U} = \frac{1}{C} + \omega \frac{d}{d\omega}\left(\frac{1}{C}\right) \tag{5.4}$$

that is, group velocity can be calculated from phase velocity (but not conversely). Strictly speaking, a group velocity curve does not provide additional information on the medium compared with that given by phase velocity, yet both are extensively used in geophysical interpretation analyses. Among other things, this is due to the fact that, not infrequently, group velocity alone can be measured, and also, that the uncertainties in U and C measurements are not necessarily connected through a relation like (5.4).

We now turn from this discussion of a filtering medium to consideration of signal characteristics. Suppose a signal is given in the form of a real-valued function, $w(t)$. We define the spectrum of the signal as

$$K(\omega) = |K(\omega)|e^{i\psi(t)} = \begin{cases} \sqrt{\dfrac{2}{\pi}} \displaystyle\int_{-\infty}^{\infty} w(t)e^{-i\omega t}\, dt, & \omega > 0 \\ \dfrac{1}{\sqrt{2\pi}} \displaystyle\int_{-\infty}^{\infty} w(t)\, dt, & \omega = 0 \\ 0, & \omega < 0 \end{cases} \tag{5.5}$$

where $K(\omega)$, $\psi(\omega)$ are spectral amplitude and phase, respectively.

The above definition allows a complex-valued signal, $W(t)$, to be constructed that corresponds to $w(t)$ and is related to $K(w)$ through the ordinary Fourier transform on an infinite axis:

$$\begin{aligned} W(t) &= |W(t)|e^{i\varphi(t)} = \frac{1}{\sqrt{2\pi}} \int_{-\infty}^{\infty} K(\omega)e^{i\omega t}\, d\omega \\ K(\omega) &= |K(\omega)|e^{i\psi(\omega)} = \frac{1}{\sqrt{2\pi}} \int_{-\infty}^{\infty} W(t)e^{-i\omega t}\, dt \end{aligned} \tag{5.6}$$

where $|W(t)|$, $\varphi(t)$ are time-related (or instantaneous) amplitude and phase. In many cases the functions $|W(t)|$ and $\varphi(t)$ have the visualizable meanings of 'envelope amplitude' and 'carrier phase'. This is true, in particular, for narrow-band signals ($D_\omega/\langle\omega\rangle \ll 1$), where D_ω and $\langle\omega\rangle$ are the width and dominant frequency of the signal spectral band.

From (5.5) and (5.6) it follows that $w(t)$ and $W(t)$ are connected by $w(t) = \text{Re}\, W(t)$, that is, the complex-valued signal contains the whole of the information about the real-valued one; $W(t)$ is sometimes called the analytic signal. While we are working within the adopted model of a linear stationary medium, $w(t)$ and $W(t)$ are solutions to the same differential equations.

Once an analytic signal has been defined, the notions of amplitude and phase can be used both in the frequency and time domain. Two characteristics will be serviceable (a prime means differentiation with respect to frequency, a dot that with respect to time):

$$\tau(\omega) = -\psi'(\omega) \quad \text{(group time)},$$

$$\Omega(t) = \dot{\varphi}(t) \quad \text{(instantaneous frequency)}.$$

There is a simple relation that connects phase and group velocities with the spectral characteristics of a signal for a laterally homogeneous medium. After passing through a filtering medium, a wave has the phase spectrum

$$\psi(\omega) = -\xi(\omega)r + \psi_s(\omega) = -\frac{\omega}{C(\omega)}r + \psi_s(\omega) \tag{5.7}$$

where $\psi_s(\omega)$ is the spectral phase at the source (Levshin, 1973; Frez and Schwab, 1976). Differentiation of (5.7) yields

$$\tau(\omega) = -\psi'(\omega) = \frac{r}{U(\omega)} - \psi_s'(\omega). \tag{5.8}$$

This relation explains the term 'group time', $\tau(\omega)$ being that particular characteristic of the signal which is used to determine group velocity. For the same reason we shall sometimes employ the phrase 'spectral dispersion curve of the signal' for the function $\tau(\omega)$ and, proceeding on analogy, shall call $\Omega(t)$ 'time dispersion curve of the signal'.

Before $C(\omega)$ and $U(\omega)$ are determined from (5.7) and (5.8), the source phase has to be eliminated. This is achieved either by measuring $\psi(\omega)$ and $\tau(\omega)$ at two or more stations or by using the theoretical concepts relating to source mechanism (see 1.2). Seismic sources usually act during a time that is short compared with the typical values of $\tau(\omega)$, enabling $-\psi_s'(\omega)$ in (5.8) to be replaced by a constant ('source time') to within the required accuracy. Group velocity can thus be found from known epicenter parameters, even though $\tau(\omega)$ has been measured at a single station. We shall often set $\psi_s'(\omega) \equiv 0$ in what follows.

Thus, to study the velocity characteristics of a medium one must have measurements of $\psi(\omega)$ and $\tau(\omega)$ for a signal. Amplitude problems involve measurements of the spectral amplitude $|K(\omega)|$.

Characterization of surface wave signals. The routine aspect of the processing consists in functional transformations. Some of these, such as filtering, are signal-oriented. This kind of transformation works the more effectively, the better we know our signal and, generally speaking, the noise. Below we list some well-known general properties of surface wave signals (mainly in relation to teleseismic records).

'Finiteness' in frequency and time. A surface wave involves a limited interval,

both on the seismogram and in the spectrum, signal amplitude being small and hidden in the noise outside that interval. The behaviour of a model signal there can be chosen arbitrarily, provided the amplitude does not exceed some threshold. In particular, both functions, $|W(t)|^2$ and $|K(\omega)|^2$, can always be assumed to vanish at $\pm\infty$ and to be integrable. Processing results should not significantly depend on intervals where signal amplitude is small and cannot be reliably determined.

Deterministic character. Apart from some special cases, surface waves make up the largest, easily identifiable portion of a seismic record. Noise plays a subordinate role. For this reason a surface wave signal can be treated as a deterministic function with relatively smooth curves of $|K(\omega)|$, $\tau(\omega)$. The figures in 7.3 provide some notion of what these functions look like. Intervals of unstable behaviour in $|K(w)|$ or $\tau(\omega)$ frequently encountered in seismogram spectra are usually caused by noise rather than by some property of the signal.

The dispersion of $\tau(\omega)$ associated with group velocity dispersion is the most striking feature of the signals discussed. As a result, a surface wave shows no definite front, but has a broad spectral band and is long in duration. The frequency axis may be divided into three or four intervals with the dispersion curve $U(\omega)$ for teleseismic waves monotonic within each of these (see Figure 6.4). Frequently the entire signal may be recorded within a signal interval of this kind, i.e., $\tau(\omega)$ is a monotonic. In such cases a useful model is $\tau'(\omega) = \text{constant}$, here called a 'linearly dispersed signal'.

Polarization. Two well-pronounced types of surface wave polarization typical of laterally homogeneous isotropic media (Love and Rayleigh waves, see 1.4) occur in seismograms. Considerable departures from the theoretical polarization directions are not infrequent, however. These are not so large as to cause serious difficulty in wave identification, yet may reach 50° (see 7.3). Values around 10° are more typical.

Noise. Stationary noise is small compared with surface waves on most seismograms. Local impulsive nonstationary noise is the main source of trouble. This kind of noise may be largely generated by the surface wave itself. Because it is nonstationary, regular noise models are useless in most of the cases. The common practice is to treat 'everything that is unlike the signal' as noise, but interference between different identified surface waves occurs frequently enough too.

Signal parameterization. The general properties of signals closely related to processing techniques in use are here described by a parameterization that is widely employed in physics, namely, by using moments and quadratic norms.

We define the normalized amplitudes

$$A(t) = \frac{|W(t)|}{\sqrt{\int_{-\infty}^{\infty} |W(t)|^2 \, dt}} \qquad B(\omega) = \frac{|K(\omega)|}{\sqrt{\int_{-\infty}^{\infty} |K(\omega)|^2 \, d\omega}} . \tag{5.9}$$

Then (Parceval's equality)

$$\int_{-\infty}^{\infty} A^2(t) \, dt = \int_{-\infty}^{\infty} B^2(\omega) \, d\omega = 1 \tag{5.10}$$

and, from the Fourier transform (5.6)

$$B(\omega)e^{i\psi(\omega)} = \frac{1}{\sqrt{2\pi}} \int_{-\infty}^{\infty} A(t)\, e^{i[\varphi(t) - \omega t]}\, \mathrm{d}t. \tag{5.11}$$

Treating $A^2(t)$ and $B^2(\omega)$ as 'energy distributions', we define parameters that characterize 'mean' properties of the signal: the typical time

$$\langle t \rangle = \int_{-\infty}^{\infty} tA^2(t)\, \mathrm{d}t \tag{5.12}$$

the mean instantaneous frequency

$$\langle \Omega \rangle = \int_{-\infty}^{\infty} \Omega(t)A^2(t)\, \mathrm{d}t, \tag{5.13}$$

the duration D_t

$$D_t^2 = 2\int_{-\infty}^{\infty} (t - \langle t \rangle)^2 A^2(t)\, \mathrm{d}t \tag{5.14}$$

the 'interval' where the instantaneous frequency D_Ω varies

$$D_\Omega^2 = 2\int_{-\infty}^{\infty} (\Omega(t) - \langle \Omega \rangle)^2 A^2(t)\, \mathrm{d}t \tag{5.15}$$

the typical frequency

$$\langle \omega \rangle = \int_{-\infty}^{\infty} \omega B^2(\omega)\, \mathrm{d}\omega \tag{5.16}$$

the mean group time

$$\langle \tau \rangle = \int_{-\infty}^{\infty} \tau(\omega)B^2(\omega)\, \mathrm{d}\omega \tag{5.17}$$

the spectral bandwidth D_ω

$$D_\omega^2 = 2\int_{-\infty}^{\infty} (\omega - \langle \omega \rangle)^2 B^2(\omega)\, \mathrm{d}\omega \tag{5.18}$$

the 'interval' of signal group time D_τ

$$D_\tau^2 = 2\int_{-\infty}^{\infty} (\tau(\omega) - \langle \tau \rangle)^2 B^2(\omega)\, \mathrm{d}\omega. \tag{5.19}$$

It is understood that $A(\pm\infty) \to 0$, $B(\pm\infty) \to 0$ in the above expressions and that all integrals are bounded. It has been mentioned that 'the practical finiteness' of time intervals and spectral bandwidths of the signals always justify these assumptions.

We now consider a few relations that involve 'mean' signal properties.

'*Group-time relation*'. We differentiate both sides of (5.11) with respect to the frequency

$$(B' - iB\tau)e^{i\psi} = -\frac{i}{\sqrt{2\pi}}\int_{-\infty}^{\infty} tA\exp[i(\varphi - \omega t)]\,\mathrm{d}t. \tag{5.20}$$

Now form the product of (5.20) and the complex conjugate of (5.11), and integrate it over ω. Using $\int_{-\infty}^{\infty}\exp[i(t-s)\omega]\,\mathrm{d}\omega = 2\pi\delta(t-s)$ and $\int_{-\infty}^{\infty} B'B\,\mathrm{d}\omega = B^2|_{-\infty}^{\infty} = 0$, we get

$$\langle\tau\rangle = \int_{-\infty}^{\infty}\tau(\omega)B^2(\omega)\,\mathrm{d}\omega = \int_{-\infty}^{\infty} tA^2(t)\,\mathrm{d}t = \langle t\rangle. \tag{5.21}$$

The typical signal time $\langle t\rangle$ equals $\langle\tau\rangle$, the group time averaged over the power spectrum. Note that $\langle\tau\rangle$ and $\langle t\rangle$ have been obtained by averaging with different weighting functions.

Equation (5.21) explains how a spectral quantity, group velocity $U(\omega)$, can be the 'velocity of propagation in space and time'. A wave propagating in a dispersive medium varies in shape from point to point, so the concept 'velocity of propagation' needs a special definition. We take for the 'time at which a wave is at point r', the value of $\langle t\rangle$ obtained from the seismogram recorded at that point. Then, putting $\psi_s(\omega) \equiv 0$, we shall have from (5.21) and (5.8):

$$\frac{1}{\langle U\rangle} = \frac{\langle t\rangle}{r} = \int_{-\infty}^{\infty}\frac{B^2(\omega)}{U(\omega)}\,\mathrm{d}\omega. \tag{5.22}$$

In the sense indicated, a wave travels in a dispersive medium at a slowness $1/\langle U\rangle$ that equals the mean (with respect to the signal power spectrum) 'group slowness' of the medium.

The quantity $1/(\mathrm{d}\langle t\rangle/\mathrm{d}r)$ may occasionally be treated as the 'envelope velocity', in the first place for narrow-band signals. According to (5.22), this is equal to the group velocity at some frequency within the signal bandwidth. This treatment of group velocity is physically meaningful and therefore rather helpful; indeed, it underlies a number of group velocity measurement techniques. One should bear in mind, nevertheless, that what we convey by the above is an approximate statement that means just as much as is implied by (5.21). For this reason the relevant measurement techniques are all subject to systematic error (see 5.3).

Proceeding in an analogous manner to (5.21), one can derive the result that the typical signal frequency (5.16) is equal to the mean instantaneous signal frequency (5.13)

$$\langle\Omega\rangle = \int_{-\infty}^{\infty}\Omega(t)A^2(t)\,\mathrm{d}t = \int_{-\infty}^{\infty}\omega B^2(\omega)\,\mathrm{d}\omega = \langle\omega\rangle. \tag{5.23}$$

Thus, when we measure the apparent frequency Ω on a seismogram, we also learn an approximate value of that frequency $\langle\omega\rangle$ around which the signal spectrum concentrates. This last statement when applied to a narrow-band signal provides a still more graphic interpretation of group velocity: 'envelope velocity' is approximately equal to the group velocity at the apparent frequency of the signal.

We conclude by citing an example in which (5.21) yields a somewhat unexpected physical result. Suppose a narrowband wave is travelling in a dispersive medium subject to attenuation, the latter increasing with frequency and $U(\omega)$ decreasing. As the wave propagates, the long-period portion of the signal spectrum will gain in importance, hence the effective interval of integration in (5.22) will shift towards longer periods. $\langle U\rangle$ will consequently grow. An observer that is 'envelope velocity'-oriented will seem to notice an acceleration, although the medium remains laterally homogeneous. This bias in spectral bandwidth can certainly be eliminated by measuring $\langle\Omega\rangle$ too, but the example demonstrates that caution is needed in velocity determinations based on time characteristics of a signal.

Duration and spectral bandwidth. Uncertainty relation. Using (5.11) and (5.12), we rewrite (5.20) in the form

$$[B' - iB\cdot(\tau-\langle\tau\rangle)]e^{i\psi} = -\frac{i}{\sqrt{2\pi}}\int_{-\infty}^{\infty}(t-\langle t\rangle)A\exp[i(\varphi-\omega t)]\,\mathrm{d}t \tag{5.24}$$

and apply Parceval's equality to this. The result is

$$D_t^2 = D_\tau^2 + 2\int_{-\infty}^{\infty}B'^2(\omega)\,\mathrm{d}\omega. \tag{5.25}$$

The duration of a signal is thus controlled by two different properties of its spectrum to be considered below.

When we were discussing (5.21), we began to talk about the connection between $\tau(\omega)$ and the times t where the signal occurs. The first term on the right-hand side of (5.25) is an expression of that connection, as follows. The statement $\langle t\rangle = \langle\tau\rangle$ and that derived from (5.25), $D_t > D_\tau$, mean that ('on average') the time interval t occupied by a signal at least overlaps that where $\tau(\omega)$ varies within the relevant bandwidth. Note that $D_\tau = 0$ only holds when the phase spectrum is linear ($\tau(\omega)$ = constant). Any amount of dispersion in $\tau(\omega)$ invariably leads to an increase in the duration D_t. Taking the important case of a linearly dispersed signal ($\tau'(\omega) = \tau'$ = constant), we have

$$D_t = |\tau'|D_\omega. \tag{5.26}$$

For surface waves D_τ is significantly dependent on the velocity dispersion in the medium. Using (5.8), (5.19), (5.22) and assuming $\psi_s'(\omega)\equiv 0$, we get

$$D_\tau^2 = r^2\int_{-\infty}^{\infty}\left(\frac{1}{U(\omega)}-\frac{1}{\langle U\rangle}\right)^2 B^2(\omega)\,\mathrm{d}\omega \tag{5.27}$$

that is, D_τ grows in proportion to the distance from the source.

The second term on the right-hand side of (5.25) is only controlled by the amplitude spectrum of the signal. We can evaluate this term by making use of the identities

$$1 = \int_{-\infty}^{\infty} B^2\,d\omega = -2\int_{-\infty}^{\infty} B'B\omega\,d\omega = -2\int_{-\infty}^{\infty} (\omega - \langle\omega\rangle)BB'\,d\omega.$$

Application of Cauchy's inequality to the last of these integrals yields

$$2D_\omega^2 \int_{-\infty}^{\infty} B'^2(\omega)\,d\omega \geqslant 1. \tag{5.28}$$

Substituting (5.28) into (5.25), we get

$$D_t^2 \geqslant D_\tau^2 + 1/D_\omega^2. \tag{5.29}$$

The second term on the right-hand side of (5.25) is thus responsible for the ordinary uncertainty relation which provides a lower bound on $D_\omega D_t$. The expression (5.29) is the complete form of the uncertainty relation. The equality in (5.28) and (5.29) occurs when $B' \sim (\omega - \langle\omega\rangle)B$, that is, for a signal having a Gaussian amplitude spectrum, $B(\omega) = c_1 \exp[c_2(\omega - \langle\omega\rangle)]$. We must require $c_2 < 0$ for $B(\omega)$ to vanish at infinity. Among all signals having arbitrary but identical phase spectra, the Gaussian signal has the minimum product $D_\omega D_t$.

For wave propagation in a nondissipative medium, $B(\omega)$ and hence $\int_{-\infty}^{\infty} B'^2(\omega)\,d\omega$ do not depend on the distance to the source. Then, using (5.27), we get

$$D_t = \sqrt{ar^2 + b} \approx D_\tau + b/2ar \tag{5.30}$$

where a and b are two positive constants. $D_t \to D_\tau$ at large distances from the source, that is, signal duration is often controlled by dispersion alone.

Relations that connect D_ω and D_Ω are derived in a similar fashion:

$$D_\omega^2 = D_\Omega^2 + 2\int_{-\infty}^{\infty} A'^2(t)\,dt \tag{5.31}$$

$$D_\omega^2 = D_\Omega^2 + 1/D_t^2. \tag{5.32}$$

Dimensionless parameters I and q. Uncertainty. The integral properties of signals discussed above are conveniently described in terms of two dimensionless parameters:

$$I^2 = 2D_\omega^2 \int_{-\infty}^{\infty} B'^2(\omega)\,d\omega \tag{5.33}$$

$$q^2 = \frac{D_\tau^2}{2\int_{-\infty}^{\infty} B'^2(\omega)\,d\omega}. \tag{5.34}$$

Then (5.25) and the uncertainty relation (5.29) take the dimensionless form

$$D_\omega^2 D_t^2 = I^2(q^2 + 1), \qquad I^2 \geqslant 1. \tag{5.35}$$

The quantity $D_\omega D_t$ will be termed signal uncertainty. It attains a minimum equal to $D_\omega D_t = 1$ for a Gaussian signal ($I = 1$) with a linear phase spectrum ($\tau(\omega) =$ constant, $q = 0$). The value of uncertainty thus characterizes 'nonoptimality' of a signal in the sense of $D_\omega D_t$. The parameters I and q express two possible reasons for the 'nonoptimality'.

The parameter I describes 'nonoptimality' of an amplitude spectrum. It is readily verified that the substitution $|K(\omega)| \to |K(a\omega)|$, where a is a constant, does not affect I. In that sense I is independent of the spectral bandwidth of a signal; I is close to one for signals that are 'similar' to Gaussian ones. In actual practice, most spectral peaks when considered separately have Is that are close to one. In contrast to this, signals involving several spectral peaks are always nonoptimal in the sense of I being greater than one. A simple example will illustrate this.

Consider two signals with $\tau(\omega) \equiv 0$ (their amplitude spectra are shown in Figure 5.3a): one of these has a single spectral peak, while the other consists of several non-overlapping spectral peaks of the same form, but different heights. For these signals the ratio of the respective integrals $\int_{-\infty}^{\infty} |K|'^2 \, d\omega$ is equal to that of the energies $\int_{-\infty}^{\infty} |K|^2 \, d\omega$ (in Figure 5.3 the energies are chosen equal for convenience), $D_\tau = 0$, consequently the parameters $D_t = 2 \int_{-\infty}^{\infty} B'^2 \, d\omega$ for both signals are equal. From the figure one can see that the values of D_ω are significantly different, and the Is differ in the same ratio.

The parameter q characterizes the relative effects of phase and amplitude spectrum on signal duration. The statement $q = 0$ is equivalent to $\tau(\omega) =$ constant. It is only when $q > \sim 1$ that nonlinearity in the phase spectrum seriously affects signal shape. We shall occasionally call signals with $q > 0$ 'dispersed' signals (weakly dispersed, when $q \ll 1$ and strongly so when $q \gg 1$). In the case of a linear $\tau(\omega)$

$$q = \frac{\tau' D_\omega^2}{I} . \tag{5.36}$$

Below we assume q to have the same sign as τ' for such signals. (I is always assumed positive).

It is often convenient to calculate q separately for each spectral peak. In that

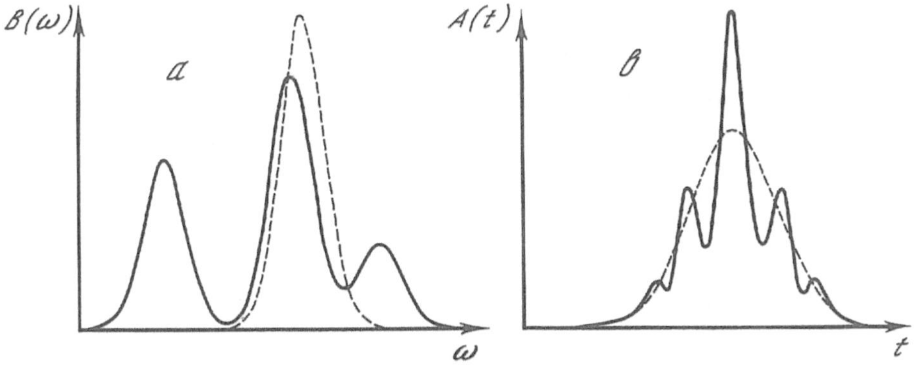

Fig. 5.3. Parameter I different from unity: an example. (a) spectral and (b) time amplitudes of two signals. The dashed line is for $I = 1$, the solid line is for $I \approx 3$.

case $I \approx 1$, and when $q \gg 1$, we shall have from (5.33) through (5.35):

$$D_\omega D_t \approx D_\omega D_\tau \approx |q| \tag{5.37}$$

that is, the q for any individual spectral peak of a strongly dispersed signal is approximately its uncertainty. Indeed, from (5.35) it follows that dispersed signals are always nonoptimal in the sense of $D_\omega D_t$.

We have mentioned that the most characteristic feature of surface wave signals is the dispersion $\tau(\omega)$. The statement can now be given a more precise meaning. The typical values for teleseismic surface waves are $q^2 > \sim 10$. As D_ω is nearly independent of the distance to the source, while D_τ is directly proportional to it according to (5.27), it follows that the q for surface waves grows in about direct proportion to r. Surface wave signals are thus rather nonoptimal in the sense of uncertainty. That fact as we shall see largely determines techniques of surface wave processing.

The q of a surface wave train can be roughly estimated visually from the seismogram. When a signal is strongly dispersed, we usually have $D_\omega \approx D_\Omega$ in (5.32). Hence, substituting D_Ω into (5.37) and remembering how D_Ω and D_t were defined, we can evaluate q around a peak of the envelope from

$$|q| \approx \Delta t \cdot \Delta\Omega \tag{5.38}$$

where Δt, $\Delta\Omega$ are the intervals of variation for t and Ω within the time interval where the envelope amplitude $|W(t)|$ is greater than about 0.6 times the peak value. Note that a signal may be strongly dispersed only around some of the envelope peaks.

We now discuss what special properties follow from the high value of uncertainty for surface wave signals.

Locally narrow-band signal. Stationary phase approximation. If a signal has $q^2 \ll 1$ and $I \approx 1$, then any transformation that diminishes D_ω will, according to the uncertainty relation (5.35), lead to an increase in D_t. Such signals and, hence, statements like that just quoted, frequently occur in physics, but surface waves present a different situation. When $q^2 \gg 1$, both D_ω and D_t can be diminished simultaneously. We are going to demonstrate this for the simplest case of a linearly dispersed signal ($\tau' =$ constant) with a sufficiently broad band D_ω and $I \approx 1$ (this last condition means that a single spectral peak is being considered).

Let us pass that signal through a filter with a real-valued frequency response $H(\omega)$ ($\arg H(\omega) \equiv 0$) that is symmetric about some frequency ω^H. Signal parameters at the filter output are distinguished by placing the $\tilde{}$ sign above a letter. As $H(\omega)$ is real-valued, we have $\tilde{\tau}' = \tau'$, and (5.25) for a filtered signal becomes

$$\tilde{D}_t^2 = \tilde{D}_\tau^2(1 + \tilde{q}^2) \approx (\tau' D_\omega^H)^2 + \left(\frac{I^H}{D_\omega^H}\right)^2 \tag{5.39}$$

$\tilde{D}_\omega^H$ and $\tilde{I}$ have been replaced by the filter parameters D_ω^H and I^H, which would not incur serious errors when $D_\omega^H \ll D_\omega$ i.e., when the filter is narrow compared with the bandwidth of the original signal, and $\tilde{B}(\omega) \approx \text{constant} \cdot H(\omega)$.

Let us narrow the filter bandwidth D_ω^H, while keeping its shape unchanged (i.e.,

$I^H =$ constant). Then (5.39) yields

$$\tilde{D}_t^2 \approx \tilde{D}_\tau^2 \approx (\tau' D_\omega^H)^2 \quad \text{if} \quad q^2 \gg 1 \tag{5.40}$$

that is, $\tilde{D}_t$ falls off in proportion to $\tilde{D}_\omega \approx D_\omega^H$. The filtered signal has the additional properties

$$\langle \tilde{\Omega} \rangle = \langle \tilde{\omega} \rangle \approx \omega^H \tag{5.41}$$

$$\langle \tilde{t} \rangle = \langle \tilde{\tau} \rangle \approx \tau(\omega^H) \tag{5.42}$$

which follow from the fact that $H(\omega)$ is symmetric about ω^H and $\tau(\omega)$ is a linear function. We see that a filtered signal has the shape of a wave train whose instantaneous frequency is close to ω^H, concentrates around the time instant $t = \tau(\omega^H)$, and has smaller duration and spectral bandwidth compared with the original signal.

Several filters with different ω^H can separate out an original signal into a sequence of relatively short, narrow-band trains that arrive at different times. Loosely speaking, each narrow spectral band of a strongly dispersed signal is significant only within a limited interval of time that is shorter than the total duration. Similarly, one may say that a strongly dispersed signal has a time-dependent, relatively narrow spectral band around any instant of time t. (If the $\tau(\omega)$ curve is not monotonic, there may be several spectral bands of this kind corresponding to a given time moment.) A signal of this kind can appropriately be called a locally narrow-band signal. While being broadband as a whole, it behaves like a narrow-band one in the vicinity of a given instant of time. In particular, the functions $|W(t)|$ and $\Omega(t)$ have the easily visualizable sense of envelope amplitude and carrier frequency in the case of a locally narrow-band signal.

The conclusions just made are not mathematically rigorous, but acquire a well-defined meaning asymptotically within the stationary phase approximation: this last asserts that if $\tau'(\omega) \neq 0$ within a frequency interval and

$$\psi(\omega) = N\eta(\omega) + \eta_0(\omega), \qquad N \to \infty \tag{5.43}$$

(whence $q \to \infty$), then there exists an interval of t where, apart from $O(1/N)$, the relation connecting $K(\omega)$ and $W(t)$ is a functional rather than an integral one:

$$|W(t)| = \sqrt{|\dot{\Omega}(t)|}\,|K(\Omega(t))| \tag{5.44}$$

$$\varphi(t) = \psi(\Omega(t)) + \Omega(t) \cdot t + \frac{\pi}{4} \operatorname{sign}[\dot{\Omega}(t)] \tag{5.45}$$

where $\Omega(t)$ is obtained by solving

$$t = \tau(\Omega). \tag{5.46}$$

It is easy to see that $\Omega(t)$ as given by (5.46) is the instantaneous frequency of the signal $W(t)$:

$$\dot{\varphi}(t) = [t - \tau(\Omega(t))] \cdot \dot{\Omega}(t) + \Omega(t) = \Omega(t). \tag{5.47}$$

From (5.46) and (5.47) it follows that the spectral and time dispersion curves of a signal, $\tau(\omega)$ and $\Omega(t)$, are mutually inverse functions. This fact is frequently employed in visual measurements of seismograms: $\Omega(t)$ is measured directly in the time domain instead of $\tau(\omega)$. The large parameter in (5.43) for the case of surface waves is, according to (5.47), the distance r, that is, the degree of approximation available in (5.44), (5.45) improves with increasing distance to the source. Still, the accuracy provided by the stationary phase approximation is poor for most teleseismic surface wave records (especially in the distance range around 3 to 5 thousand kilometers), being below the modern requirements on the measurement of phase and group velocities. Nevertheless, that approximation remains a useful tool for making rough estimates and qualitative conclusions as to the behaviour of a dispersed signal.

Formulas (5.44) and (5.45) can be qualitatively interpreted as follows: 'there is a single spectral component at each instant of time on the record'. However, whatever the accuracy of the stationary phase approximation (however large the value of q is), it does not follow that $\tilde{D}_\omega$ and $\tilde{D}_t$ can be made simultaneously as small as we wish. To see this, we turn to (5.39) again.

As D^H_ω decreases, $\tilde{D}_t$ will eventually come to be significantly dependent on the second term in (5.39). It will eventually predominate for small D^H_ω, i.e.,

$$\tilde{D}_t = \frac{\tilde{I}}{\tilde{D}_\omega}\sqrt{1+\tilde{q}^2} \approx \frac{I^H}{D^H_\omega} \quad (\text{if } \tilde{q}^2 \ll 1). \tag{5.48}$$

The uncertainty relation now becomes $\tilde{D}_t \sim 1/\tilde{D}_\omega$ and D_t increases with D^H_ω decreasing.

The function $\tilde{D}_t(D^H_\omega)$ is not monotonic, so it has a minimum which can conveniently be found by rewriting (5.25) in the form

$$\tilde{D}_t^2 \approx \tau' I^H(\tilde{q} + 1/\tilde{q}). \tag{5.49}$$

The factor in front of the brackets is independent of ω^H, i.e., $\tilde{D}_t$ is the least when $|\tilde{q}| = 1$. In that case

$$\tilde{D}^2_\omega = (D^H_\omega)^2 = \frac{I^H}{|\tau'|} \tag{5.50}$$

$$\tilde{D}_t^2 = 2|\tau'|I^H \tag{5.51}$$

$$\tilde{D}_\omega \tilde{D}_t = \sqrt{2}\, I^H. \tag{5.52}$$

If $I^H \approx 1$, then the minimum value of $\tilde{D}_t$ is $\approx \sqrt{q/2}$ times as small as the original D_t, and the corresponding $\tilde{D}_\omega$ is $\approx \sqrt{q}$ times as small as D_ω. Filtering (with a real-valued $H(\omega)$) can thus separate a signal into $\sim \sqrt{q}$ signals of minimum duration.

Frequency-time representation. When one deals with a research problem, where everything cannot be formalized, one would like to look at the most characteristic features of the object of study — the signal in this case — in order to understand whether it is consistent with one's theoretical notions or, if it is not, what are the principal discrepancies. For this reason an appropriate representation of data is very important for successful processing.

Short impulsive waves that are characterized by a single arrival time can frequently be studied in the time domain. Long-continued quasi-harmonic oscillations like free oscillations of the Earth are much more conveniently studied in spectral form. Surface waves are rather difficult to process. Their principal feature, the dispersion $\tau(\omega)$, is described by a function, rather than a single parameter. A visual picture of such a signal requires a function of two variables. This is the frequency—time representation which has the remarkable property of separating signals in accordance with their dispersion curves. In the present section we briefly review this representation, discussing it in more detail in Section 5.3.

We have in fact used the idea of a frequency—time representation when discussing special properties of dispersed signals. If a signal is passed through a system of parallel relatively narrow-band filters $H(\omega - \omega^H)$ with varying central frequency ω^H, then each resulting signal will, according to (5.42), concentrate around the time $t = \tau(\omega^H)$. (This conclusion is not related to the above requirement $q \gg 1$, and is true for any signal with a smoothly varying $\tau(\omega)$). The combination of all signals at the output of all the filters $H(\omega - \omega^H)$ will now be treated as a (complex) function of two variables, ω^H and t:

$$S(\omega^H, t) = \int_{-\infty}^{\infty} H(\omega - \omega^H) K(\omega) e^{i\omega t} \, d\omega \tag{5.53}$$

which is what we call the frequency—time representation of a signal. The signal analysis based on (5.53) will be abbreviated further on as FTAN (frequency—time analysis) (Levshin *et al.*, 1972).

A contour map of $|S(\omega^H, t)|$ called a FTAN map is used for visual representation. For ω fixed, $|S(\omega^H, t)|$ is the signal envelope at the output of the relevant filter. For this reason, corresponding to each input signal is a 'mountain ridge' (increased values) in the FTAN map extending along the dispersion curve $t(\omega^H) = \tau(\omega^H)$ (Figure 5.4).

The introduction of $S(\omega^H, t)$ enables us to employ a convenient terminology based, in particular, on the concept of the frequency—time region of a signal. This is understood to be that part of the (ω^H, t)-plane occupied by the relevant ridge. It can be seen from Figure 5.4 that the picture of a frequency—time region gives a

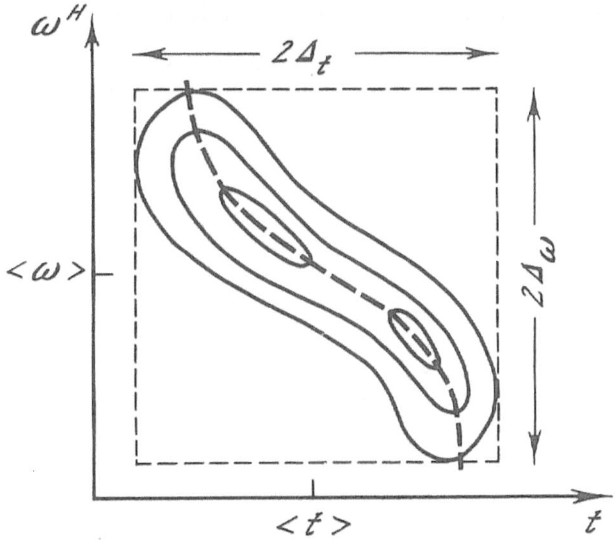

Fig. 5.4. Diagrammatic representation of the FTAN map. Dashed line is a dispersion curve $t(\omega^H) = \tau(\omega^H)$.

much clearer notion of a dispersed signal than the parameters $\langle t \rangle$, D_t, $\langle \omega \rangle$, D_ω can. The statement 'the energy of a signal concentrates (on the (ω^H)-plane) around its dispersion curve' acquires a definite meaning in terms of $|S(\omega^H, t)|$. From this one can also see that the large uncertainty $D_\omega D_t$ of a dispersed signal is in some sense fictitious: the area of the relevant rectangle on the (ω^H, t)-plane considerably exceeds that of the frequency—time region.

There is an important feature of the frequency—time representation that makes it essentially different from the more usual spectral or time representation. The function $S(\omega^H, t)$ is not a property of the original signal alone, but also involves the filter characteristic $H(\omega - \omega^H)$ chosen by the investigator. Different choices of $H(\omega - \omega^H)$ will transform one and the same signal to different $S(\omega^H, t)$ functions. An analogy can be drawn with measurements in quantum mechanics where the result is also affected by the 'influence of the experimentalist'. It does not follow nevertheless that the calculation of $S(\omega^H, t)$ loses some information contained in the signal (that might have occurred if a single filter had been used instead of a set of filters). When the shape of $H(\omega - \omega^H)$ is known, the function $W(t)$ or $K(\omega)$ can in principle be recovered from $S(\omega^H, t)$.

A complete frequency—time representation involves two functions: $S(\omega^H, t)$ proper and the law of variation of the filter shape. When we use a frequency—time representation, we are in fact dealing with a whole class of signal representations that differ in filter choice. This gives rise to a peculiar difficulty, the question at issue being the choice of that representation which is the most relevant to the processing problem in hand (optimal FTAN filtering). We note that the spectrum and the record can be regarded as two extreme cases in the class of frequency—time representations: $K(\omega)$ is for infinitesimally narrow filters $H(\omega - \omega^H) = \delta(\omega - \omega^H)$, $W(t)$ is for infinitely broad filters $H(\omega - \omega^H) = 1/\sqrt{2\pi}$.

The uses of FTAN are not confined to a convenient representation of the data, it can be useful in practically all problems arising in surface wave processing. The advantages for surface wave identification are obvious. Also:

(1) A smaller area of the frequency—time region compared with the uncertainty $D_\omega D_t$ provides extra possibilities for separation of dispersed signals (Figure 5.5).

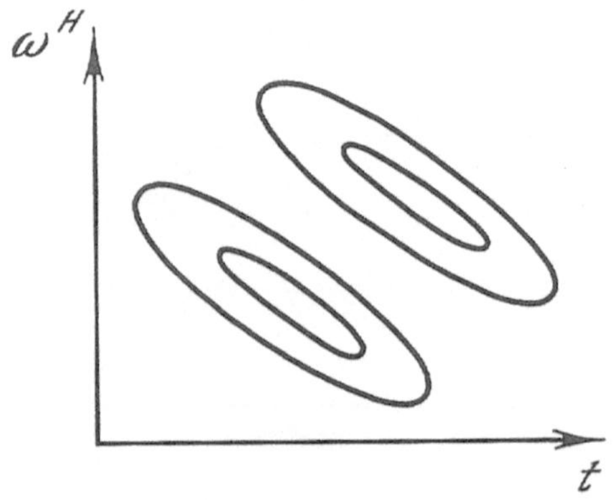

Fig. 5.5. Separation of signals in a FTAN map. These signals are indistinguishable in the time and the frequency domain.

(2) For the same reason the portion of noise that falls into the region $(\langle \omega \rangle \pm D_\omega, \langle t \rangle \pm D_t)$, but lies far from the signal dispersion curve, separate from it. Consequently, the frequency—time representation has higher overall signal-to-noise ratio than that for the original record or its spectrum. This conclusion is also applicable to stationary noise distributed uniformly over the (ω^H, t)-plane.
(3) The axis of the ridge corresponding to the signal on the (ω^H, t)-plane is remarkable, not only because it passes near the dispersion curve $t = \tau(\omega^H)$, but also because the values of $|S(\omega^H, t)|$ arg $S(\omega^H, t)$ at the ridge axis enable $|K(\omega)|$ and $\psi(\omega)$ to be evaluated to some accuracy. FTAN can thus serve as a method (perhaps, a rather rough one) for measuring nearly every spectral characteristic of interest.

To sum up, FTAN is in some sense a 'versatile tool' in surface wave processing. One has to pay for versatility in terms of quality, however. In the next section we shall see that FTAN does not make use of all available resources and has obvious disadvantages in every processing problem. For this reason it should be looked upon as a procedure useful in model building problems of surface wave processing. Nonetheless, results obtained through FTAN are frequently quite satisfactory and do not need further refinement.

We shall revert to a detailed discussion of FTAN in 5.4, but until then we are going to look at what can be done to improve FTAN performance.

'Floating' filtering. Phase equalization. FTAN yields models (values of the characteristics) of the signal and noise that are sufficiently detailed, even though rough. The processing from now on enters the routine stage which requires efficiency rather than visualizability and versatility. The working technique should resolve two questions: (1) separation of the wave signal from nonstationary noise, and (2) measurement of the spectral characteristics of a wave signal using a method that is stable under stationary noise.

We shall skip the latter question. Noise is of minor importance for teleseismic surface waves, and the random error of $K(\omega)$ is usually insignificant when the signal has been separated from nonstationary noise. Otherwise one should turn to statistical methods in use for signal identification upon a noisy background.

We now consider the most conventional method for signal identification as applied to surface waves, namely, linear filtering. The term 'filter' is frequently employed in a narrow sense for a transformation whose parameters are invariant under a time shift (frequency filtering). That is the procedure we have used until now. It is however convenient to use that term in a broader sense when applied to processing of dispersed signals. We write the general form of a linear transformation (the ~ sign above a letter denotes a result of filtering)

$$\tilde{K}(\omega) = \int_{-\infty}^{\infty} F(\omega, \lambda) K(\lambda)\, \mathrm{d}\lambda. \tag{5.54}$$

The requirements on $F(\omega, \lambda)$ are best expressed in the frequency—time language. The filter described by (5.54) should separate, without distortions as far as possible, the part of the plane in which the signal frequency—time region lies (Figure 5.6). (The region where $|S(\omega^H, t)| \neq 0$ cannot be finite. What is meant is that the values of $|S(\omega^H, t)|$ are negligibly small outside the separate region).

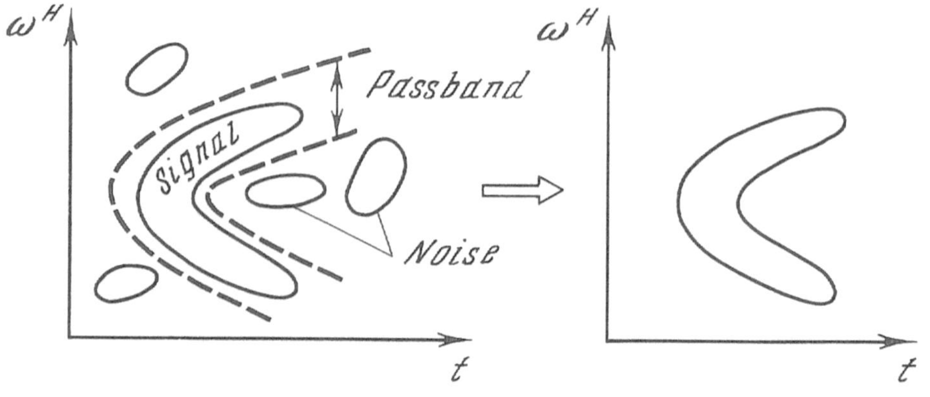

Fig. 5.6. Diagrammatic representation of a floating filter in action.

Loosely speaking, one can imagine this operation as a 'frequency filter whose parameters vary in time'. The filter band is 'floating' along the dispersion curve. Because it is around this curve that the signal energy is concentrated, it is not significantly distorted, while the noise located far from the signal dispersion curve does not pass through a floating filter. Below we discuss the construction of a simple floating filter whose time section width is independent of frequency. This extra requirement is technically convenient, because it enables (5.54) to be computed within the framework of ordinary spectral analysis, in particular, using the fast Fourier transform.

Ordinary time-invariant bandpass filtering separates a band paralled to the t-axis on the (ω^H, t)-plane. Similarly, a time window $G(t)$ applied to a record to give $W(t) = G(t)W(t)$ separates a band parallel to the ω^H-axis. The most that one can achieve with the two operations combined is to separate a 'rectangular' region on the (ω^H, t)-plane whose sides are parallel to the axes. In contrast to this, the 'band' of a floating filter may be located in any way relative to the ω^H and t-axes. Nonetheless, floating filtering consists in a sequence of frequency filters and time windows, provided the 'bandwidth along t' is kept the same for all the ω^H.

The most important thing in this operation is phase equalization (otherwise called phase-consistent filtering). The dispersion curve of a signal (denoted $\hat{\tau}(\omega)$) is known approximately from FTAN results, so transforming the spectral phase of the whole record

$$\begin{aligned}\tilde{K}(\omega) &= K(\omega)\exp[-i\hat{\psi}(\omega)]\\ \hat{\psi}(\omega) &= -\left(\int_0^\omega \hat{\tau}(\eta)\,\mathrm{d}\eta + c_1\omega + c_2\right)\end{aligned} \tag{5.55}$$

makes the signal of interest weakly dispersed ($\tilde{q} \ll 1$). The constants c_1 and c_2 do not affect the envelope shape $|\tilde{W}(t)|$, only altering the initial phase of the resulting signal and shifting it to a convenient instant of time, for instance, to the midpoint of the record ($\langle\tilde{t}\rangle \approx c_1$ for the signal of interest). Note that the use of (5.55) may increase the slope of the noise dispersion curve. Generally speaking, one can roughly imagine that corresponding to phase equalization is a deformation of the

(ω^H, t)-plane that transforms a curve $t = \hat{\tau}(\omega^H)$ + constant into the straight line t = constant parallel to the ω^H-axis. This means that the distance between the axes of the frequency—time regions for different signals as measured along the t-axis is not changed, but the region of the signal of interest is stretched approximately parallel to the ω^H-axis. A frequency—time region of this kind is quite well separated by using a time window (as well as a bandpass frequency filter which it is more convenient to apply prior to phase equalization). If now we wish to recover the original signal shape, we should only have to use the inverse procedure of phase equalization, that is, to add the same function $\hat{\psi}(\omega)$ to the signal phase spectrum.

The procedure of floating filtering thus separates into four successive steps (Figure 5.7 shows diagrammatically the resulting FTAN maps):

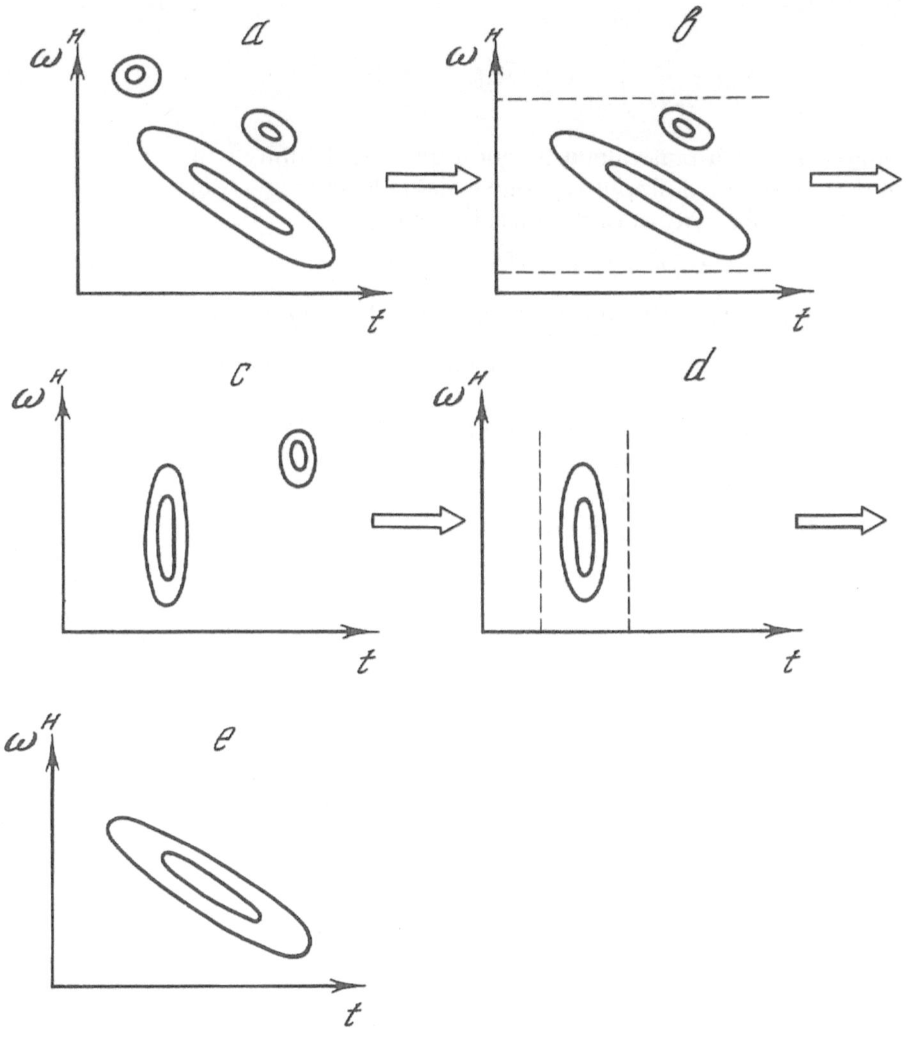

Fig. 5.7. Implementation of a floating filter. (a) FTAN map of the original signal; (b—e) FTAN maps showing the results of the sequence of operations that makes up floating filtering.

(1) bandpass filtering (Figure 5.7b)

$$\tilde{K}_1(\lambda) = H(\lambda) \cdot K(\lambda) \tag{5.56}$$

where $H(\lambda)$ is a real-valued function that equals unity within the signal band and falls off to zero outside it;

(2) phase equalization (Figure 5.7c) and the transition to the time domain

$$\tilde{W}_2(t) = \frac{1}{\sqrt{2\pi}} \int_{-\infty}^{\infty} \tilde{K}_1(\lambda) e^{i[\lambda t - \hat{\psi}(\lambda)]}\, d\lambda. \tag{5.57}$$

As a result, an originally strongly dispersed surface wave signal becomes a short violent pulse;

(3) the use of a time window $G(t)$ (Figure 5.7d) and the reversion to the spectral representation

$$\tilde{K}_3(\omega) = \frac{1}{\sqrt{2\pi}} \int_{-\infty}^{\infty} G(t) \tilde{W}_2(t) e^{-i\omega t}\, dt \tag{5.58}$$

where $G(t)$ is a real-valued function that equals unity within the time interval of the transformed (impulsive) signal and falls off to zero outside it.

(4) the inverse phase transformation (Figure 5.7e)

$$\tilde{K}(\omega) = \tilde{K}_3(\omega) e^{i\hat{\psi}(\omega)} \tag{5.59}$$

the record assumes the original form, but with the noise eliminated.

Combining (5.56) through (5.59), we finally get the kernel $F(\omega, \lambda)$ in (5.54)

$$\begin{aligned} F(\omega, \lambda) &= \frac{1}{\sqrt{2\pi}} H(\lambda) H_t(\omega - \lambda) e^{i[\hat{\psi}(\omega) - \hat{\psi}(\lambda)]} \\ H_t(\omega) &= \frac{1}{\sqrt{2\pi}} \int_{-\infty}^{\infty} G(t) e^{i\omega t}\, dt. \end{aligned} \tag{5.60}$$

Actually it is of course unnecessary to compute $F(\omega, \lambda)$. It is just the possibility of partitioning (5.54) into simple operations which makes the above version of floating filtering an efficient processing tool.

Apart from $\hat{\psi}(\omega)$, the principal parameter of a floating filter is the bandpass width along t or equivalently, the width D_t^G of the time window $G(t)$. When D_t^G is decreased, that diminishes the noise-related error ('random error'), but enhances the filter-related distortion ('systematic error'). In other words, in actual practice it is advisable to make the window as broad as the noise allows. The optimal compromise value is record-specific.

Indeed, a surface wave usually involves a rather broad frequency range with noise distributed unevenly within it. Hence it is frequently not convenient to keep D_t^G the same at all frequencies. This difficulty is simply obviated by separating the original signal by bandpass filtering (5.56) into several ranges and applying a floating filter to each. From the example given in Figure 5.3 it follows that $\tilde{D}_t$ will not be significantly affected by separating each spectral peak of the signal into a frequency range of its own.

Floating filtering has advantages over FTAN largely due to the fact that it can achieve a more effective compression of dispersed signals along t. The main compression tool is narrow-band real-valued filtering for FTAN and phase equalization for floating filtering. We have seen (5.51) that the latter can decrease D_t by the factor of $\sim \sqrt{q/2}$. Even so, the uncertainty in the resulting signal remains at least $\sqrt{2}$ times the optimal value (3.52). That means that FTAN does not utilize all resources for signal compression. Phase equalization applied to a signal compresses it by a factor of $\sim q$; if $I \approx 1$ for the original signal, then the uncertainty of the compressed signal is close to the optimal. As a result, the overall random and systematic errors of floating filtering results are below the FTAN values.

We conclude by remarking that the above classification into FTAN and floating filtering is to some degree arbitrary. Both are based on the same technique, linear filtering (in a broad sense), both implement the same idea of processing, a gradual adaptation of the filter to fit the signal and noise. The apparent difference consists, in the first place, in the representation of results, while the advantages of floating filtering over FTAN are related to the fact that more a priori information is put into the former procedure. There are other ways to do this, in particular, the use of phase equalization within FTAN (Dziewonski, Mills and Bloch, 1972; Barmin, Levshin and Starovoit, 1984). We will not discuss these techniques, because the principles on which they are based are similar to those underlying our procedures.

5.3. FREQUENCY-TIME ANALYSIS

The frequency—time representation has been in use in analog form for at least four decades (Ewing *et al.*, 1959), while its digital form was introduced into seismology in 1969 (Dziewonski, Bloch and Landisman, 1969). This version with some slight modifications has been actively exploited for more than 15 years, and we too will largely focus on this. In order to gain a better understanding of some assumptions underlying the technique we are going to look at the relevant definition from a more general standpoint.

Alternative definitions. A definition of the frequency—time representation of a signal $W(t)$ that is symmetric about the ω^H- and t^H-axes is

$$S(\omega^H, t^H) = \int_{-\infty}^{\infty} G^*(\omega^H, t^H, t)\, W(t)\, \mathrm{d}t$$

$$= \int_{-\infty}^{\infty} H^*(\omega^H, t^H, \omega)\, K(\omega)\, \mathrm{d}\omega \tag{5.61}$$

$$\langle t(G)\rangle(\omega^H, t^H) \equiv t, \qquad \langle \omega(H)\rangle(\omega^H, t^H) \equiv \omega^H$$

where the notation $\langle t(G)\rangle$, $\langle \omega(H)\rangle$ implies that the relevant mean values have been obtained from $G(\omega^H, t^H, t)$ and its Fourier transform $H(\omega^H, t^H, \omega)$ for fixed ω^H and t^H. The expression (5.61) means that the original signal is projected onto a two-parameter family of 'trial' functions, each of these being 'concentrated' around the frequency ω^H and the time t^H.

The definition (5.61) is flexible enough, since it allows an independent choice of G for any point on the (ω^H, t^H)-plane, but it would consume too much computer time in this general form. The situation simplifies when the parameters ω^H, t^H are no longer treated symmetrically, G or H being assumed to be invariant with respect to a shift in one of the two parameters. This leads to two alternative definitions:

(1) the spectral shape of the 'trial' signal depends on t^H only:

$$S_t(\omega^H, t^H) = \int_{-\infty}^{\infty} H^*(t^H, \omega - \omega^H) K(\omega)\, d\omega$$

$$= \int_{-\infty}^{\infty} G_t^*(t^H, t) e^{-i\omega^H t} W(t)\, dt \tag{5.62}$$

where $G_t(t^H, t)$ is the Fourier transform of $H(t^H, \omega)$. When G_t is real, S_t is called the 'instantaneous spectrum' of the signal $W(t)$;

(2) the shape of the 'trial' signal depends on ω^H only:

$$S_\omega(\omega^H, t^H) = \int_{-\infty}^{\infty} G^*(\omega^H, t - t^H) W(t)\, dt$$

$$= \int_{-\infty}^{\infty} H_\omega^*(\omega^H, \omega) e^{i\omega t^H} K(\omega)\, d\omega \tag{5.63}$$

where $H_\omega(\omega^H, \omega)$ is the Fourier transform of $G(\omega^H, t)$. The operation (5.63) is a multichannel frequency filtering (whose response is $H_\omega^*(\omega^H, \omega)$) and is consistent with the frequency–time representation as defined in the preceding section.

Requirements 1 and 2 are mutually exclusive, so the functions S_t and S_ω as given by (5.62) and (5.63) are different. The simplest case would be for them to differ by the phase factor $e^{i\omega^H t^H}$. This will occur when we put $G_t(t^H, t) = G_t(t - t^H)$ in (5.62) or $H_\omega(\omega^H, \omega) = H_\omega(\omega - \omega^H)$ in (5.63). In particular, the function

$$S_0(\omega^H, t^H) = \int_{-\infty}^{\infty} H_\omega^*(\omega - \omega^H) e^{i\omega t^H} K(\omega)\, d\omega \tag{5.64}$$

is known in radio engineering by the name of 'cross-uncertainty function'.

The definition (5.63) is the one that is most widely encountered. This is simpler to use in analog form than (5.62), but with digital computations (5.62) and (5.63), as well as (5.64), all need about the same amount of labor. Physical considerations usually require, however, that the resolution be controlled along the frequency rather than time. For instance, one may recall that frequency intervals of equal length within different ranges of ω are responsible for widely different intervals of depths to which surface waves penetrate.

The choice of $H_\omega^*(\omega^H, \omega)$ in (5.63) may be guided by typical properties of the signal to be processed. As a rule, preliminary surface wave processing uses filters possessing the following simple properties: no phase distortion (a real-valued H_ω) and the best resolution. From (5.51) it follows that signal duration at the filter output, H_ω, is proportional to I^H, that is, the Gaussian filter is optimal in the sense of signal resolution. It is this kind of filter which will be considered below.

To sum up, the frequency—time representation of a signal given by its spectrum $K(\omega)$ is understood as the following complex function on the (ω^H, t)-plane

$$S(\omega^H, t) = \frac{1}{\sqrt{2\pi}\beta(\omega^H)} \int_{-\infty}^{\infty} \exp\left(-\frac{(\omega-\omega^H)^2}{2\beta^2(\omega^H)}\right) K(\omega) e^{i\omega t} \, d\omega. \quad (5.65)$$

Since other definitions will no longer be considered, all superfluous indices have been omitted in (5.65).

The real-valued Gaussian filter used in (5.65) will be called the FTAN filter. It is fully specified by the two parameters

$$\langle \omega \rangle^H = \omega^H \quad (5.66)$$

$$D_\omega^H = \beta. \quad (5.67)$$

Model signals. The properties of FTAN are conveniently illustrated by using as examples parametric signals in analytic form modelling a surface wave record. We take two signals ($K_g(\omega)$ and $K_l(\omega)$) whose frequency—time representations can be calculated in explicit form.

Both signals have a quadratic phase spectrum (linear dispersion curve $\tau(\omega)$)

$$\psi_g(\omega) = \psi_l(\omega) = \psi_0 - \tau_0 \cdot (\omega - \omega_0) - \frac{\tau'}{2} \cdot (\omega - \omega_0) \quad (5.68)$$

where ψ_0, τ_0, τ' are constants.

The amplitude spectrum of a Gaussian model signal K_g is

$$|K_g(\omega)| = K_0 \exp\left(-\frac{(\omega-\omega_0)^2}{2\sigma^2}\right). \quad (5.69)$$

From (5.68), (5.69) it follows that the following relations hold for K_g:

$$\begin{aligned} &\langle \omega \rangle = \omega_0, \quad \langle t \rangle = \tau_0, \quad D_\omega = \sigma, \\ &I = 1, \qquad q = \tau'\sigma^2, \quad D_t^2 = (1+q^2)/\sigma^2. \end{aligned} \quad (5.70)$$

The amplitude $|K_g(\omega)|$ is different from zero for $\omega < 0$, that is, the corresponding time function $W_g(t)$ is not an 'analytic signal', so that there is no real-valued signal $w(t)$ corresponding to it. If we assume $D_\omega \ll \omega_0$, however, that circumstance can be disregarded, and $K_g(\omega)$ will satisfactorily model an individual spectral peak in a surface wave.

We are going to demonstrate the local properties of the frequency—time representation by using K_l, a linear model signal whose amplitude spectrum around a frequency ω^H has the form

$$|K_l(\omega)| = |K(\omega^H)| + |K|'(\omega^H) \cdot (\omega - \omega^H). \quad (5.71)$$

We also assume the phase spectrum $\psi_l(\omega)$ (5.63) to be expanded at $\omega_0 = \omega^H$.

Because the frequency response of a Gaussian FTAN filter falls off rapidly, the results of frequency—time analysis do not significantly depend on how $|K_l(\omega)|$ behaves outside the filter passband of width $\sim \beta$.

The frequency—time representations of the model signals are

$$S_g(\omega^H, t) = \frac{K_0}{\sqrt{(1 + \beta^2/\sigma^2) + ip}} \times \exp\left(i(\psi_0 + \omega_0 t) - \frac{(v - is)^2}{2(1 + \beta^2/\sigma^2 + ip)} - \frac{s^2}{2} \right) \tag{5.72}$$

$$S_l(\omega^H, t) = \frac{|K(\omega^H)|}{\sqrt{1 + ip}} \left(1 + \frac{i\bar{K}v^H}{1 + ip} \right) \times \exp\left\{ i[\psi(\omega^H) + \omega^H t] - \frac{v^2}{2(1 + ip)} \right\} \tag{5.73}$$

where $\beta = \beta(\omega^H)$ and

$$v = \beta \cdot (t - \tau_0), \quad (\text{in (5.73) } v = \beta \cdot (t - \tau(\omega^H))),$$
$$s = \frac{\omega^H - \omega_0}{\beta}, \quad p = \tau'\beta^2, \quad \bar{K} = \frac{|K|'(\omega^H)}{|K(\omega^H)|}\beta. \tag{5.74}$$

Below we quote only final results for the model signals. Lander (1974, 1975, 1978) discusses the properties of $S_g(\omega^H, t)$ and those of a signal similar to $S_l(\omega^H, t)$ in more detail.

Optimal filtering. The function $S(\omega^H, t)$ is significantly dependent on the choice of FTAN filters or, to be more exact, on the function $\beta(\omega^H)$. We should have two problems in view when choosing $\beta(\omega^H)$: separation of signal and noise, and a satisfactory estimation of signal characteristics. The requirements that these two problems impose on $\beta(\omega^H)$ are largely contradictory. However, practical processing of teleseismic surface waves shows that when FTAN is regarded as preliminary processing (where a high accuracy of estimation is not needed), one should give preference to the problem of signal separation. The criteria of optimal filtering will be understood in the narrow sense just indicated.

Since we do not know noise models a priori, optimality criteria are only based on the properties of the signal of interest, consisting in minimizing either the area of the frequency—time region or some section of it.

First, we consider optimal filtering for a Gaussian model signal, simplifying our task by restricting ourselves to the condition $\beta(\omega^H) = \beta = \text{constant}$. Let us find the value of β that minimizes the area of the signal frequency—time region. From (5.72) it follows that the contours of $|S_g(\omega^H, t)|$ are ellipses whose areas are proportional to

$$\frac{\sigma^2}{\beta^2}\left[\left(1 + \frac{\beta^2}{\sigma^2}\right)^2 + p^2\right] \tag{5.75}$$

which attains a minimum when ((5.70) has here been used)

$$\beta_{0s}^2 = \frac{1}{\sqrt{\tau'^2 + 1/\sigma^4}} = \frac{\sigma^2}{\sqrt{1+q^2}} = \frac{D_\omega}{D_t}. \tag{5.76}$$

All quantities on the right-hand sides of (5.76) are for the original signal; β_{0s} provides a simple convenient estimate of the optimal β for any values of q.

The values of β for which the frequency—time region of a signal is the most compressed either in time or frequency are different from β_{0s}. The minimum width of the section across the region along w^H is attained when

$$\beta_{0\omega}^2 = \frac{1}{|\tau'|} \frac{q^2 - |q|}{q^2+1} \tag{5.77}$$

while the minimum section width along t (that is, the $\tilde{D}_t$ of the signal passed through a FTAN filter) is attained when

$$\beta_{0t}^2 = \frac{1}{\tau'} \frac{q}{|q|-1} \tag{5.78}$$

the last expression being true independent of the condition $\beta(\omega^H)$ = constant. When $\beta(\omega^H) = \beta_{0t}$, we shall have for a FTAN-filtered signal

$$\tilde{q}_{0t}(\omega^H) = \tau' \Bigg/ \left(\frac{1}{\beta_{0t}^2} + \frac{1}{\sigma^2}\right) \equiv 1. \tag{5.79}$$

From (5.76) through (5.78) one can see that β_{0s} exists for any signal parameters, but the width of the frequency-time region along ω^H or t only attains a minimum when $|q| > 1$. A weakly dispersed signal ($|q| < 1$) has the best time resolution as $\beta \to \infty$ and the best frequency resolution as $\beta \to 0$. When $\beta \to \infty$, however, the frequency—time representation is (under an appropriate normalization) identical with the time representation, while when $\beta = 0$, it is identical with the spectral representation. We thus arrive at the important conclusion that FTAN should only be used to separate strongly dispersed signals. FTAN resolution for $|q| < 1$ signals is poor compared with that of a spectrum or a time record.

We used the general relation (5.39) in the preceding section to obtain a filter width that minimizes $\tilde{D}_t$ for strongly dispersed signals. Formula (5.50) when rewritten in FTAN notation becomes

$$\beta_0^2(\omega^H) = 1/|\tau'(\omega^H)|. \tag{5.80}$$

From (5.76) through (5.78) it follows that, when $q \to \infty$, all optimal β's corresponding to different criteria are tending to that value. Thus, all resolution criteria for a strongly dispersed signal yield a single value of β_0 (5.80).

Formula (5.80) can also be used locally without $\tau(\omega)$ being linear in a wide frequency range. It is sufficient that the condition $\tau'(\omega)$ = constant holds within the filter width, that is (all quantities are at ω^H):

$$\eta = \left|\frac{\tau''\beta_0}{2\tau'}\right| \ll 1. \tag{5.81}$$

When both quantities were written in (5.80) as functions of ω^H, (5.81) was assumed to be true. The condition $q \gg 1$ must be replaced by a local one:

$$\bar{K} \ll 1 \tag{5.82}$$

where $\bar{K}$ is obtained from (5.74) for $\beta = \beta_0$.

(5.80) is the main working formula for most of the surface wave frequency range. Exceptions are vicinities of extrema of $\tau(\omega)$ where (5.81) breaks down. Construction of optimal filters there requires an examination of the general relation (5.25) for more complex $\tau(\omega)$. Cara (1973) assumes $\bar{K} = 0$ to derive formulas of optimal FTAN filters (ones minimizing $\tilde{D}_t$) for an arbitrary function $\tau(\omega)$ represented as a series in powers of ω. We quote the results for a quadratic $\tau(\omega)$.

When $|K(\omega)| \approx$ constant, in a vicinity of ω^H of width $\sim\beta$ (i.e., $\bar{K} \ll 1$), and for

$$\tau(\omega) = \tau(\omega^H) + \tau'(\omega^H)\cdot(\omega - \omega^H) + \frac{\tau''(\omega^H)}{2}(\omega - \omega^H)^2 \tag{5.83}$$

we seek to determine the optimal value of β_0 (that minimizing $\tilde{D}_t$) based on

$$\tau'\beta_0^4 + \frac{\tau''^2}{2}\beta_0^6 = 1. \tag{5.84}$$

Then the output of the optimal filter will be

$$\tilde{q}_0^2 = (1 + \eta^2/2)\cdot(1 + \eta^2) \tag{5.85}$$

where η is the parameter from (5.81). q_0 is not necessarily unity in the quadratic approximation as was the case for the linear approximation, but remains of the same order. q_0 varies between 1/2 and 1, depending upon the relation between the linear and quadratic term in the expansion of $\tau(\omega)$. At an extremum of $\tau(\omega)$, (5.84) and (5.85) become

$$\beta_0^2 = \sqrt[3]{\frac{2}{|\tau''|}}, \qquad \tilde{q}_0^2 = 1/2. \tag{5.86}$$

The choice of optimal filter is made automatic by expressing β_0 in terms of the velocities. For example, if linear portions of $\tau(\omega)$ are taken, (5.80) is transformed (using (5.8)) into

$$\beta_0^2(\omega^H) = U^2(\omega^H) \Big/ \left| r\frac{du}{d\omega}(\omega^H)\right| \tag{5.87}$$

that is, the FTAN optimal filter width decreases with epicentral distance increasing. No excessive accuracy is actually needed in the choice of β_0. The available information on average group velocities of surface waves in the Earth is sufficient for a satisfactory choice of β_0 before processing. $U(\omega)$ and its derivatives are fixed beforehand, the only input parameter to calculate β_0 being the epicentral distance r.

Measurement problem. The general problem of measuring the parameters of the original signal on the basis of $S(\omega^H, t)$ requires a linear integral equation to be solved. If $S(\omega^H, t)$ is given in a sufficiently large region, the equation is solved with satisfactory stability under low noise. For example, the inversion formula for the simplest case $\beta(\omega^H) = \text{constant}$ is

$$W(t) = \frac{1}{\sqrt{2\pi}} \int_{-\infty}^{\infty} S(\omega^H, t)\, d\omega^H. \tag{5.88}$$

Equation (5.65) is much harder to invert when $S(\omega^H, t)$ is known with satisfactory accuracy in some part of the signal frequency—time region only.

Let us discuss simpler measurement techniques that are actually used and are in keeping with the preliminary nature of FTAN. We are going to measure, instead of spectral parameters of the original signal, suitably chosen parameters of $S(\omega^H, t)$, that is, time signal parameters at the FTAN output. This substitution, time parameters in place of spectral ones, gives rise to systematic errors. The parameters of $S(\omega^H, t)$ used for the measurement will be called estimates in what follows and distinguished by putting an ^ above the relevant letter. The ~ sign denotes signal parameters at the FTAN filter output as before. Many different FTAN estimates have been proposed; we shall list those in general use. The formulas for systematic errors quoted below are mainly based on the model signals (5.72), (5.73). These formulas adequately describe the qualitative behavior of the errors, but are not accurate enough to be used as corrections in actual situations.

Group time estimates. The three estimates of group time to be discussed are constructed as follows. Two parameters, $\hat{\tau}$ and $\hat{\omega}$, are chosen for each signal at the FTAN filter output (that is, for a section of $S(\omega^H, t)$ along t). The point $\hat{\tau}(\hat{\omega})$ is regarded as an estimate for a point of the dispersion curve $\tau(\omega)$.

Estimate $\hat{\tau}(\hat{\omega}) = \langle \tilde{t} \rangle(\langle \tilde{\omega} \rangle)$. Expanding $\tau(\omega)$ at the point $\langle \tilde{\omega} \rangle$ into a power series involving a quadratic term and recalling that $\langle \tilde{\omega} \rangle = \langle \tilde{\Omega} \rangle$, we shall have

$$\langle \tilde{t} \rangle = \langle \tilde{\tau} \rangle = \tau(\langle \tilde{\Omega} \rangle) + \frac{\tau''(\langle \tilde{\Omega} \rangle)}{4} \cdot D_{\omega}^2. \tag{5.89}$$

It follows that for a linearly dispersed signal ($\tau'' = 0$), $\langle \tilde{t} \rangle(\langle \tilde{\Omega} \rangle)$ is not biased. When $\bar{K} \ll 1$, we have $\tilde{D}_{\omega} \approx D_{\omega}^H = \beta$, and the error $\delta\tau$ of the estimate is given by

$$\delta\tau = \tau(\langle \tilde{\Omega} \rangle) - \langle \tilde{t} \rangle(\langle \tilde{\Omega} \rangle) = - \frac{\tau''\beta^2}{4} \underset{\beta=\beta_0}{=} - \frac{\tau''}{4|\tau'|}. \tag{5.90}$$

The last equality in (5.90) holds for the optimal filter (5.80) assuming the measurements to be made away from extrema of the dispersion curve.

Calculation of $\langle \tilde{t} \rangle$ and $\langle \tilde{\Omega} \rangle$ requires a long noise-free time interval. For this reason use is made of estimates that, even though less accurate, are also less sensitive to noise; these are based on parameters taken at the point where $|\tilde{W}(t)|$ is a maximum (where $|S(\omega^H, t)|$ has a conditional maximum in t). Such parameters will be marked with the subscript m; in particular, t_m is the instant of time where $|\tilde{W}(t)|$ is a maximum.

Estimate $\hat{\tau}(\hat{\omega}) = t_m(\omega^H)$. This is the most widely used and simplest group time estimate. If there is a linearly dispersed signal whose spectral amplitude $\tilde{B}(\omega)$ is symmetric about ω^H, then $t_m = \langle \tilde{t} \rangle$, $\omega^H = \langle \tilde{\omega} \rangle$, so that this estimate is identical with the preceding. Since a Gaussian filter is symmetric about ω^H, the bias of the estimate $t_m(\omega^H)$ is largely controlled by $\bar{K}$ (5.74) in intervals where $\tau(\omega)$ is linear. Examination of the linear model signal (5.73) shows that the error has the form

$$\delta\tau_\omega = \tau(\omega^H) - t_m(\omega^H) = -\beta\tau'\bar{K} = -\tilde{q}\,\frac{|K|'}{|K|} \underset{\beta=\beta_0}{=} -\frac{|K|'}{|K|}. \quad (5.91)$$

A fairly simple method to get a rough idea of the bias is by examining the Gaussian model signal (5.72). $|\omega^H - \omega_0| = \sigma|K|'/|K| = 1/\sigma$ for this signal, that is, the error $\delta\tau_\omega$ is of the order of the inverse of the spectral peak width.

Figure 5.8 shows a typical behavior of $t_m(\omega^H)$. Note that the bias changes sign in passing through a peak along the ridge of $|S(\omega^H, t)|$. This causes a spurious inflexion to appear around the saddle point.

Estimate $\hat{\tau}(\hat{\omega}) = t_m(\Omega_m)$. When one measures at an edge of the signal frequency band, the filter parameter ω^H may grossly misrepresent the properties of the signal at the filter output, as demonstrated by Figure 5.9. In such a case the estimate $t_m(\omega^H)$ has an indefensibly large error. The error can be lowered by choosing $\hat{\omega}$ as the instantaneous frequency at time t_m:

$$\hat{\omega} = \tilde{\Omega}_m = \frac{\partial}{\partial t} \arg S(\omega^H, t_m). \quad (5.92)$$

The estimate $t_m(\tilde{\Omega}_m)$ for the Gaussian model signal is unbiased over the entire frequency band, while the error for the signal (5.73) is of third order in $\bar{K}$

$$\delta\tau_\Omega = \tau(\tilde{\Omega}_m) - t_m(\tilde{\Omega}_m) \approx \frac{\beta\tau'}{1+q^2}\,\bar{K}^3$$
$$= \beta^2\left(\frac{|K|'}{|K|}\right)^3 \underset{\beta=\beta_0}{=} \frac{1}{|\tau'|}\left(\frac{|K|'}{|K|}\right)^3. \quad (5.93)$$

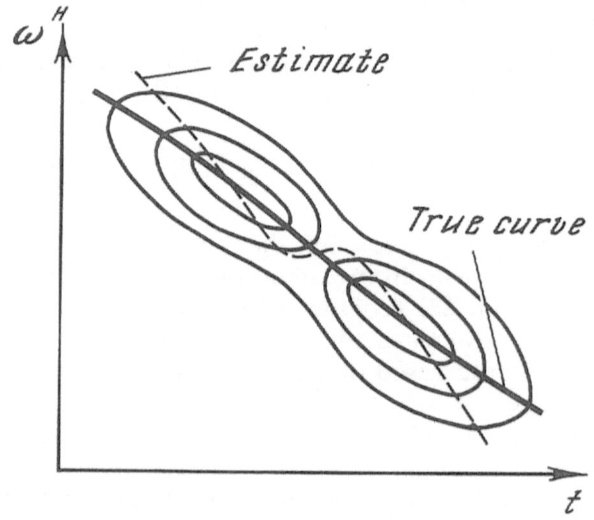

Fig. 5.8. Typical departure of FTAN estimates (dashed line) from the true signal dispersion curve (solid line).

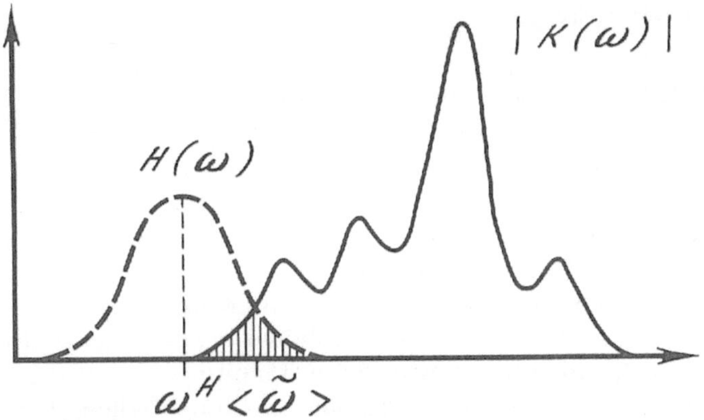

Fig. 5.9. Explanation of a typical bias of the estimate $t_m(\omega^H)$ at the edge of the spectral band of a signal.

The last estimate is thus to be preferred to $t_m(\omega^H)$ over the entire frequency band under measurement.

Note that, although the estimate $t_m(\tilde{\Omega}_m)$ is written in a manner similar to (5.46), it has nothing to do with the stationary phase approximation, which is usually inapplicable to signals at the output of FTAN filters, because optimal filtering gives $\tilde{q} \sim 1$.

It follows from the formulas for systematic errors that the measurement accuracy for $\tau(\omega)$ when $\beta(\omega^H)$ is fixed is the higher the smaller is the characteristic value of τ' for the signal. For this reason the dispersion curves of different signals that are present in the same FTAN map may be measurable with different accuracies.

Amplitude and phase estimates. These are constructed by using S_m, a function of a single variable obtained from $S(\omega^H, t)$ at the line where $|S|$ has maxima along t. One can choose different quantities to serve as the argument of S_m, in particular, ω^H, t, or $\tilde{\Omega}_m$ which are taken at the same point on the plane where S_m is. The formulas given below, (5.94) through (5.100), are for a Gaussian model signal, hence they approximate the case of a separate spectral peak along a linear portion of the dispersion curve.

S_m as a function of $\tilde{\Omega}_m$ (5.92) is the most convenient to use in calculating amplitude and phase estimates. In that case

$$|K_g(\tilde{\Omega}_m)| = \sqrt{(1+\beta^2/\sigma^2)^2 + \tau'^2\beta^4}\, |S_m(\tilde{\Omega}_m)|$$

$$= \frac{\beta \tilde{D}_t}{\sqrt[4]{1+\tilde{q}^2}}\, |S_m(\Omega_m)| \tag{5.94}$$

$$\psi_g(\tilde{\Omega}_m) = \arg S_m(\tilde{\Omega}_m) - \tilde{\Omega}_m t_m + \tfrac{1}{2} \operatorname{arctg} \tilde{q}. \tag{5.95}$$

The right-hand sides of (5.94) and (5.95) involve only quantities that can be found from $S(\omega^H, t)$. The expression can be regarded as estimating spectral amplitude and phase. The estimates are especially simple for the optimal FTAN filter ($\beta =$

β_{0t}). Assuming $q \gg 1$, we get:

$$|\hat{K}(\tilde{\Omega}_m)| = \sqrt[4]{2}\,\sqrt{1+\beta_{0t}^2/\sigma^2}\,|S_m(\tilde{\Omega}_m)| \approx \sqrt[4]{2}\,|S_m(\tilde{\Omega}_m)| \tag{5.96}$$

$$|\hat{\psi}(\tilde{\Omega}_m)| = \arg S_m(\tilde{\Omega}_m) - \tilde{\Omega}_m t_m + \frac{\pi}{8}\,\text{sign}\,\tau'. \tag{5.97}$$

Note that the last term in (5.97) equals half the analogous term in the stationary phase approximation (5.45).

The accuracy given by the above estimates of group time, amplitude and phase is enhanced with decreasing β. This is as it should be, considering that the spectral representation of a signal can be regarded as a limiting case of the frequency-time representation for δ-like, infinitesimally narrow filters. When $\beta < \beta_0$, however, filter narrowing spreads the signal $\tilde{W}(t)$, and so is limited by the presence of noise. Here again, we encounter the conflict of systematic and random errors.

Estimates of 'mean' signal parameters. From (5.96) it follows that the optimal filter choice allows $\langle \omega \rangle$ and D_ω to be estimated directly from similar parameters of $S_m(\tilde{\Omega}_m)$.

If $S_m(\omega^H)$ is used, then (denoting its second central moment as χ_ω^2) D_ω is to be estimated by

$$\hat{D}_\omega^2 = \chi_\omega^2 - \beta^2. \tag{5.98}$$

Taking also $S_m(t)$ and denoting its second central moment as χ_t^2, we can write the estimates of D_t and q for the original signal as

$$\hat{D}_\omega^2 = (1 + \chi_t^2\chi_\omega^2)/(\chi_\omega^2 - \beta^2) \tag{5.99}$$

$$|\hat{q}| = \chi_\omega\chi_t. \tag{5.100}$$

Relation between the parameters of $S(\omega^H, t)$. We conclude by discussing a few general relations that clarify the nature of $S(\omega^H, t)$. This is not an arbitrary function of two variables. It follows from the definition (5.65) that the parameters of $S(\omega^H, t)$ satisfy certain differential relations. We discuss the relations for the case $\beta(\omega^H)$ = constant alone where they are the simplest.

We rewrite (5.65) as (change of variable $x = \omega/\beta$)

$$S(\omega^H, t) = |S|e^{i\tilde{\varphi}}$$
$$= e^{-(\omega^H/\beta)^2/2}\frac{1}{\sqrt{2\pi}}\int_{-\infty}^{\infty} K(\beta x)e^{-x^2}e^{i(\beta t - i\omega^H/\beta)x}\,\mathrm{d}x. \tag{5.101}$$

When β = constant, it follows that

$$\ln(S(\omega^H, t)e^{(\omega^H/\beta)^2/2}) = \ln|S(\omega^H, t)| + (\omega^H/\beta)^2/2 + i\tilde{\varphi}(\omega^H, t)$$
$$= J(\omega^H, t) = J(\beta t - i\omega^H/\beta) \tag{5.102}$$

is an analytic function of the complex variable $z = \beta t - i\omega^H/\beta$. Writing down the Cauchy–Riemann conditions for (5.102), we get relations that hold at any point of the (ω^H, t)-plane (assuming $|S(\omega^H, t)| \neq 0$):

$$\frac{\partial \tilde{\varphi}}{\partial t} = \tilde{\Omega} = \omega^H + \frac{\beta^2}{|S|}\frac{\partial |S|}{\partial \omega^H} \tag{5.103}$$

$$\frac{\partial \tilde{\varphi}}{\partial \omega^H} = -\frac{1}{\beta^2 |S|}\frac{\partial |S|}{\partial t} \tag{5.104}$$

whence

$$\frac{1}{\beta^2}\frac{\partial^2 \tilde{\varphi}}{\partial t^2} + \beta^2 \frac{\partial \tilde{\varphi}}{(\partial \omega^H)^2} = 0 \tag{5.105}$$

$$\frac{1}{\beta^2}\frac{\partial^2 \ln |S|}{\partial t^2} + \beta^2 \frac{\partial^2 \ln |S|}{(\partial \omega^H)^2} = -1. \tag{5.106}$$

From (5.103), (5.104) it follows that the functions $\tilde{\varphi}(\omega^H, t)$ and $\ln |S(\omega^H, t)|$ can be recovered one from the other, apart from a constant, in any (singly connected) region (where $|S| \neq 0$). An FTAN (amplitude) map thus contains the whole of information. on a signal, apart from a single value of initial phase. Of course, that does not mean that $\tilde{\varphi}(\omega^H, t)$ is actually useless to calculate. For instance, such a useful quantity as $\tilde{\Omega}_m$ is much simpler to find from $\tilde{\varphi}(\omega^H, t)$ than from $|S(\omega^H, t)|$.

The fact that $S(\omega^H, t)$ only differs from an analytic function by a known factor $\exp[(\omega^H/\beta)^2/2]$ means that the frequency—time representation can in principle be recovered from the values of $S(\omega^H, t)$ (amplitude and phase) on a segment on the (ω^H, t)-plane. In particular, all values of $S(\omega^H, t)$ within the frequency—time region of a signal are determined by the values of $S(\omega^H, t_m(\omega^H))$ at the ridge line.

We now list a few useful statements that follow from the statements connecting $\tilde{\varphi}$ and $|S|$ (recall that these are hold the case $\beta(\omega^H) = \text{constant}$):

(1) $\tilde{\Omega} = \omega^H$ at the line where $|S|$ has extrema along ω (a corollary of (5.103));
(2) the line where $|S|$ has extrema along t is also that where $\tilde{\varphi}$ has extrema along ω^H (a corollary of (5.104)). Hence the contours of $\tilde{\varphi}(\omega^H, t)$ and $|S(\omega^H, t)|$ are mutually orthogonal at these points (are parallel to the ω^H- and t-axes respectively);
(3) the last property yields two relations involving the function $\arg S_m$ that has been used for phase estimation:

$$\frac{d}{dt} \arg S_m(t) = \tilde{\Omega}_m \tag{5.107}$$

$$\frac{d}{d\tilde{\Omega}_m} [\arg S_m(\tilde{\Omega}_m) - \tilde{\Omega}_m t_m] = -t_m. \tag{5.108}$$

The last equality means that when $\tilde{q}(\omega^H) = \text{constant}$, in particular for optimal FTAN filters, the group time estimate $t_m(\omega_m)$ and the estimate (5.95) are connected by a relation similar to that which exists between the estimated functions $\psi(\omega)$ and $\tau(\omega)$, namely, (5.8). Consequently the same relation connects the systematic errors of these quantities.

5.4. LINEAR POLARIZATION ANALYSIS

Definition. A three-component record is described by a real-valued vector function $\mathbf{w}(t)$ with the components

$$\| w_x(t)\, w_y(t)\, w_z(t) \|. \tag{5.109}$$

Transformation (5.5) turns (5.109) into a triad of complex spectra

$$\| K_x(\omega)\, K_y(\omega)\, K_z(\omega) \| \tag{5.110}$$

which can for each ω be regarded as an element of a complex linear space, i.e., a complex vector $\mathbf{K}(\omega)$. One can consider, on the same complex space, the analytic vector signal $\mathbf{W}(t)$ obtained from (5.110) by Fourier-transforming the components. The real-valued space whose element is (5.109) (let us call it the physical space) is immersed in the complex space, being identical with the set of all its real-valued vectors.

$\mathbf{K}(\omega)$ describes the polarization, amplitude and phase spectral properties of a signal. Definition of polarization proper is based on the proposition: *two signals have identical polarizations, if their vector spectra* $\mathbf{K}(\omega)$ *are linearly dependent for any* ω.

From this it follows that all vectors that differ by a complex scalar factor, i.e., belong to the same subspace of one dimension, are equivalent as regards polarization. For this reason the mathematical object that describes the polarization properties of a signal is taken to be the one-dimensional space to which the complex vector $\mathbf{K}(\omega)$ belongs. Below, one-dimensional linear subspaces of a complex space are called straigt lines and two-dimensional ones are called planes.

Parameterization of a straight line. Polarization vector. A straight line is fully specified by the parameters of the unit vector that belongs to it. A vector that differs from the unit vector by a scalar factor $\exp[-i\gamma]$ (γ is real) belongs to the same subspace. To have a one-to-one correspondence, the unit vector should satisfy additional conditions which are chosen from the following considerations.

Let $\mathbf{K} = \mathbf{a} + i\mathbf{b}$ and $\mathbf{K}e^{-i\gamma} = \mathbf{c} + i\mathbf{d}$ be complex vectors ($\mathbf{a}$, $\mathbf{b}$, $\mathbf{c}$, $\mathbf{d}$ are real-valued vectors). Then

$$\begin{aligned} \mathbf{c} &= \mathbf{a}\cos\gamma + \mathbf{b}\sin\gamma \\ \mathbf{d} &= -\mathbf{a}\sin\gamma + \mathbf{b}\cos\gamma. \end{aligned} \tag{5.111}$$

When γ is varied, both $\mathbf{c}$ and $\mathbf{d}$ describe the same ellipse in the physical space, while remaining the conjugate radii of the ellipse. There exists such $\gamma = \psi$ that $\mathbf{c}$ and $\mathbf{d}$ coincide with the principal axes of the ellipse. The value of ψ is found from the orthogonality condition for $\mathbf{c}$ and $\mathbf{d}$:

$$\psi = \frac{1}{2}\tan^{-1}\frac{2(\mathbf{a}, \mathbf{b})}{(\mathbf{a}, \mathbf{a}) - (\mathbf{b}, \mathbf{b})}. \tag{5.112}$$

The vectors of the principal axes are obtained by substituting $\gamma = \psi$ into (5.111).

Thus, there exists a definite ellipse in the physical space corresponding to any complex vector. The reduction of the ellipse to the principal axes corresponds to the proposition that a complex vector $\mathbf{K}$ can be represented as

$$\mathbf{K} = z\mathbf{P} \tag{5.113}$$

where z is a scalar, while $\mathbf{P}$ has the special form

$$\begin{gathered}\mathbf{P} = \mu_1\mathbf{r}_1 + i\mu_2\mathbf{r}_2 \\ \mu_1, \mu_2 \geqslant 0, \quad \mu_1^2 + \mu_2^2 = 1, \quad |\mathbf{r}_1| = |\mathbf{r}_2| = 1, \quad (\mathbf{r}_1, \mathbf{r}_2) = 0\end{gathered} \tag{5.114}$$

where μ_1, μ_2, $\mathbf{r}_1$ and $\mathbf{r}_2$ are real-valued scalars and vectors, the vectors $\mu_1\mathbf{r}_1$ and $\mu_2\mathbf{r}_2$ are the principal axes of the relevant ellipse and are found from $\mathbf{K} = \mathbf{a} + i\mathbf{b}$ by the procedure indicated above with the normalization condition $\mu_1^2 + \mu_2^2 = 1$. This means that $|\mathbf{P}| = 1$.

A vector $\mathbf{P}$ of the form (5.114) will be assumed to be a parameter of a complex straight line. Then the vector spectrum of an arbitrary signal can be written as

$$\mathbf{K}(\omega) = \mathbf{P}(\omega)K(\omega). \tag{5.115}$$

We shall call $\mathbf{P}(\omega)$ the 'polarization vector' and $K(\omega)$ the 'scalar spectrum' of a signal: $|K| = |\mathbf{K}|$, while $\psi = \arg K$ is calculated from (5.112). This definition of the phase is equivalent to the measurement of it along that component of the physical space which is identical with the principal axis of the relevant ellipse.

Note. The conditions in (5.114) define four unit vectors $\mathbf{P}$ that differ as a result of successive multiplication by i, corresponding to the existence of four principal semi-axes. Strictly speaking, a definition of $\mathbf{P}$ must involve a condition that allows one of the four to be chosen. Such a condition is easy to formulate in specific practical problems; for instance, when we deal with Rayleigh waves, the 'most principal' semi-axis may be taken to be that closest to the 'upward' direction.

Examples. Take an orthogonal frame of reference whose first axis is identical with the direction of wave propagation, while the third points vertically downward; a Rayleigh wave (in a laterally homogeneous isotropic medium) has the polarization vector

$$\mathbf{P}(\omega) = \|\mu_1(\omega)\; 0\; i\mu_2(\omega)\|$$

while the Love wave vector is

$$\mathbf{P}(\omega) = \|0\; 1\; 0\|.$$

Complex space. Below we use the ordinary notions on a unitary vector space, in particular, the Hermite scalar product

$$\langle \mathbf{K}_1, \mathbf{K}_2 \rangle = (\mathbf{a}_1 + i\mathbf{b}_1, \mathbf{a}_2 - i\mathbf{b}_2). \tag{5.116}$$

Here and below, we denote real-valued vectors by bold lower case letters. Since the physical space is immersed in the complex space, any result for the former can be obtained within the framework of the latter. For example, the spectrum of a signal along the real-valued component $\mathbf{e}$ is equal to $\langle \mathbf{K}(\omega), \mathbf{e} \rangle$.

A vector product is defined as

$$\langle \mathbf{K}_1 \times \mathbf{K}_2 \rangle = [(\mathbf{a}_1 - i\mathbf{b}_1) \times (\mathbf{a}_2 - i\mathbf{b}_2)]. \tag{5.117}$$

This definition preserves the properties of vector and mixed products that are known for real spaces. In particular, the result of $\langle \mathbf{K}_1 \times \mathbf{K}_2 \rangle$ is orthogonal to each of the factors in the sense of (5.116). The following relation holds:

$$\langle \mathbf{K}_1 \times \langle \mathbf{K}_2 \times \mathbf{K}_3 \rangle \rangle = \mathbf{K}_2 \cdot \langle \mathbf{K}_1, \mathbf{K}_3 \rangle^* - \mathbf{K}_3 \cdot \langle \mathbf{K}_1, \mathbf{K}_2 \rangle^*.$$

We emphasize that the form $\langle \mathbf{K}_1, \mathbf{K}_2 \rangle$ implies complex conjugation of the second factor, while the form $\langle \mathbf{K}_1 \times \mathbf{K}_2 \rangle$ implies that of both. For this reason there is no complex conjugation of $\mathbf{K}$ in, say, the explicit form $\langle \mathbf{K}, \mathbf{K}^* \rangle = (\mathbf{a} + i\mathbf{b}, \mathbf{a} + i\mathbf{b})$.

The angle between 'directions' $\mathbf{K}_1$ and $\mathbf{K}_2$ (denoted henceforward as $\alpha(\mathbf{K}_1, \mathbf{K}_2)$) is defined by

$$\cos \alpha(\mathbf{K}_1, \mathbf{K}_2) = \frac{|\langle \mathbf{K}_1, \mathbf{K}_2 \rangle|}{|\mathbf{K}_1| \cdot |\mathbf{K}_2|} \tag{5.118}$$

$$\sin \alpha(\mathbf{K}_1, \mathbf{K}_2) = \frac{|\langle \mathbf{K}_1 \times \mathbf{K}_2 \rangle|}{|\mathbf{K}_1| \cdot |\mathbf{K}_2|}. \tag{5.119}$$

The angle always lies within the interval $(0, \pi/2)$.

Vector notation provides a concise way of writing for the parameters of the scalar spectrum and polarization vector (5.114) (as well as for the relevant ellipse):

$$|K| = |\mathbf{K}|, \qquad \arg K = \tfrac{1}{2} \arg \langle \mathbf{K}, \mathbf{K}^* \rangle \tag{5.120}$$

$$\mu_1 = \cos \tfrac{1}{2} \alpha(\mathbf{K}, \mathbf{K}^*), \qquad \mu_2 = \sin \tfrac{1}{2} \alpha(\mathbf{K}, \mathbf{K}^*) \tag{5.121}$$

$\mathbf{P}$ fixes, in addition to the geometry of the ellipse, the sense of rotation along it in a spectral component of the signal (from $\mathbf{r}_2$ towards $\mathbf{r}_1$). The positive sense of rotation axis (in the physical space) is identical with the vector

$$i\langle \mathbf{K} \times \mathbf{K}^* \rangle = 2|\mathbf{K}|^2 \mu_1 \mu_2 [\mathbf{r}_1 \times \mathbf{r}_2] \tag{5.122}$$

which is orthogonal to the ellipse plane. From (5.122) it follows that $|\langle \mathbf{K} \times \mathbf{K}^* \rangle|$ is proportional to the ellipse area. In the important particular case μ_1 (or μ_2) $= 0$ (we shall refer to it as 'real-valued polarization'), the ellipse degenerates to a straight segment of zero area. The condition for real-valued polarization is

$$\frac{|\langle \mathbf{K} \times \mathbf{K}^* \rangle}{|\mathbf{K}|^2} = \sin \alpha(\mathbf{K}, \mathbf{K}^*) = 2\mu_1 \mu_2 = 0. \tag{5.123}$$

The parameter $2\mu_1\mu_2$ for an arbitrary straight line takes on values between 0 and 1, reaching unity when $\mu_1^2 = \mu_2^2 = 1/2$ (circular polarization). Formulas (5.121) allow each of μ_1 and μ_2 to be found from $2\mu_1\mu_2$.

Apart from vectors, straight lines and planes will also be considered as independent objects. A straight line is denoted by the letter of the corresponding vector (usually in the form (5.114)) with a left superscript 1 (${}^1\mathbf{P}$). For a plane we employ the letter of the vector orthogonal to it, adding the superscript 2 (${}^2\mathbf{P}$). Equivalence of straight lines or planes is denoted by the $\sim$ symbol. This also marks the fact that vectors belong to the same straight line.

With a fixed basis, consisting of real-valued vectors all straight lines can be classified into two types: $C1$ is a straight line that does not contain real-valued vectors; $R1$ is one that contains at least one, hence an infinity of real-valued vectors. Such straight lines will be called 'real-valued'. The condition for a straight line to belong to $R1$ is (5.123), where $\mathbf{K}$ is an arbitrary vector on the line.

One can similarly distinguish planes of two types. One of these, $C2$, contains a single real-valued straight line. The vector $\mathbf{e}$ that belongs to it is given by

$$\mathbf{e} \sim \langle \mathbf{P} \times \mathbf{P}^* \rangle \sim [\mathbf{r}_1 \times \mathbf{r}_2] \tag{5.124}$$

where $\mathbf{P} = \mu_1 \mathbf{r}_1 + i\mu_2 \mathbf{r}_2$ is a vector (in the form (5.114)) orthogonal to the plane.

A plane of the other type, $R2$, contains a single plane of the physical space. A plane of the $R2$ type will also be called 'real-valued'. The condition for a plane to belong to $R2$ expressed in terms of the orthogonal vector is (5.123).

Note that the above classification of subspaces is, generally speaking, not invariant under the choice of an arbitrary complex basis. This is true both as regards the operation of complex conjugation and the straight line parameterization in the form (5.114). We shall assume, however, that all the expressions involve objects of the complex space as written in a basis that belongs to the physical space, the situation of real measurements. This restriction removes the difficulties just indicated.

Elliptically and linearly polarized signals. Techniques of polarization analysis are largely based on the model of a signal with the polarization vector independent of frequency:

$$\mathbf{K}(\omega) = \mathbf{P}K(\omega). \tag{5.125}$$

The above classification of straight lines ($C1$ and $R1$) corresponds to signals of the type (5.125) having elliptic and linear polarization. The latter can be written in the form

$$\mathbf{K}(\omega) = \mathbf{r}K(\omega) \tag{5.126}$$

where $\mathbf{r}$ is the unit real-valued vector. Although (5.126) is a particular case of (5.125), it would be more convenient as regards the terminology to treat elliptically and linearly polarized signals as two independent types. The reason for this lies in the fact that when a signal has a definite polarization type, that provides an important piece of prior information useful in signal processing. Accordingly, we shall take it for granted that the condition $\langle \mathbf{P} \times \mathbf{P}^* \rangle \neq 0$ is incorporated in the definition of an elliptically polarized signal (5.125).

The particle path $\mathbf{w}(t) = \operatorname{Re} \mathbf{W}(t)$ of an elliptically polarized signal in the physical space is generally different from an ellipse. However, the path lies on a plane that is orthogonal to the real-valued vector $i\langle \mathbf{P} \times \mathbf{P}^* \rangle$; there are two orthogonal directions on the plane (identical with the $\mathbf{r}_1$ and $\mathbf{r}_2$ of $\mathbf{P}$), such that the projections of $\mathbf{K}(\omega)$ on these are, apart from a constant factor, the real and imaginary part of one and the same analytic signal. With the linear polarization (5.126), motion in the physical space occurs along the straight line fixed by $\mathbf{r}$.

A harmonic signal with components $K_m \exp[i(\omega t + \psi_m)]$ always has the form (5.125); indeed, according to (5.111), its path in the physical space is an ellipse. For this reason, if a wave recorded by a seismometer is a narrow-band signal, the observed path is close to an ellipse. As time goes on, the ellipse is slowly deforming for a locally narrow-band signal, a frequent case with surface waves.

The special importance attaching to elliptic and linear polarization is not due to the mere fact alone that these are simple to study. The model (5.125) frequently quite adequately represents real signals. The velocities of elastic waves in the Earth (P, S, Rayleigh and Love waves etc.) are intimately related to their polarization. For this reason the division of a record into waves based on their typical velocities is largely equivalent to the division into signals whose polarization vectors are weakly dependent on frequency.

Model for linear polarization analysis. (1) Each wave that is present on a record is assumed to be either elliptically or linearly polarized:

$$\mathbf{K}(\omega) = \sum_j \mathbf{K}_j(\omega) = \sum_j \mathbf{P}_j K_j(\omega). \tag{5.127}$$

(2) The scalar spectra $K_j(\omega)$ are linearly independent. Hence the important corollary:

If the total spectrum $\mathbf{K}(\omega)$ belongs to some linear subspace, then each of the interfering signals also belongs to it.

The model (5.127) is an extension of the notions of elliptically and linearly polarized signals. If (5.125) is a degenerate case of three-dimensional motion along a complex straight line, then motions are possible within the framework of (5.127) for which $\mathbf{K}(\omega)$ at any frequency belongs to a fixed plane. Linear polarization analysis must investigate all degenerate situations of this kind.

If a real record is to be adequately modelled, (5.127) must incorporate the noise too. For the sake of brevity, however, we shall demonstrate the main ideas of polarization analysis from the deterministic standpoint, quoting whenever necessary final results obtained by incorporating random noise in (5.127). Kanareikin, Pavlov and Potekhin (1966) provide a detailed account of numerous statistical aspects arising in polarization analysis.

Covariance matrix. The most popular tool for studying subspaces is the covariance matrix $\mathbf{C}$ calculated directly from the record or from the spectrum:

$$\mathbf{C} = \frac{\int_{-\infty}^{\infty} \mathbf{K}(\omega)\mathbf{K}^{T*}(\omega)\,\mathrm{d}\omega}{\int_{-\infty}^{\infty} |\mathbf{K}(\omega)|^2\,\mathrm{d}\omega} = \frac{\int_{-\infty}^{\infty} \mathbf{W}(t)\mathbf{W}^{T*}(t)\,\mathrm{d}t}{\int_{-\infty}^{\infty} |\mathbf{W}(t)|^2\,\mathrm{d}t}. \tag{5.128}$$

The normalization in (5.128) is chosen so as to make the trace of $\mathbf{C}$ equal to unity, $\mathbf{K}^{T*}$, $\mathbf{W}^{T*}$ being the complex conjugate row vectors.

Properties of $\mathbf{C}$ that are invariant under a change of orthogonal (and real-valued) reference frame are in a one-to-one correspondence with the parameters of subspaces. Denote: $\lambda_1 \geqslant \lambda_2 \geqslant \lambda_3$ are the eigenvalues of $\mathbf{C}$; they are nonnegative and $\lambda_1 + \lambda_2 + \lambda_3 = 1$. $\mathbf{G}$, $\mathbf{V}$ are the eigenvectors of $\mathbf{C}$ for λ_1 and λ_3, respectively (in the form (5.114)):

$$\mathbf{G} = \eta_1 \mathbf{g}_1 + i\eta_2 \mathbf{g}_2, \qquad \mathbf{V} = \nu_1 \mathbf{v}_1 + i\nu_2 \mathbf{v}_2. \tag{5.129}$$

Table 5.I lists equivalent properties of the covariance matrix for each subspace type.

Thus, to determine the parameters of the particular subspace to which $\mathbf{K}(\omega)$ belongs, one must perform two operations:

(1) calculate the eigenvalues and eigenvectors of $\mathbf{C}$ and find the dimension of the subspace from the λ_m;
(2) transform the relevant eigenvector ($\mathbf{G}$ when rank $\mathbf{C} = 1$, and $\mathbf{V}$ when rank $\mathbf{C} = 2$) to the form (5.114) and determine the subspace type from the value of $2\eta_1\eta_2$ or $2\nu_1\nu_2$.

When we know beforehand that the subspace sought is of dimension one, that

TABLE 5.I

$\mathbf{K}(\omega)$ belongs to subspace	Subspace type	Equivalent properties of $\mathbf{C}$	
${}^1\mathbf{P}$	$C1$	rank $\mathbf{C} = 1$ (or $\lambda_1 = \lambda_2 = 0$)	$2\eta_1\eta_2 = 0$
${}^2\mathbf{P}$	$R1$	$\mathbf{G} \sim \mathbf{P}$	$2\eta_1\eta_2 = 0$
${}^1\mathbf{Q}$	$C2$	rank $\mathbf{C} = 2$ (or $\lambda_3 = 0, \lambda_1, \lambda_2 = 0$)	$2\nu_1\nu_2 = 0$
${}^2\mathbf{Q}$	$R2$	$\mathbf{V} \sim \mathbf{Q}$	$2\nu_1\nu_2 = 0$
to the entire subspace		rank $\mathbf{C} = 3$	

is, the signal has elliptic or linear polarization, a single operation will suffice. For in that case

$$\text{Re}\,\mathbf{C} = \text{Re}\,\mathbf{P}\mathbf{P}^{T*} = \mu_1^2\mathbf{r}_1\mathbf{r}_1^T + \mu_2^2\mathbf{r}_2\mathbf{r}_2^T. \tag{5.130}$$

whence the desired quantities μ_1^2, μ_2^2, $\mathbf{r}_1$, $\mathbf{r}_2$ are seen to be the eigenvalues and eigenvectors of the matrix Re $\mathbf{C}$. The third eigenvalue is in that case zero.

In actual practice, because of the noise or because the polarization departs from (5.127), $\mathbf{C}$ is nondegenerate, and its properties as listed on the right in Table 5.I are stated as approximate requirements. For example, we put $\lambda_m = 0$ when $\lambda_m < \lambda_0$, where λ_0 is a threshold chosen beforehand. Results obtained in this way are stable enough under small departures from (5.127). The reason is that $\mathbf{C}$ plays the same part with respect to the polarization parameters of a signal as $\langle\omega\rangle$, D_ω etc. do with respect to its scalar parameters. In other words, the covariance matrix yields the 'mean location of the linear subspace' to which the signal belongs.

To see this, consider an arbitrary unit vector $\mathbf{Y}$ and write the quadratic form using the normalized spectral amplitude $B(\omega)$ (5.9)

$$\mathbf{Y}^T\mathbf{C}\mathbf{Y} = \int_{-\infty}^{\infty} |\langle\mathbf{P}(\omega), \mathbf{Y}\rangle|^2 B^2(\omega)\, d\omega$$

$$= \int_{-\infty}^{\infty} \cos^2\alpha(\mathbf{P}(\omega), \mathbf{Y}) B^2(\omega)\, d\omega. \tag{5.131}$$

This quantity is extremal at the eigenvectors of $\mathbf{C}$ and is equal to the relevant eigenvalues. The vector $\mathbf{G}$ (5.129) fixes a straight line, and $\mathbf{V}$ fixes a plane onto which most of signal energy is projected. In this sense, the straight line and the plane obtained from $\mathbf{G}$ and $\mathbf{V}$, respectively, are 'mean over the signal'.

From (5.131) one readily evaluates the typical (for the signal) angle of departure from the 'mean' straight line ${}^1\mathbf{G}$

$$D_\alpha^G = \sin^{-1}\sqrt{\lambda_2 + \lambda_3} \tag{5.132}$$

and that from the 'mean' plane

$$D_\alpha^V = \sin^{-1}\sqrt{\lambda_3}. \tag{5.133}$$

The averaging properties of the covariance matrix allow it to be used as a tool for calculating smoothed polarization parameters of any signal (including one with $\mathbf{P}(\omega) \neq$ constant). The ranges of integration in (5.128) are then replaced by finite ones, the equality separating into two independent determinations of the spectral and temporal covariance matrix.

It is more convenient in processing of dispersed surface waves to use $\mathbf{C}$ as determined through the frequency—time representation, this being calculated for each component with the same $\beta(\omega)$. In that case the vectorial frequency—time representation for any signal of form (5.125) is

$$\mathbf{S}(\omega^T, t) = \mathbf{P}S(\omega^H, t) \tag{5.134}$$

that is, the transition from $\mathbf{K}(\omega)$ to $\mathbf{S}(\omega^H, t)$ does not distort the polarization vector. When $\mathbf{P}(\omega)$ = constant, polarization parameters as measured from $\mathbf{S}(\omega^H, t)$ are subject to systematic errors.

The covariance matrix is given in terms of $\mathbf{S}(\omega^H, t)$ by

$$\mathbf{C}(\omega_\varepsilon^H, t_\varepsilon) = \frac{\displaystyle\int_{\varepsilon(\omega_\varepsilon^H, t_\varepsilon)} \mathbf{S}(\omega^H, t)\mathbf{S}^{T*}(\omega^H, t)\,\mathrm{d}\omega^H\,\mathrm{d}t}{\displaystyle\int_{\varepsilon(\omega_\varepsilon^H, t_\varepsilon)} \mathbf{P}\mathbf{P}^{T*}|S(\omega^H, t)|^2\,\mathrm{d}\omega^H\,\mathrm{d}t} \tag{5.135}$$

where $\varepsilon(\omega_\varepsilon^H, t_\varepsilon)$ is a 'scanning window', a limited region of the (ω^H, t)-plane located around the point $(\omega_\varepsilon^H, t_\varepsilon)$. Matrix $\mathbf{C}$ is thus treated as a functionn fixed on the frequency—time plane. At each point $(\omega_\varepsilon^H, t_\varepsilon)$, this function possesses all the properties listed above in relation to $\mathbf{S}(\omega^H, t)$ considered within the window $\varepsilon(\omega_\varepsilon^H, t_\varepsilon)$.

The techniques of polarization analysis to be discussed below are based on the properties of the covariance matrix, not on the procedure used to calculate it. Even so, the results do significantly depend on this and, in particular, on the choice of ε.

Analysis of one-dimensional subspaces. This simplest version of (5.127) is the one most frequently used; the assumption here is that a single elliptically or linearly polarized signal is dominant on the record segment under study. There is extensive literature devoted to this question (Flinn, 1965; Means, 1972; Kats and Mikhailova, 1977; Galmakov and Sitnikov, 1984; Galperin, 1984), so we shall only briefly touch on some of its aspects.

Since (5.125) is in good agreement with real waves, the criteria of that model can be used in signal identification and separation. Identification is based on a decision function L whose value is a measure of how much the straight-line parameters deviate from the model.

For example, if an elliptically or linearly polarized signal is to be identified, L may be chosen to be the following combination (λ_1 being the greatest eigenvalue of $\mathbf{C}$)

$$L_E = \frac{3\lambda_1 - 1}{2\lambda_1} \tag{5.136}$$

which estimates the ratio of signal energy to the total energy occurring within the windom ε. The value of (5.136) varies from zero (when $\lambda_1 = \lambda_2 = \lambda_3 = 1/3$) to one (when $\lambda_1 = 1$). Scanning the time–frequency area of interest with the window, we represent L_E as a function of $(\omega_\varepsilon^H, t_\varepsilon)$. Values of it close to unity correspond to regions where polarization is close to the model. The polarization vector is estimated by the eigenvector **G** (5.129) calculated at the point where $L_E(\omega_\varepsilon^H, t_\varepsilon)$ is a maximum.

To separate signals of the form (5.125), one identifies the $(\omega_\varepsilon^H, t_\varepsilon)$-plane with the (ω^H, t)-plane and finds the product

$$\tilde{\mathbf{S}}(\omega^H, t) = L(\omega^H, t) \cdot \mathbf{S}(\omega^H, t). \tag{5.137}$$

This operation is called polarization filtering, usually involving the projection of $\tilde{\mathbf{S}}(\omega^H, t)$ onto a direction that is either chosen beforehand or estimated from **C**. The filter (5.137) is nonlinear, hence causes signal distortions. This effect may be diminished for wanted signals by taking L as shown in Figure 5.10. One chooses one or several parameters X to show how much the polarization measured from **C** is close to the model, as well as the relevant threshold values X_0. $L \approx 1$ for signals similar to the model one, so they are passed by (5.137) without significant distortions. Below we will construct L using a simple function such as $1/(1 + (X/X_0)^n)$, $n > 1$. Here the presence of the free parameter n allows one to choose L as close to the step function as desired.

The function L_E (5.136) in the above example can be converted to the form indicated above as follows. X is taken to be ρ, which characterizes the noise-to-signal energy ratio

$$\rho = \frac{1 - L_E}{L_E} = \frac{1 - \lambda_1}{3\lambda_1 - 1} \tag{5.138}$$

and varies between 0 and ∞. Let ρ_0 be a threshold value for ρ. Then

$$L_1(\rho) = \frac{1}{1 + (\rho/\rho_0)^{n_1}}, \tag{5.139}$$

for an appropriate n_1 approximately corresponds to the function shown in Figure 5.10. For low noise ($\rho \ll \rho_0$) we have $L_1 \approx 1$; when the signal-to-noise ratio is

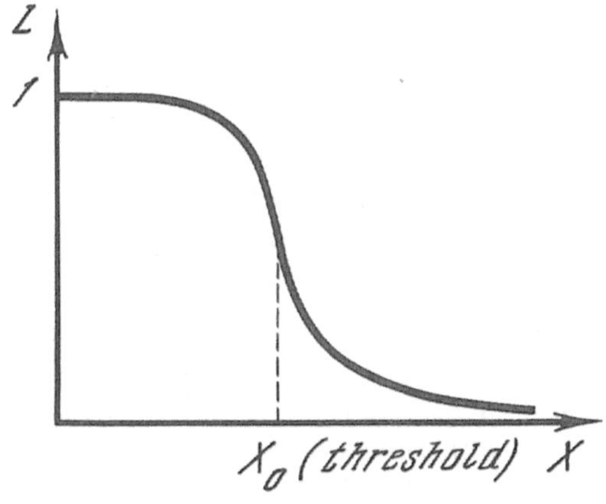

Fig. 5.10. Typical shape of the decision function.

above the threshold ($\rho \gg \rho_0$), then $L_1 \ll 1$. The filter (5.137) mainly distorts signals with $\rho \approx \rho_0$.

We now discuss a few examples of the decision functions.

(1) For identification of linearly (by no means elliptically) polarized signals (for instance, Love waves) one can make use of the same functions (5.139), (5.138) with the argument λ_1 replaced by the greatest eigenvalue of Re **C**. It is the appropriate eigenvector taken at the point where L is a maximum which is assumed to be an estimate of the desired polarization vector.

(2) For identification of a signal having a specific polarization vector **P**, a factor depending on the following parameter is inserted into the decision function

$$\theta = \frac{\lambda_1}{1-\lambda_1}\tan^2\alpha(\mathbf{P},\mathbf{G}) \tag{5.140}$$

where **G** is the eigenvector of (5.129) which corresponds to λ_1. Parameter θ varies between 0 and ∞ and equals one when $\alpha(\mathbf{P}, \mathbf{G}) = D_\alpha^G$ (5.132). The decision function is like (5.139)

$$L_2 = L_1 \frac{1}{1+(\theta/\theta_0)^{n_2}} \tag{5.141}$$

where θ_0 is the threshold value for θ.

(3) Suppose it is required to identify a fundamental mode Rayleigh wave arriving at an arbitrary azimuth that is not known beforehand. The identification criteria will be that the polarization ellipse plane should be vertical

$$\alpha(\mathbf{e}_z, \langle\mathbf{P}\times\mathbf{P}^*\rangle) = \pi/2 \tag{5.142}$$

($\mathbf{e}_z$ is the vertical unit vector) and that μ_2/μ_1 should be close to unity: $1/c < \mu_2/\mu_1 < c$ or the equivalent requirement

$$\tan^2\alpha(\mathbf{P},\mathbf{P}^*) = \frac{|\langle\mathbf{P}\times\mathbf{P}^*\rangle|^2}{|\langle\mathbf{P},\mathbf{P}^*\rangle|^2} > \frac{4}{(c-1/c)^2} = \gamma_0 \tag{5.143}$$

where γ_0 is a threshold value for $\gamma = \tan^2(\mathbf{G}, \mathbf{G}^*)$. The other threshold value θ_0 (for (5.142)) is found by introducing a parameter θ in the sense similar to that of (5.140):

$$\theta = \frac{\lambda_1}{1-\lambda_1}\operatorname{cotan}^2\alpha(\mathbf{e}_z, \langle\mathbf{G}\times\mathbf{G}^*\rangle). \tag{5.144}$$

The final form of decision function

$$L_3 = L_1 \frac{1}{1+(\theta/\theta_0)^{n_{31}}}\,\frac{1}{1+(\gamma_0/\gamma)^{n_{32}}} \tag{5.145}$$

performs signal identification based on an approximate validity of three requirements: (5.152), (5.153), and the agreement with (5.125).

We conclude by a few words on choosing the scanning window size. The best signal-to-random-noise ratio is obtained for an ε that approximately coincides with the time—frequency region of the signal. The latter can be treated as an 'upper bound' on ε. On the other hand, values of $\mathbf{S}(\omega^H, t)$ at neighbouring points of the (ω^H, t)-plane are strongly interdependent. It is therefore necessary for effective identification that at least one of the linear dimensions of ε, either along ω^H or t, should exceed quantities of the order of β or $1/\beta$, respectively. Otherwise the eigenvalue λ_1 would be close to unity irrespective of how well (5.125) fits the signal. This requirement determines a 'lower bound' on ε. It is valid both in the study of one-dimensional subspaces and in problems of interfering signal identification and measurement. As to the rest, one is to be guided by the consideration that increased ε generally inflates the systematic errors of **P** estimates and diminishes the errors due to random noise.

Identification of interfering signals. The potential of linear polarization analysis in determining the number of signals that are present in some, say, frequency range is rather limited. To see this, one should consider that the expansion of any signal on a fixed basis $\mathbf{E}_1, \mathbf{E}_2, \mathbf{E}_3$ of the complex space

$$\mathbf{K}(\omega) = \mathbf{E}_1 K_1(\omega) + \mathbf{E}_2 K_2(\omega) + \mathbf{E}_3 K_3(\omega) \tag{5.146}$$

is just the representation of the signal as a sum of three signals that have elliptic (or even linear) polarization. Therefore the result of interference for more than two signals cannot in principle be distinguished from the general case and identified by polarization methods. Even so, the only remaining version of (5.127), interference between two signals with different polarization vectors, is important enough for practice. We proceed to deal with this.

A sum of two signals of the form (5.125) belongs to a fixed plane that contains both polarization vectors. The identification criteria for this situation (within the window ε) are

$$\lambda_3 = 0, \qquad \lambda_2 > 0. \tag{5.147}$$

In practice one constructs a decision function which checks whether both requirements (5.147) are satisfied; for instance (l_0 is a threshold),

$$L_4 = 1 \Bigg/ \left[1 + \left(\frac{1}{l_0} \frac{\lambda_1 \lambda_3}{\lambda_2^2} \right)^{n_4} \right]. \tag{5.148}$$

(5.148) is incapable of distinguishing between planes of the $C2$ and $R2$ type. If only signals that belong to $R2$ are to be identified, a factor should be inserted into (5.148) to check the condition $2\nu_1\nu_2 = 0$. We assume $\nu_2 \leqslant \nu_1$ for the sake of definiteness. In such a case the real-valued straight line that is the closest (in the sense of the angle) to **V** passes through the vector $\mathbf{v}_1$ (5.129), and

$$\tan \alpha(\mathbf{V}, \mathbf{v}_1) = \nu_1/\nu_2. \tag{5.149}$$

Comparing this quantity to the typical value of the tangent of the error D_α^v (5.133), one can write down a decision function to distinguish $R2$-type planes, for example, in the form (m_0 is a threshold)

$$L_5 = L_4 \Bigg/ \left[1 + \left(\frac{1}{m_0} \frac{\nu_2^2}{\nu_1^2} \frac{\lambda_3}{1 - \lambda_3} \right)^{n_5} \right]. \tag{5.150}$$

The plane that contains both interfering signals is estimated by ${}^2\mathbf{V}$ (or ${}^2\mathbf{v}_1$, if the plane is assumed to be of the $R2$ type).

The fact that the total signal belongs to a plane of a certain type provides some information on each of the interfering signals.

If the plane is identified as $R2$, then any of the interfering signals is orthogonal to the real-valued vector $\mathbf{v}_1$. In that case, though, we also have $(\mathrm{Re}\,\mathbf{W}_j(t), \mathbf{v}_1) \equiv 0$ (in the time domain), that is, the original real-valued signals at any instant of time lie on the same plane of the physical space. Each may have either linear or elliptic polarization. In the latter case the same plane also contains the respective polarization ellipses.

When the plane is $C2$, at least one of the two interfering signals formally has elliptic polarization ($C2$ contains a single real-valued straight line (5.134)). However, that conclusion cannot be considered quite definite from the practical standpoint. Any ${}^2\mathbf{V}$ plane with an arbitrary parameter $2\nu_1\nu_2$ contains infinitely many straight lines that are indistinguishable from the real-valued one to within the measurement error. Thus, both interfering signals may, generally speaking, be close to linearly polarized ones.

Further conclusions as to the characteristics of interfering signals can only be made if certain prior information on these is available. It should be remembered that, as must be clear from the preceding chapters, this kind of information is scarce for laterally varying media. Actual polarization parameters of surface waves may be significantly different from what one would expect in a laterally homogeneous isotropic medium.

We now proceed to a discussion of how to measure the polarization parameters of interfering signals. We begin by examining two rather specific examples which, however, are of practical interest, and then briefly touch on more general relationships.

Interference of Love and Rayleigh waves. The model is a result of interference between a linearly (Love wave) and an elliptically (Rayleigh wave) polarized signals:

$$\mathbf{K}(\omega) = \mathbf{r}K_L(\omega) + (\mu_1\mathbf{r}_1 + i\mu_2\mathbf{r}_2)K_R(\omega). \tag{5.151}$$

Thus, the first prior assumptions to be considered are that the polarization type of each signal is known. Note that we have not yet assumed any polarization properties of Love and Rayleigh waves other than those usual for a laterally homogeneous medium, and shall supply such information only when necessary.

If the observed $\mathbf{K}(\omega)$ lies on a ${}^2\mathbf{V}$ plane of the $C2$ type, the assumptions we have made suffice to find the Love wave polarization vector. As $\mathbf{r}$ belongs to a plane that contains a single real-valued straight line, it follows that

$$\mathbf{r} = [\mathbf{v}_1 \times \mathbf{v}_2] \tag{5.152}$$

where, as before, $\mathbf{V} = \nu_1\mathbf{v}_1 + i\nu_2\mathbf{v}_2$ is the eigenvector from (5.129).

The Rayleigh wave polarization vector cannot be determined without making additional assumptions. A clear understanding of the restrictions to which it is subject can be gained by utilizing a special kind of transformation which is also helpful in practice. This linear transformation on an orthogonal basis consisting of $\mathbf{v}_1, \mathbf{v}_2, \mathbf{r}$ has the form

$$\tilde{\mathbf{K}}(\omega) = \begin{Vmatrix} i & 0 & 0 \\ 0 & 1 & 0 \\ 0 & 0 & 1 \end{Vmatrix} \mathbf{K}(\omega). \tag{5.153}$$

The notions of a real-valued straight line and plane are not invariant under

(5.153). This transformation converts a vector $\mathbf{V}$ into a vector $\tilde{\mathbf{v}}$ that lies on a real-valued straight line:

$$\mathbf{V} = \nu_1\mathbf{v}_1 + i\nu_2\mathbf{v}_2 \rightarrow \tilde{\mathbf{v}} = i(\nu_1\mathbf{v}_1 + \nu_2\mathbf{v}_2) \tag{5.154}$$

while the relevant ${}^2\mathbf{V}$ plane becomes a ${}^2\tilde{\mathbf{v}}$ plane of the $R2$ type. It is to this plane that the transformed Rayleigh wave belongs, as well as the Love wave which is not affected by (5.153) at all.

The practical significance of (5.153) is due to another assumption relating to (5.151): we assume that one of the principal axes of Rayleigh wave polarization ellipse ($\mathbf{r}_2$ for definiteness) is known beforehand and is orthogonal to the Love wave $\mathbf{r}$ vector. This assumption is quite natural within the framework of a laterally homogeneous isotropic medium where one of the principal axes of a Rayleigh wave ellipse is known beforehand, while Love waves are polarized in the horizontal plane. However, the direction $\mathbf{r}_1$ and the values of μ_1, μ_2 are assumed to be so far unknown. Then, expressing $\mathbf{V}$ in terms of the vector product of $\mathbf{r}$ and $\mu_1\mathbf{r}_1 + i\mu_2\mathbf{r}_2$, we obtain

$$\mathbf{v}_1 = \mathbf{r}_2. \tag{5.155}$$

However, $\mathbf{r}_2$ has been assumed to be known (vertical). The transformation (5.153) can then be applied to the record before the covariance matrix, hence $\mathbf{V}$ is calculated.

In practice, (5.153) reduces to a change of $\pi/2$ in the phase of the vertical component spectrum. Afterwards the desired plane is assumed beforehand to have the $R2$ type and is found by using the Re $\mathbf{C}$ matrix. The eigenvector corresponding to the least eigenvalue of Re $\mathbf{C}$ is taken to be an estimate of $\tilde{\mathbf{v}}$. The angle it makes with the vertical determines the values of ν_1 and ν_2:

$$\nu_1 = \cos\alpha(\tilde{\mathbf{v}}, \mathbf{r}_2), \qquad \nu_2 = \sin\alpha(\tilde{\mathbf{v}}, \mathbf{r}_2). \tag{5.156}$$

The direction of Love wave polarization is estimated by the line of intersection between the ${}^2\tilde{\mathbf{v}}$ plane and the horizontal plane:

$$\mathbf{r} \sim [\tilde{\mathbf{v}} \times \mathbf{r}_2]. \tag{5.157}$$

This procedure is valuable in that it requires less parameters to be estimated than in the calculation of $\mathbf{V}$, hence is more stable under noise.

According to (5.155), (5.153) transforms a Rayleigh wave into a linearly polarized signal with the vector

$$\tilde{\mathbf{p}} = \mu_1\mathbf{r}_1 + \mu_2\mathbf{r}_2. \tag{5.158}$$

As $\tilde{\mathbf{p}}$ belongs to the ${}^2\tilde{\mathbf{v}}$-plane, its unknown parameters are connected by the relation

$$\mu_1 \sin\alpha(\mathbf{r}_1, \mathbf{r})/\mu_2 = \nu_2/\nu_1. \tag{5.159}$$

Note that $\alpha(\mathbf{r}_1, \mathbf{r})$ is the angle between the Love wave $\mathbf{r}$ vector and the plane of initial Rayleigh wave polarization ellipse. From (5.159) it follows that

$$\mu_2/\mu_1 \leqslant \nu_1/\nu_2. \tag{5.160}$$

More exact statements about the parameters of Rayleigh wave polarization vector cannot, according to (5.159), be made, unless additional information is provided on the values of either the angle $\alpha(\mathbf{r}_1, \mathbf{r})$ or the ratio of ellipse axes μ_2/μ_1. For example, the full assumption of a laterally homogeneous isotropic medium, i.e., $\alpha(\mathbf{r}_1, \mathbf{r}) = \pi/2$, yields all the unknown parameters:

$$\mu_1\mathbf{r}_1 + i\mu_2\mathbf{r}_2 = \nu_1\mathbf{v}_2 + i\nu_2\mathbf{r}_1 \tag{5.161}$$

On the other hand, for a fundamental mode Rayleigh wave we usually have $\mu_2/\mu_1 > k \sim 0.5$, and this (when μ_2/μ_1 is not known) imposes certain restrictions on the possible values of $\alpha(\mathbf{r}_1, \mathbf{r})$: $\sin \alpha(\mathbf{r}_1, \mathbf{r}) \geqslant k\nu_1/\nu_2$. Since the angle is allowed to vary between 0 and $\pi/2$, we must specify that, corresponding to each value of $\mu_2/\mu_1 \leqslant \nu_1/\nu_2$, there are two vectors $\mathbf{r}_1$ that are symmetric about the $\mathbf{v}_1$-axis on the horizontal plane, hence there are two Rayleigh wave polarization vectors.

Interference between two Rayleigh waves. This situation is modelled by

$$\mathbf{K}(\omega) = (\mu_{11}\mathbf{r}_{11} + i\mu_{21}\mathbf{r}_{21})K_1(\omega) + (\mu_{12}\mathbf{r}_{12} + i\mu_{22}\mathbf{r}_{22})K_2(\omega) \tag{5.162}$$

and can be studied in a similar fashion to the preceding.

Suppose the principal axes $\mathbf{r}_{21}$ and $\mathbf{r}_{22}$ are identical and vertical. Then the $\mathbf{v}_1$-axis (5.129) too is identical with these, the vector $\mathbf{V}$ being orthogonal to the plane that contains $\mathbf{K}(\omega)$:

$$\mathbf{V}_1 = \mathbf{r}_{21} = \mathbf{r}_{22}. \tag{5.163}$$

Now apply (5.153) to the record. Each of the interfering Rayleigh waves will then become a linearly polarized signal belonging to the ${}^2\tilde{\mathbf{v}}$-plane. The parameters of each polarization vector satisfy (5.159), (5.160), where $\alpha(\mathbf{r}_{1j}, \mathbf{r})$ is the angle which $\mathbf{r}_{1j}$ makes with the line of intersection between ${}^2\tilde{\mathbf{v}}$ and the horizontal plane.

Searching for the polarization vectors needs additional information on either $\alpha(\mathbf{r}_{1j}, \mathbf{r})$ or μ_{2j}/μ_{1j}. For instance, appealing to the model of a laterally homogeneous medium again, we can assume $\mu_{21}/\mu_{11} = \mu_{22}/\mu_{12}$ (the value of this ratio is not known beforehand). In that case, knowing $\mathbf{r}_{11}$, i.e., the propagation direction of one of the waves (the wave of interest), we can use (5.159) to determine that of the other (noise):

$$\mathbf{r}_{12} = 2(\mathbf{r}_{11}, [\mathbf{r} \times \mathbf{r}_{21}])\,[\mathbf{r} \times \mathbf{r}_{21}] - \mathbf{r}_{11}$$

and the ratio of ellipse axes

$$\frac{\mu_{21}}{\mu_{11}} = \frac{\mu_{22}}{\mu_{12}} = \frac{\nu_2}{\nu_1 \sin \alpha(\mathbf{r}_{11}, \mathbf{r})}$$

whence both polarization vectors are fully determined.

General restrictions. Below we give a few relations associated with the models (5.151), (5.162) which are independent of the prior assumptions made above.

The parameters of any vector $\mathbf{P} = \mu_1\mathbf{r}_1 + i\mu_2\mathbf{r}_2$ that belongs to a ${}^2\mathbf{V}$-plane of $C2$ type are related through

$$\frac{\mu_1\mu_2}{\nu_1\nu_2} = \frac{\sin^2 \alpha(\mathbf{P}, \mathbf{v}_r)}{|([\mathbf{r}_1 \times \mathbf{r}_2], \mathbf{v}_r)|} = \frac{\sin^2 \alpha(\mathbf{P}, \mathbf{v}_r)}{\sin \gamma} \tag{5.164}$$

where $\mathbf{v}_r$ fixes the only straight line that lies on the plane, and γ is the angle (in the physical space) between $\mathbf{v}_r$ and the ellipse plane corresponding to $\mathbf{P}$.

A $C2$-type plane contains, apart from a real-valued straight line ($2\mu_1\mu_2 = 0$), also two straight lines corresponding to circular polarization ($2\mu_1\mu_2 = 1$) which are given by the vectors ($\nu_2 \leqslant \nu_1$ for definiteness)

$$\mathbf{v}^c_{1,2} = \frac{1}{\sqrt{2}}\left(\frac{\nu_2}{\nu_1}\mathbf{v}_1 \pm \sqrt{1-\nu_2^2/\nu_1^2}\mathbf{v}_r + i\mathbf{v}_2\right) \tag{5.165}$$

the two lines coalescing into one only when $\nu_1 = \nu_2$. Any other value of $2\mu_1\mu_2$ (except zero and unity) produces infinitely many straight lines in the ${}^2\mathbf{V}$-plane. This fact distinguishes the general case from the above examples in which one or two straight lines correspond to a value of μ_2/μ_1.

If a vector is assumed to lie on a real-valued straight line, but actually has a 'weak ellipticity' $\delta\mu_2 \ll 1$, then the estimate we got for $\mathbf{v}_r$ is usually (when $2\nu_1\nu_2 \sim 1$) close to the true vector $\mathbf{P}$. This statement is ensured by an inequality derived from (5.164), namely,

$$\sin\alpha(\mathbf{P}, \mathbf{v}_r) \leqslant \min[\mu_2/\nu_2, \mu_1/\nu_1],$$

i.e., the estimate differs from the true vector by $\delta\alpha \leqslant \delta\mu_2/\nu_2$ (it has been assumed that $\mu_2/\mu_1 < \nu_2/\nu_1$). The measurements of parameters of straight lines that are known to be real-valued beforehand are unstable only for planes that are 'close' to real-valued ones ($\nu_2 \ll 1$).

If a $C2$-type plane is due to two signals that are close to linearly polarized ones, then their polarization vectors are close to $\mathbf{v}_r$ and, hence, to each other. This condition distinguishes $C2$ from $R2$. The latter can be due to linearly polarized signals with an arbitrary angle between them.

Polarizational separation of signals. Consider the problem of determining scalar spectra $K_j(\omega)$ on the basis of known (or measured) polarization vectors $\mathbf{P}_j$ within the framework of (5.127). It is understood that the $K_j(\omega)$ are significantly correlated and comparable in amplitude, that is, can be separated neither by frequency-time methods nor by polarization filtering.

The idea of polarization filtering consists in the total record being projected onto a direction where the selected 'signal of interest' might dominate the rest ('noise'). With any number of interfering signals, the problem requires knowledge of the covariance matrix of scalar spectra $K_j(\omega)$. Information of this kind is not available in surface wave processing, and polarizational separation is effective only when interference occurs between three signals at most.

Suppose the model (5.127) involves three signals having the polarization vectors $\mathbf{P}_1$, $\mathbf{P}_2$, $\mathbf{P}_3$. A conjugate set of vectors $\mathbf{E}_1$, $\mathbf{E}_2$, $\mathbf{E}_3$ is constructed on the basis of these, each of the $\mathbf{E}$ being orthogonal to a pair taken from the $\mathbf{P}_j$, for instance,

$$\mathbf{E}_1 = \langle\mathbf{P}_2 \times \mathbf{P}_3\rangle/\langle\mathbf{P}_1, \langle\mathbf{P}_2 \times \mathbf{P}_3\rangle\rangle. \tag{5.166}$$

This can be done if the $\mathbf{P}_j$ are linearly independent. $K_1(\omega)$ can be separated by projecting $\mathbf{K}(\omega)$ onto $\mathbf{E}_1$. The coefficient of the vector product in (5.166) ensures a correct measurement of the amplitude and phase of $K_1(\omega)$.

If only two signals interfere, then the signal-to-random-noise ratio is maximized by choosing as $\mathbf{P}_1$ the $\mathbf{V}$ in (5.129), which is the eigenvector of $\mathbf{C}$ corresponding to the least eigenvalue.

The practically important thing is that the vector $\mathbf{P}_1$ enters (5.166) through a scalar coefficient only. This means that one need not know the polarization vector of the signal itself in order to separate $K_1(\omega)$, apart from an unknown scalar factor, knowledge of noise polarization being sufficient. Conversely, if the $\mathbf{P}_j$ of the noise are unknown, a measurement of the polarization vector parameters is no guarantee for separation of the relevant scalar spectrum. Note that the frequency-independent constant factor does not affect the measurement of group times (and phases too, provided it is real-valued).

These considerations are directly relevant to the interference between Love and Rayleigh waves. We have seen that it is only the Love wave polarization vector that can be measured in the general case. This means that only the Rayleigh wave scalar spectrum is easy to find by polarizational separation (apart from a constant factor). The latter circumstance is familiar to practical seismologists who study Rayleigh waves as recorded on the vertical component.

5.5. SPATIAL ANALYSIS OF SURFACE WAVES

The presence of lateral inhomogeneities, as follows from theoretical treatments (see Chapters 2 and 3) and experimental data (Sikharulidze, 1978; Bungum and Capon, 1974; Capon, 1970; Levshin and Berteussen, 1979), gives rise to curved surface wave paths and to the appearance of delayed signals, in addition to the main surface wave train.

Such anomalies in spatial wave structure are best studied by means of spatially distributed seismic arrays (Lander and Levshin, 1982). Two principal techniques are available, space–time and frequency–time methods of analysis.

Space–time analysis (*STAN*). This is done in the time domain and may be regarded as a two-dimensional analogue of controlled directional reception of the second kind, since it is based on summation of traces with time delays determined by the parameters of the phase surface, that is, the following function of three variables is sought

$$P(t, \alpha, C) = \frac{1}{\Delta T} \int_{t-\Delta T/2}^{t+\Delta T/2} \left[\sum_j w_j u_j(s - \tau_j(\alpha, C)) \right]^2 \mathrm{d}s \qquad (5.167)$$

where P is the total signal power, t is the time; α and C are the parameters of the plane phase surface for which the summation is carried out, α being the arrival azimuth, C the phase velocity through the array area; ΔT is the width of the moving time window used for power estimation, j the trace number, W_j the weight of $u_j(t)$, the jth trace observation; τ_j is the time delay for jth trace. Since a surface wave is characterized by definite ranges of t, α, and C, such signals stand out in $P(t, \alpha, C)$ diagrams as local regions of increased values. If an array has a diameter (aperture) of about 100 km (exactly the NORSAR dimensions), then no time summation can provide adequate resolution in C. For this reason diagrams of $P_{C_0}(t, \alpha)$ cross-sections are the most convenient for analysis, C_0 being the mean phase velocity within the frequency range of interest.

An example illustrating the use of azimuth-time diagrams is given in Figure 5.11. Seismic waves are being analysed as recorded at NORSAR (to be more specific, by its long-period vertical seismometers), the theoretical arrival azimuth

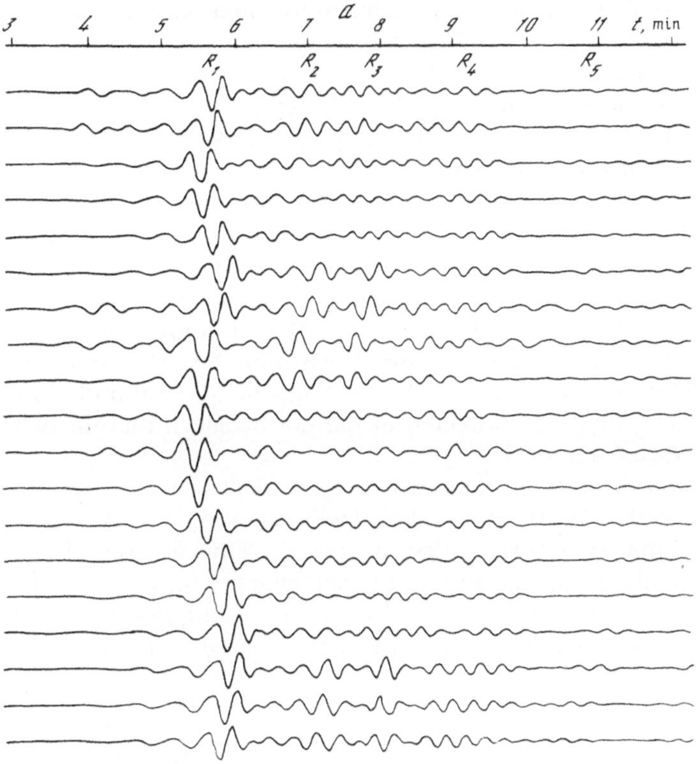

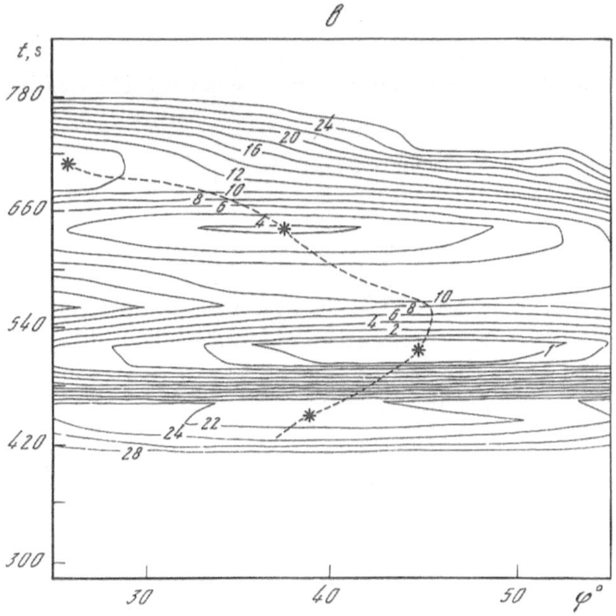

Fig. 5.11. Space-time analysis of records at NORSAR. (*a*) portions of record showing the fundamental mode of Rayleigh waves *R*1 and the coda (*R*2—*R*5 waves). The time *t* is measured from the beginning of the record; (*b*) azimuth-time diagram $P_c(t, \alpha)$, numerals at contours showing the decrease in power in dB relative to the maximum. Stars mark the maximum of the main and later arrivals, the dashed line showing the variation of the maximum in time.

being 41.7°. The phase velocity C was taken to equal 3.5 km/sec, the time window was $\Delta T = 40$ sec.

Apart from the main signal R_1 which has noticeable dispersion, the record contains several intensive later signals with delays of 1 to 3 min. Examination of the diagram shows that the main signal has an arrival azimuth appreciably larger that the theoretical one, and which varies smoothly with time. Later signals have local maxima of $P_C(t, \alpha)$ that are clearly identified in the diagram and have significantly different arrival azimuths compared with the main signal. An analysis of time delays and azimuth departures for the later arrivals identifies these arrivals as surface waves reflected at near-vertical crustal discontinuities: a boundary between two blocks in the Barents Sea, the circumference of the Swalbard Platform, the continental slope in the Norwegian Sea (Levshin and Berteussen, 1979). The same method was used by Berteussen, Levshin and Ratnikova (1982) to identify a wave that was reflected at the northeastern margin of the Dnieper—Donetsk basin (see Chapter 7, 7.1).

Space—frequency analysis. This is based on spectral transformations of records (Aki and Richards, 1980; Pisarenko, 1973; Capon, 1969). The cross-spectral power density matrix $\hat{f}_{ij}(\omega)$ is first estimated. This is done by separating the interval to be analysed on the record of each channel into M identical data blocks of N samples each after preliminary filtering by a bandpass filter that is symmetric about a selected frequency ω. Spectral density at ω is then calculated for each block:

$$S_{jn} = \frac{1}{\sqrt{N}} \sum_{m=1}^{N} u_j((m-1)\Delta t + (n-1)N\Delta t)\, e^{-i(m-1)\Delta t} \tag{5.168}$$

where Δt is the sampling interval, $n = 1, \ldots, M$.

Spectral power density is then estimated by between-blocks averaging:

$$\hat{f}_{jl}(\omega) = \frac{1}{M} \sum_{n=1}^{M} S_{jn}(\omega) S^*_{ln}(\omega). \tag{5.169}$$

Knowing $\hat{f}_{jl}$, one can estimate the space—frequency spectral power $P(\omega, \mathbf{k})$, where $\mathbf{k}$ is a vector wavenumber. The direction of $\mathbf{k}$ with respect to the cardinal points characterizes the arrival azimuth and $\omega/|\mathbf{k}|$, the phase velocity of the wave packet travelling through the array. The standard estimate $\tilde{P}(\omega, \mathbf{k})$ is taken to be

$$\hat{P}(\omega, \mathbf{k}) = \frac{1}{K^2} \sum_{j,l=1} w_j w_l^* f_{jl}(\omega)\, e^{-i(\mathbf{k}, (\mathbf{x}_j - \mathbf{x}_l))} \tag{5.170}$$

where the w_j are weights which control the window shape along the wavenumber axis (usually $w_j \equiv 1$); $\mathbf{x}_j$ is the radius vector which characterizes the location of the jth seismometer in the array.

$\hat{P}(\omega, \mathbf{k})$ is an unbiased consistent estimator of spectral density. This means, in particular, that for a plane harmonic wave $\exp[i(\omega_0 t - (\mathbf{k}_0, \mathbf{r}))]$, the estimate $\tilde{P}(\omega, \mathbf{k})$ approaches the delta function $\delta(\omega - \omega_0, \mathbf{k} - \mathbf{k}_0)$ as $M \rightarrow \infty$. In that case the power diagram (at a fixed ω) has a sharp peak at $\mathbf{k} = \mathbf{k}_0$.

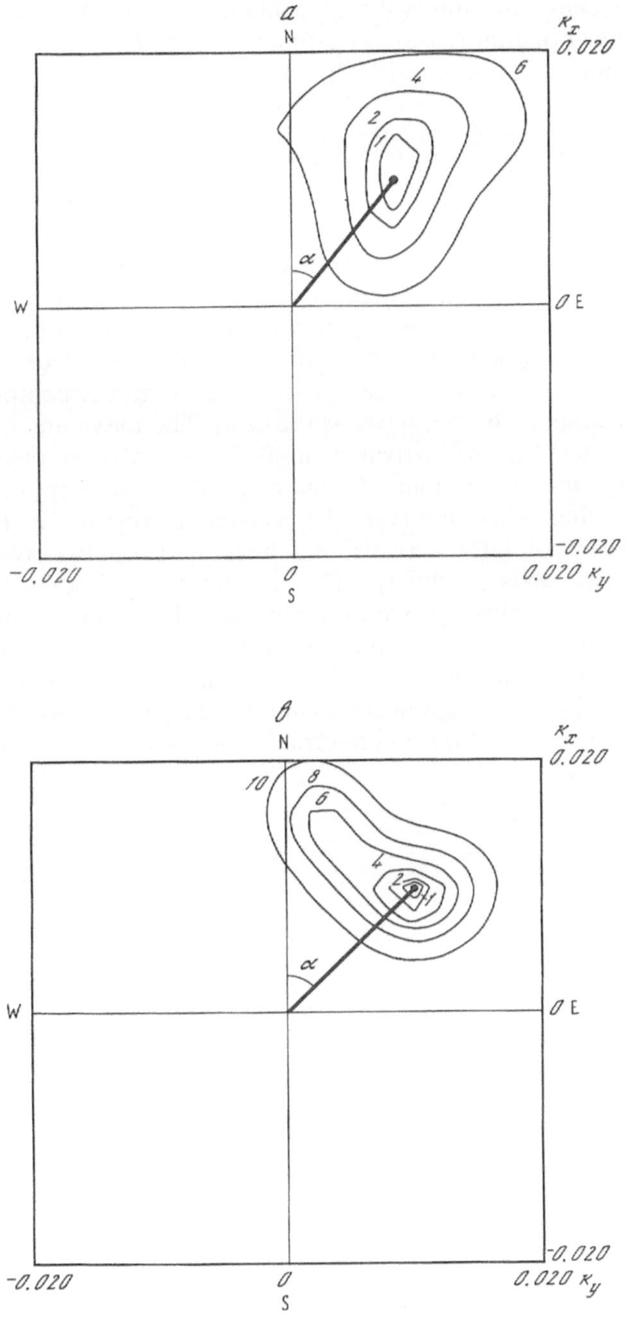

Fig. 5.12. Diagrams of space-time spectral power density $P'(\omega, \mathbf{k})$ for time intervals (*a*) 220—360 sec and (*b*) 360—500 sec. The solid circle indicates the maximum of P', numerals at contours give the decrease in P' relative to the maximum in dB.

However, because the interval that contains the signal is finite, $\hat{P}(\omega, \mathbf{k})$ has a lower resolution. For this reason an estimator $\tilde{P}'(\omega, \mathbf{k})$ has been proposed having enhanced resolution compared with $\hat{P}(\omega, \mathbf{k})$:

$$P'(\omega, \mathbf{k}) = \left\{ \sum_{j,l=1}^{K} \hat{f}_{jl}^{-1}(\omega) e^{i(\mathbf{k}, (\mathbf{x}_j - \mathbf{x}_l))} \right\}^{-1}. \tag{5.171}$$

A general theory of such estimators can be found in Pisarenko (1973).

As an exercise in the use of this estimator for surface wave analysis, we examine the $P'(\omega, \mathbf{k})$ diagrams shown in Figure 5.12. They were obtained from the records of Figure 5.11a and correspond to two nonoverlapping intervals of 140 sec at frequency 0.05 Hz. Each interval is divided into two blocks. A diagram is a contour map of P' on the coordinate plane of wavenumber components k_x, k_y (the x-axis points to the north, the y-axis to the east). The maximum of $P'(\omega, \mathbf{k})$ for the first interval occurs at the arrival azimuth $\alpha = 45°$ and phase velocity $C =$ 3.51 km/s. This interval contains the main signal whose amplitude dominates the record. As to the other interval, the maximum region of $P'(\omega, \mathbf{k})$ spreads northwest because of later arrivals that have significantly lower azimuths with similar values of phase velocity. The results from both analyses are fairly consistent; the time analysis provides a considerably better resolution in time than does the spectral one, whereas the spectral analysis has a better resolution in azimuth and especially in phase velocity. One can conclude that different methods of signal treatment provide significant complementary information.

Nolet and Panza (1974) discuss alternative approaches in the analysis of spatial array observations.

6. METHODS FOR QUANTITATIVE INTERPRETATION OF OBSERVATIONS

6.1. A REVIEW OF RECENT APPROACHES TO THE INVERSION OF SURFACE WAVE DATA

Measured characteristics of surface waves (phase and group velocities, amplitude and phase spectra) are used in seismological interpretation to solve various problems, such as the distribution of velocity and attenuation along the vertical and on the horizontal plane, and the study of earthquake mechanisms. Since this monograph is concerned with surface wave fields in laterally varying media, we shall mainly discuss those aspects of interpretation which are associated with lateral inhomogeneities. These include lateral variations proper in the upper earth structure, as well as the effects due to lateral inhomogeneity, and which should be taken into account in solving other problems (for example, for earthquake mechanism determinations).

Two problems of inversion based on the data of surface wave velocity dispersion are to be distinguished: one concerns the construction of so-called local dispersion curves from the data along paths for which surface wave velocities may significantly vary; the other consists in deriving a velocity structure based on a local dispersion curve thus obtained. The latter is an inverse problem proper formulated within the framework of laterally homogeneous media; as has been mentioned, we shall not discuss it in any detail, the more so because methods available for solving it can be found in the literature (Brune and Dorman, 1963; Brune, 1969; Knopoff, 1972; Levshin, 1973; Nolet, 1976; Aki and Richards, 1980).

The problem of mapping lateral crustal and upper mantle inhomogeneities based on the differences in surface wave dispersion along different paths has a long history. It was as long ago as the nineteen-fifties that significant differences between continental and oceanic crustal structure had been found based on surface wave group velocities over oceanic and continental paths. A simple procedure was used for this purpose which, by the way, may still be valuable. It consists in selecting paths that traverse tectonically homogeneous areas. This could naturally be done only for large regions with well-defined structural differences, and even then, only averaged results could be obtained for sufficiently long paths. The use of a differential method (based on data from two closely spaced stations) to obtain phase and group velocities yields dispersion curves and (based on these) vertical distributions of density and seismic wave velocities for smaller areas. This approach has the drawback of being applicable to data for those areas of the Earth where there are seismic stations. One should also note that the velocities derived by the differential method are subject to considerable uncertainty.

It is possible to use much more extensive observational material by incorporating data over 'mixed' paths, i.e., ones traversing regions of different structures. If the ratio between segments of a path corresponding to different regions varies with the path, then the velocities in each region can be estimated from travel times for such paths. This possibility is based on the assumption that the medium is laterally homogeneous within each region, the boundaries between them are known, while the effects of refraction and phase distortion at boundaries can be neglected. The simplest version of this technique is based on successive elimination of paths (Levshin and Berteussen, 1979; Berteussen, Levshin, and Ratnikova, 1982).

Suppose there are only two regions (1 and 2) and two paths that traverse them in different ways, of lengths l_1 and l_2. The portions of region 1 in paths 1 and 2 are q_1 and $q_2 (q_1 \neq q_2)$. If the travel times (group or phase ones) at a frequency ω along paths 1 and 2 are equal to t_1 and t_2, respectively, these readily yield the velocities v_1 and v_2 (group or phase ones, respectively) in each region at this frequency:

$$
\begin{aligned}
v_1 &= (q_1 - q_2) \Big/ \left[\frac{t_1}{l_1} (1 - q_2) - \frac{t_2}{l_2} (1 - q_1) \right], \\
v_2 &= (q_2 - q_1) \Big/ \left[\frac{t_1}{l_1} q_2 - \frac{t_2}{l_2} q_1 \right].
\end{aligned}
\tag{6.1}
$$

This 'path elimination' technique is readily extended to handle the case of many regions. If the number of paths is greater than the number of regions (blocks), one should make use of statistical procedures in order to incorporate the presence of random errors in the observations.

Suppose there are N blocks and J paths, and let Δ_{nj} be the distance along the jth path in the nth block; ε_j is the random error involved in the measurement of travel time t_j along the jth path. In that case

$$
t_j = \sum_{n=1}^{N} \Delta_{nj} v_n^{-1} + \varepsilon_j, \qquad 1 \leqslant j \leqslant J \tag{6.2}
$$

where v_n is the unknown velocity in the nth block. The method most frequently used to determine the v_n is a least-squares technique to minimize the sum of squares of the residuals between observed and theoretical travel times

$$
\sum_{j=1}^{J} \left(t_j - \sum_{n=1}^{N} \Delta_{nj} v_n^{-1} \right)^2,
$$

or some analogues of it that incorporate knowledge of random errors in the measurement of time.

This approach to the problem of group velocity estimation in tectonically different blocks (platforms, shields, oceans, tectonic regions) has been employed by Toksöz and Ben-Menahem (1963), Santo and Sato (1966), Dziewonski (1971), Wu (1972), Tatham (1975), Okal (1977), Mills (1978), Leveque (1980) as the main source of information on differences in block structure.

An essential drawback of this method some aspects of which are examined in Section 6.2 is that, firstly, there are no constraints on the behavior of the unknowns: for poorly sampled blocks these may fall outside the limits of physically possible velocity values; and secondly, there are no constraints on the absolute residuals: for some paths these may exceed those acceptable in real situations. The method of quadratic programming considered in Section 6.3 is free from these drawbacks. The two methods are compared for efficiency in Section 6.4 on a model problem that is close enough to the real one in complexity.

In some cases the task may be formulated in a different way: previous regional studies have provided velocities in the blocks, the configuration of the boundaries between these being unknown. This problem is also solved by minimizing the sum of residuals; however, the unknown parameters are now, not the velocities v_n, but parameters of block boundaries that control the distances Δ_{nj} the wave travels in the blocks. This formulation was used by Santo (1961).

A drawback of these approaches is the assumption of rigid constraints (either on boundary configurations or on the velocities within blocks), which does not allow all information to be extracted about the lateral inhomogeneities that surface wave data contain.

Sato and Santo (1969) were the first to try to determine lateral inhomogeneities over the surface of the Earth without appealing to extraneous information, on the basis of surface wave velocities measured along various intersecting paths. They suggested expanding the reciprocal of group velocity $U(\theta, \varphi)$ (for a fixed period $T = 2\pi/\omega$) into spherical harmonics

$$U^{-1}(\theta, \varphi) = \sum_{n=0}^{N} \sum_{m=0}^{n} P_n^m(\cos\theta)\,(A_{mn}\cos m\varphi + B_{mn}\sin m\varphi), \tag{6.3}$$

where θ, φ are spherical coordinates, $P_n(x)$ are associated Legendre polynomials. The expansion can be used to represent phase or group travel time along a given path as a linear function of the unknown coefficients A_{mn}, B_{mn}, provided the path $\mathfrak{L}_j$ is a great-circle arc:

$$t_j = \int_{\mathfrak{L}_j} \frac{\mathrm{d}s}{V} = \int_{\mathfrak{L}_j} \sum_{m,n} (A_{mn}\cos m\varphi + B_{mn}\sin m\varphi)\, P_n^m(\cos\theta)\,\mathrm{d}s$$

$$= \sum_{m,n} (A_{mn}\Phi^j_{mn} + B_{mn}\Psi^j_{mn}) \tag{6.4}$$

where V is phase or group velocity;

$$\Phi^j_{mn} = R_0 \int_{(\theta_{0j},\,\varphi_{0j})}^{(\theta_{1j},\,\varphi_{1j})} P_n^m(\cos\theta)\cos m\varphi\,\mathrm{d}\gamma$$

$$\Psi^j_{mn} = R_0 \int_{(\theta_{0j},\,\varphi_{0j})}^{(\theta_{1j},\,\varphi_{1j})} P_n^m(\cos\theta)\sin m\varphi\,\mathrm{d}\gamma \tag{6.5}$$

the integration in (6.5) being performed along the great circle ($d\gamma$ is an element of arc) from the source with coordinates θ_{0j}, $\varphi_{0\varphi}$ to the station (θ_{1j}, $\varphi_{1\varphi}$). The values of Φ_{mn}, Ψ_{mn} can thus be calculated for each path, so that the A_{mn}, B_{mn} are found by minimizing

$$\sum_j \left[t_j - \sum_{m,n} (A_{mn}\Phi^j_{mn} + B_{mn}\Psi^j_{mn}) \right]^2 .$$

Avetisyan and Yanovskaya (1973), Nakanishi and Anderson (1982, 1983, 1984), Anderson (1984) used this method to investigate the distribution of Rayleigh wave phase and group velocities for a range of periods, revealing changes in upper mantle lateral inhomogeneities with depth.

The method can be utilized both for the Earth as a whole and for individual areas. It seems evident that, when the data along paths over the entire Earth are available, the expansion (6.3) is natural and the only one that can be used. When the velocity distribution is investigated over a limited area, different representations may be used to fit it. For example, if the area is comparatively small, then a suitable choice of the pole of a spherical set of coordinates can place the area in a vicinity of the equator, the coordinates θ, φ would be close to rectangular ones, hence $V^{-1}(\theta, \varphi)$ can be fitted by an ordinary polynomial in θ, φ. The choice of the fitting function is generally governed both by the dimensions of the area under study and by our theoretical notions of what the behavior of velocity may be like.

Thus, although this method does not use the constraints that are imposed in the above approach on the number of blocks and their configuration, it too is not entirely free from the constraints imposed by our prior notions of the velocity distribution. These include, on the one hand, the kind of function to be fitted to $V^{-1}(\theta, \varphi)$ and, on the other, by the number of expansion terms we use. The latter is dictated by the amount of data available: in order to estimate the coefficients by the method of least squares, the number of unknown parameters must certainly be less than the number of paths. In this approach, however, the resolving power of data is not taken into account: usually, paths unevenly sample the area of study, so the velocity distribution can be derived in finer detail in places of greater path density that can be done in those where paths are sparse. At the same time, a solution obtained with the method yields the same degree of detail in velocity distribution everywhere over the study area. This is a drawback of the method.

The drawback can be removed, if the velocity distribution is estimated by using the generalized linear inversion method, also called the method of singular value decomposition (SVD) (Forsythe, Malcolm and Moler, 1977); it is becoming popular for dealing with various inverse problems in geophysics. Its application to the current problem is as follows.

The reciprocal of group or phase velocity $V^{-1}(\theta, \varphi)$ is represented as the sum of some mean value V_0^{-1} and a correction, $\delta(V^{-1}(\theta, \varphi)) = \kappa(\theta, \varphi)$. The study area is divided into cells: their number may be greater than the number of paths, but must be such as to provide a sufficient number of cells traversed by more than one path. The $\kappa(\theta, \varphi)$ in a cell is treated as a constant equal to $\kappa_n (n = 1, 2, \ldots, N)$. Apart from random errors, the travel time along the jth path ($j = 1, 2, \ldots, J$)

equals $\Sigma_{n=1}^{N} \Delta_{nj}(v_0^{-1} + \kappa_n)$, or

$$\delta t_j = \sum_{n=1}^{N} \Delta_{nj} \kappa_n \tag{6.6}$$

where δt_j is the difference between the observed travel time for the jth path and the time in a laterally homogeneous medium with velocity v_0. The system (6.6) can be solved when $J > N$ using the method of singular value decomposition. It should be borne in mind, however, that the solution can be obtained for those cells alone which are traversed by at least one path (with a large number of cells it may happen that some are traversed by no path at all). For cells with a solution this is minimal in norm, that is, has $\Sigma_{n=1}^{N} \kappa_n^2 = \min$. The method thus yields a solution that adequately reflects the real distribution of velocity for the case of weak lateral variation. If the number of paths is greater that the number of cells, the solution obtained by this method is identical with that by least squares, while the advantage of such a 'generalized' inversion is that one can also evaluate the resolution of the result. Some aspects of the method of singular value decomposition are discussed in Section 6.1; there is extensive literature on that subject, see, e.g., Jackson (1972), Wiggins (1972), Aki and Richards (1980), Yanovskaya and Porokhova (1983), Nolet (1987).

Another possible approach to the derivation of a velocity distribution $V(\theta, \varphi)$ based on travel times along various paths is the Backus—Gilbert method (Backus and Gilbert, 1968, 1970) which has been extended by Chou and Booker (1979), Yanovskaya (1980, 1982, 1984), Tanimoto (1985) to deal with the case of a 2-D and 3-D velocity distribution. This method does not rely on prior assumptions about the distribution of velocity such as division into blocks or the choice of the fitting function. It yields a smooth velocity distribution averaged over an area whose dimensions are determined by the number and relative location of paths in the data sample used. The degree of smoothness will obviously vary from point to point, the solution being smoother in those parts of the study area which are traversed by fewer paths compared to the parts intersected by many paths and where the paths are well separated. The method is described in more detail in 6.5 and some its applications in 6.6. The sections to follow, 6.7 to 6.9, are concerned with some specific problems arising in the interpretation of surface wave observations in laterally varying media: measurement of attenuation, apparent anisotropy, formulation of the inverse problem for seismic sources.

6.2. LEAST SQUARES AND SINGULAR VALUE DECOMPOSITION

Consider the following problem. Suppose there is some regionalization which divides the study area into N regions with J observations $t_j(\omega)$ ($j = 1, 2, \ldots, J$) of phase or group travel times, $t_j(\omega)$ being obtained from a surface wave record for the jth path in some frequency range. It is required to find dispersion curves $v_n^{-1}(\omega)$ for each of the N regions ($n = 1, 2, \ldots, N$). As a rule, $N < J$ or even $N \ll J$.

A solution of this problem for any frequency within the range involved can be sought by minimizing the sum of the squared differences between observed and theoretical travel times, that is, by minimizing, with respect to all the $v_n^{-1}(n = 1, 2, \ldots, N)$, the sum

$$\sum_{j=1}^{J} \left(t_j - \sum_{n=1}^{N} \frac{\Delta_{nj}}{v_n} \right)^2 . \tag{6.7}$$

Here Δ_{nj} is that part of the jth path occurring within the nth region; v_n is the desired regional velocity at frequency ω. Minimization of (6.7) with respect to the v_n^{-1} is a least-squares (LS) solution to the problem stated above.

The minimum of (6.7) is found from a set of normal equations (Rao, 1965). We define a matrix $\mathbf{A} = \{a_{nj}\}$ with $a_{nj} = \Delta_{nj}$ and vectors $\mathbf{x} = \{x_n\}$, $x_n = v_n^{-1}$; $\mathbf{t} = \{t_j\}$. The normal equations can be written in the form

$$\mathbf{A}^T \mathbf{A} \mathbf{x} = \mathbf{A}^T \mathbf{t} \tag{6.8}$$

where T denotes the transpose. The matix $\mathbf{H} = \mathbf{A}^T \mathbf{A}$ is symmetric, and all its eigenvalues satisfy $\lambda_1 \geqslant \lambda_2 \geqslant \lambda_3 \geqslant \ldots \geqslant \lambda_N \geqslant 0$. If $\mathbf{H}$ is positive definite ($\lambda_n > 0$), then (6.8) has a unique solution; if $\mathbf{H}$ is degenerate ($\lambda_n = 0$), there is an infinity of solutions (the solutions form a subspace of dimension one or greater). The use of the LS may give rise to a situation in which $\mathbf{H}$ is ill-conditioned, that is, is nearly degenerate. This can be the case, for instance, when either the travel time across the nth region, Δ_{nj}/V_n, is small for all paths compared with the times for the other regions (a region of small dimensions is involved) or the travel times for many paths are practically the same. In that case the result is random (affected by computer rounding errors) and is close to some solution from the subspace whose basis is formed by the eigenvectors with the least λ_n. To avoid this unpleasant situation, we should investigate the eigenvalues of $\mathbf{H}$, regarding $\mathbf{H}$ as degenerate if $\lambda_N / \Sigma_{n=1}^{N} \lambda_n$ is smaller than some threshold.

It should be noted (Aki and Richards, 1980; Forsythe, Malcolm and Moler, 1977) that in minimizing (6.7) it is preferable to use the method of singular value decomposition, if we wish to have a high computational accuracy and the simplicity in verifying that $\mathbf{H}$ is degenerate. Remembering that (6.7) is the norm $\|\mathbf{t} - \mathbf{A}\mathbf{x}\|$ of the vector $\mathbf{t} - \mathbf{A}\mathbf{x}$, consider the problem of minimizing $\|\mathbf{t} - \mathbf{A}\mathbf{x}\|$ with the constraint that $\|\mathbf{x}\|$ should be a minimum. The fact that the norm of $\mathbf{x}$ is minimal means that the unique solution with the minimal norm is taken in the space of solutions (Forsythe, Malcolm and Moler, 1977). The method of singular value decomposition yields the decomposition of $\mathbf{A}$:

$$\mathbf{A} = \mathbf{U} \Lambda \mathbf{V}^T$$

where $\mathbf{U}$ is a unitary ($\mathbf{U}^T \mathbf{U} = \mathbf{E}$) $J \times K$ matrix which consists of K columns forming a basis on the column space of $\mathbf{A}$; $\mathbf{V}$ is a unitary ($\mathbf{V}^T \mathbf{V} = \mathbf{E}$) $N \times K$ matrix consisting of K columns; it forms a basis on the row space of $\mathbf{A}$:

$$\Lambda = \left\| \begin{matrix} \sigma_1 \, 0 \ldots . \, 0 \\ 0 \\ 0 \ldots\ldots \sigma_k \end{matrix} \right\|$$

is a singular (diagonal) matrix; σ_i are singular numbers, $\sigma_1 \geqslant \sigma_2 \geqslant \ldots \geqslant \sigma_K >$

0; K is the rank of $\mathbf{A}$, $\sigma_n^2 = \lambda_n$. If the rank K equals $N (K = N)$, then the solution is unique, coincides with the LS solution, and is given by

$$\mathbf{x} = \mathbf{V} \Lambda^{-1} \mathbf{U}^T \mathbf{t}. \tag{6.9}$$

If $K < N$, the vector that minimizes $\|\mathbf{t} - \mathbf{Ax}\|$ and has the minimal norm is also given by (6.9), the matrix $\mathbf{V}\Lambda^{-1}\mathbf{U}^T = \mathbf{A}^+$ being the pseudo-inverse of $\mathbf{A}$.

The technique can be used to incorporate prior information on the velocities for a degenerate $\mathbf{A}^T\mathbf{A}(K < N)$. This is achieved by reduction of (6.6), that is, by seeking corrections $\delta(x_n)$ to the initial values $x_{0n} = v_{0n}^{-1}$ rather than the solution itself. The residual then becomes

$$\delta t_j = t_j - \sum_{n=1}^{N} \frac{\Delta_{nj}}{v_{0n}} = \sum_{n=1}^{N} \Delta_{nj}\,\delta(x_n) + \varepsilon_j.$$

The method of singular value decomposition minimizes

$$\sum_{j=1}^{J} \left(\delta t_j - \sum_{n=1}^{N} \Delta_{nj}\, \delta(x_n) \right)^2 \tag{6.10}$$

under the constraint

$$\sum_{n=1}^{N} |\Delta_{nj}\, \delta(x_n)|^2 = \min. \tag{6.11}$$

When $K < N$, (6.11) ensures that a solution x_n is found that is close enough to the initial value x_{0n}.

6.3. QUADRATIC PROGRAMMING METHODS

Section 6.2 was concerned with absolute minimization of the functional (6.7), that is, no constraints were imposed on the region of the x_n constituting the N-dimensional space. We are going to introduce three types of linear constraints on the admissible x_n-region:

$$d_n \leqslant x_n \leqslant c_n, \qquad \left| \sum_{n=1}^{N} \Delta_{nj} x_n - t_j \right| \leqslant \varepsilon_j, \quad 0 \leqslant x_n.$$

The subscript n takes on some values between one and N, j varying between one and J.

The first constraint assigns the upper and lower bounds of velocity for some regions; in particular, this may be $d_n = c_n$, meaning that one knows the velocity in the nth region: $v_n = d_n^{-1}$. The second constraint does not allow theoretical times to deviate by too much from the observations t_j; the third constraint $x_n \geqslant 0$ reflects the physical formulation of the problem — a velocity cannot be negative. Thus, the combined constraints on the x_n define a convex manifold on the space of variables which is a region of the admissible x_n. Consider a problem in quadratic programming: minimize

$$\sum_{j=1}^{J}\left(t_j - \sum_{n=1}^{N} \Delta_{nj} x_n\right)^2 \tag{6.12}$$

under the constraints

$$d_n \leqslant x_n \leqslant c_n \tag{6.13}$$

$$\left|\sum_{n=1}^{N} \Delta_{nj} x_n - t_j\right| \leqslant \varepsilon_j, \tag{6.14}$$

$$0 \leqslant x_n. \tag{6.15}$$

Note that the derivation of regional dispersion curves in this formulation incorporates the following factors: assigning prior information on the bounds of possible velocity values for some regions (6.13), the nature of errors in observed dispersion, statistical validity of the resulting solution. As will be shown below, a reasonable choice of constraints can significantly reduce dispersion curve uncertainties compared with the conventional least squares method and its modifications.

Basics of quadratic programming. Consider the following problem in quadratic programming. Minimize

$$\mathbf{c}^T\mathbf{x} + \tfrac{1}{2}\mathbf{x}^T\mathbf{H}\mathbf{x} \tag{6.16}$$

under the constraints

$$\mathbf{A}\mathbf{x} \leqslant \mathbf{b} \tag{6.17}$$

$$0 \leqslant \mathbf{x}_j. \tag{6.18}$$

Here $\mathbf{c}$, $\mathbf{x}$ are N-vectors; $\mathbf{b}$ is a p-vector; $\mathbf{A}$ is a $p \times N$ matrix; $\mathbf{H}$ is a symmetric $N \times N$ matrix. It is readily seen that the problem (6.12) through (6.15) reduces to the problem (6.16) through (6.18), the eigenvalues of $\mathbf{H}$ satisfying $\lambda_n \geqslant 0$.

Denoting the Lagrange vector multipliers for the constraints $\mathbf{Ax} \leqslant \mathbf{b}$ and $\mathbf{x}_j \geqslant 0$ by $\mathbf{u}$ and $\mathbf{v}$, respectively, and the vector of additional variables by $\mathbf{y}$, we can write the Kuhn—Tucker condition in the form (see Bazaraa and Shetty, 1979)

$$\begin{aligned} &\mathbf{Ax} + \mathbf{y} = \mathbf{b}, \qquad -\mathbf{Hx} - \mathbf{A}^T\mathbf{u} + \mathbf{v} = \mathbf{c} \\ &\mathbf{x}^T\mathbf{v} = 0, \quad \mathbf{u}^T\mathbf{y} = 0, \quad x_j, y_j, u_j, v_j \geqslant 0. \end{aligned} \tag{6.19}$$

Relations (6.19) are necessary and sufficient conditions for solving (6.16) through (6.18). When $\mathbf{H}$ is positive definite, the solution is unique, otherwise uniqueness is not guaranteed.

Denote

$$\mathbf{M} = \begin{Vmatrix} 0 & -\mathbf{A} \\ \mathbf{A}^T & \mathbf{H} \end{Vmatrix}, \quad \mathbf{q} = \begin{Vmatrix} \mathbf{b} \\ \mathbf{c} \end{Vmatrix}, \quad \mathbf{w} = \begin{Vmatrix} \mathbf{y} \\ \mathbf{v} \end{Vmatrix}, \quad \mathbf{z} = \begin{Vmatrix} \mathbf{u} \\ \mathbf{x} \end{Vmatrix}$$

and rewrite (6.19) as a problem of linear complements: find non-negative vectors

w and **z** satisfying the following relations

$$\mathbf{w} - \mathbf{Mz} = \mathbf{q}$$
$$w_j \geqslant 0, \quad z_j \geqslant 0 \quad j = 1, \ldots, p \tag{6.20}$$
$$w_j z_j = 0.$$

Here ($\mathbf{w}_j$, $\mathbf{z}_j$) is a pair of complementary variables, one of the two being basic (non-zero), while the other should be zero (non-basic variable). The procedure for finding the solution of (6.20) is described in Bazaraa and Shetty (1979). The procedure can terminate in two ways: either a solution has been found or no solution can be found. The latter is associated with the fact that (6.20) can generally have either an infinite number of solutions or no solution at all. A solution of (6.12) through (6.18) cannot be unbounded when **H** is positive definite, hence the 'no solution can be found' case only arises when the set defined by (6.17), (6.18) is empty. The algorithm converges to a solution after a finite number of steps when **H** is positive definite, and also when **H** is positive semidefinite with positive diagonal elements.

Linear programming. There is a problem in linear programming arising as a particular case of (6.16) through (6.18) for **H** equal to zero: minimize (or maximize) each of the unknown $x_n (n = 1, 2, \ldots, N)$ under the constraints (6.17), (6.18). The problem can be solved by the simplex method (Bulaevsky, Zvyagina and Yakovleva, 1977) and, less effectively, within the framework of the quadratic programming algorithm described above. The $2N$ values of $x_{n\,\min}$ and $x_{n\,\max}$ define, in a space of N dimensions, a minimal parallelepiped $V = \{\mathbf{x}; x_{n\,\min} < x_n < x_{n\,\max}\}$ that contains a polyhedral set of solutions. The solution of the quadratic programming problem is within or at the boundary of V; the LS solution is with a high probability within V in virtue of the statistical properties of measurement errors.

6.4. MODELLING EXPERIMENTAL DATA IN A BLOCK STRUCTURE

Modelling of experimental data with a view to testing various inversion procedures should be implemented so that the model problem has all the main features of a real inverse problem. These features in the construction of observed regional curves include complexities in the contours of regional blocks and great disparities in size, the location of seismic sources mostly at the periphery of the study area, limited numbers of paths, the presence of measurement errors. Like the solution of a real inverse problem, modelling begins with a selection of station—source paths and a division of the study area into regions with differences in deep structure. The last procedure is based on prior information relating to the relevant tectonic structure. Further, for each selected path assumed to coincide with a great-circle arc, one should find the lengths falling within all regions traversed by the path. This is done by the algorithm described by Barmin (1983) which can divide a geodetic line into portions. When dealing with a real problem, the next step is to estimate surface wave travel times for the paths. In a model problem, the times should also be calculated. To do this, each region is ascribed a velocity and density model, dispersion curves are calculated, and then total travel times based on these are to be found along the paths for selected periods. The possibility of random measurement error simulation should also be envisaged in the modelling.

Calculation of travel times. Suppose the study area has been divided into N regions, the hypothesis of lateral homogeneity being valid within each, and J epicenter—station paths have been defined, the paths lying wholly in the regions. A number of simplifying assumptions are made: waves propagate along great-circle arcs joining the epicenter and stations; all effects due to region boundaries are ignored. With these assumptions, the total travel time for a path equals the total travel time along the path segments within individual regions. In addition, paths that partly lie along region boundaries are excluded from consideration. If this is not done, the wave will travel at a velocity that is the greater for the two regions involved, possibly causing path distortions. Reflected waves should be excluded from consideration, because they do not travel along great circles. We conclude that the selected paths must be consistent with the regionalization, which can be done visually when the paths have been mapped.

Knowing the velocities v_n in the regions traversed by the path and the lengths of the jth path, Δ_{nj}, that lie in each of the regions, one can represent the travel time t_j for a path at a fixed period T using (6.2)

$$t_j(T) = \sum_{n=1}^{N} \frac{\Delta_{nj}}{v_n(T)} + \varepsilon_j(T).$$

The sum on the right-hand side is the theoretical travel time, $\varepsilon_j(T)$ is a random error with some distribution law which models the real error involved in travel time measurements, due both to the measurement technique and to region boundary effects. The origin of time measurement errors is fairly complex and is not sufficiently known at present. Some investigators assume the normal law of error. This is hardly true, since large errors in time determinations are usually associated with high levels of noise, but such records are as a rule excluded from consideration at the outset. For this reason the distribution density function of errors outside some bounded interval is zero.

With the above in mind, $\varepsilon_j(T)$ may be modelled as follows:

$$\varepsilon_j(T_k) = \sigma_j(T_k)\, \xi_k \tag{6.21}$$

where ξ_k is a random variable with mathematical expectation $M\xi_k = 0$ and the covariance matrix

$$\operatorname{cov}(\xi_m, \xi_n) = \begin{cases} 1 - \dfrac{|m-n|}{2s+1}, & |m-n| \leqslant s \\ 0 & |m-n| > s \end{cases}$$

Here, the ξ_k have been deliberately made dependent within a limited interval of period T, because they are certain to depend on the errors at adjacent frequencies. It is the task of FTAN to remove the purely random component. The above covariance matrix is obtained, when ξ_k is defined as

$$\xi_k = \sum_{l=k-s}^{k+s} \theta_l/(2s+1), \qquad \theta_l = \left(\sum_{j=l-p/2}^{l+p/2} \eta_j - \frac{p}{2} \right) \Big/ \sqrt{p/12} \qquad (6.22)$$

where the η_j are independent random variables distributed uniformly over [0, 1]. The value of $\sigma_j(T_k)$ characterizes the mean square deviation of ξ_k added to the true travel time. The errors as given by (6.22) are calculated for some set of discrete periods T_k.

Description of the Eurasian model. The structure of Eurasia was modelled down to depths of the order 400 km with a view to estimating the relative efficiencies of different techniques in use for constructing regional dispersion curves.

The entire surface of the Earth was divided into two parts, Eurasia and the remaining area. Sixty-one regions have been identified in Eurasia based on tectonic and geomorphic criteria. Some of the regions were afterwards combined, leaving 12 regional types in the final regionalization: (1) ancient shields, (2) oceanic areas, (3) Alpine belt, (4) island arcs, (5) aseismic platforms, (6) rifts, (7) sedimentary basins, (8) 'Tibet', (9) mountain areas, (10) 'China', (11) marginal seas, (12) inner seas.

The mountain areas include all areas having high mountains (above 3000 m altitude), apart from the Alpine belt. The oceanic areas comprise deep-sea portions adjacent to Eurasia. The Tibetan plateau and central China with its complex deep structure constitute the separate blocks 'Tibet' and 'China', given those names merely because one must call them something.

It goes without saying that this division (Figure 6.1) is rough and arbitrary, the regions identified being far from homogeneous when one recalls the real Earth. However, this summary treatment is quite acceptable in so far as the potential of the above methods is to be investigated.

Crustal and upper mantle velocity structures have been taken for the 12 regions on the basis of published evidence (Brune and Dorman, 1963; Brune, 1969; Knopoff, 1972; Der and Landisman, 1972; Dziewonski, 1971; Leveque, 1980; Mills, 1978; Souriau and Souriau, 1983). The standard Gutenberg model and Bullen A model have been used for all the regions in the depth interval 400 to 1600 km. The shear wave velocity structures are presented in Figure 6.2.

The algorithms described in Chapter 2 were applied to solve the forward problem: Rayleigh and Love wave phase and group velocities were calculated for each region in a spherical nongravitating Earth in the period range 10 to 300 sec at intervals of 10 sec. Figures 6.3 and 6.4 illustrate phase and group velocities of fundamental Rayleigh mode for the 12 regions.

Eleven real earthquake epicenters have been chosen in Eurasia and adjacent oceanic areas; 28 long-period seismograph stations have been selected from the existing seismic network (see Figure 6.1).

Out of 308 epicenter–station paths, 40 paths have been chosen traversing Eurasia (see Figure 6.1) along widely different directions. For each path, the lengths Δ_{nj} belonging to individual regions were determined and travel times $t_j(T_k)$ were calculated for the periods 50, 100, 150 and 200 sec. Ten realizations of

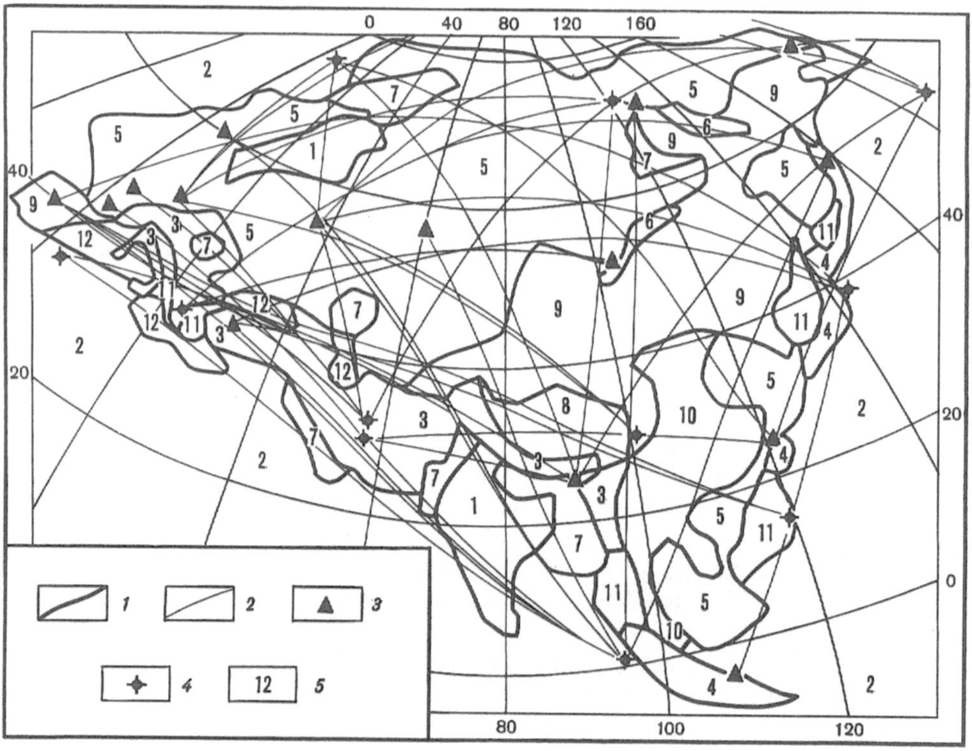

Fig. 6.1. Regionalization of Eurasia and the paths (1) region boundary, (2) path, (3) seismic station, (4) earthquake epicenter, (5) region number.

observed times t_j have been simulated that differ in the realizations of random errors.

Comparison of least squares and singular value decomposition. We are going to examine possibilities presented by LS and SVD methods for the model situation described above. An analysis of the matrix **A** by the SVD method has demonstrated that the matrix is positive definite, its singular numbers σ_i^2 lying within the range 21 to 2144. Regional dispersion curves of phase and group velocity were determined by the SVD and LS method for each of the ten variants of travel times. There is fairly satisfactory agreement between the two methods: the relative error for all the ten variants is

$$(v_{\mathrm{SVD}} - v_{\mathrm{LS}})/v_{\mathrm{LS}} \approx 10^{-5}.$$

The LS results for the first variant are shown in Table 6.I. A comparison with the original dispersion curves shows that dispersion is best determined for extended regions such as aseismic platforms, mountain areas, the Alpine belt, oceanic areas. Since the relative accuracy of raw data (travel times) fluctuates within 0 to 2 percent, depending on the simulated error and path length, the accuracy of dispersion for the regions is of the same order, about 1 percent. Determinations for island arcs and marginal seas are the least reliable, because these are traversed by few paths and the relevant path lengths (see Figure 6.1) are

TABLE 6.I.
Results of model experiments by least squares (LS) (variant 1).

region number	$T = 50$ sec			$T = 100$ sec		
	v^0 km/sec	v_{LS} km/sec	Uncertainty %	v^0 km/sec	v_{LS} km/sec	Uncertainty %
1	4.115	4.096	0.5	4.261	4.278	−0.4
2	4.022	3.997	0.6	4.132	4.115	0.4
3	3.955	3.994	0.04	4.156	4.179	−0.5
4	3.839	3.745	2.4	3.980	4.080	−2.5
5	4.006	4.004	0.04	4.099	4.101	−0.05
6	3.848	3.858	−0.3	4.027	4.273	−5.4
7	3.911	3.827	2.2	4.065	4.044	0.5
8	3.668	3.650	0.5	4.067	4.045	0.5
9	3.802	3.769	0.9	4.012	4.066	−1.3
10	3.887	3.955	−1.7	3.988	4.064	−1.9
11	3.990	3.971	0.5	4.108	3.852	6.2
12	3.997	4.180	−4.6	4.106	4.108	−0.005
	3.668*			3.980*		
	4.115**			4.261**		

region number	$T = 150$ sec			$T = 200$ sec		
	v^0 km/sec	v_{LS} km/sec	Uncertainty %	v^0 km/sec	v_{LS} km/sec	Uncertainty %
1	4.415	4.423	−0.2	4.653	4.619	0.7
2	4.299	4.326	−0.6	4.572	4.497	1.7
3	4.327	4.372	−1.0	4.595	4.651	−1.2
4	4.158	3.993	4.0	4.446	4.831	−8.7
5	4.282	4.309	−0.6	4.566	4.552	0.3
6	4.277	5.046	−18.0	4.578	4.501	1.7
7	4.276	4.275	0.01	4.575	4.547	0.6
8	4.274	4.238	0.8	4.541	4.510	0.6
9	4.265	4.241	0.6	4.587	4.591	−0.1
10	4.201	4.357	−3.7	4.519	4.421	2.2
11	4.281	3.807	11.1	4.557	4.593	−0.8
12	4.278	4.475	−4.6	4.553	4.422	2.9
	4.158*			4.446*		
	4.415**			4.653**		

* Least value v'_{min} for all the regions.
** Greatest value v'_{max}.

small (below 500 km). Errors for these areas strongly fluctuate between regions, reaching values as large as 20 percent. For example, they are 18 and 11.1 percent, respectively, for variant 1 at $T = 150$ sec. This suggests that in determining velocities for several regions greatly disparate in size, the worst results must be expected for small regions. Physically meaningless results may be obtained, like negative values of velocity. The errors for the other areas fluctuate between 2 and 5 percent.

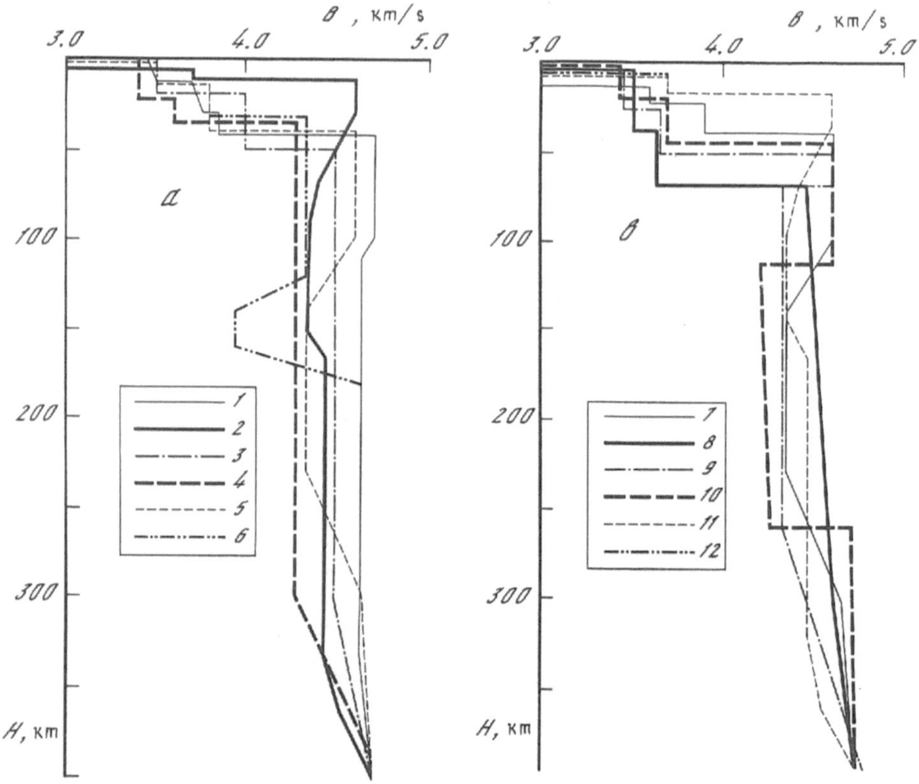

Fig. 6.2. Regional crustal and upper mantle velocity structures for regions 1—6 (a) and 7—12 (b). a: (1) ancient shields, (2) oceanic areas, (3) Alpine belt, (4) island arcs, (5) aseismic platforms, (6) intercontinental rifts; b: (7) sedimentary basins, (8) 'Tibet', (9) mountain areas, (10) 'China', (11) marginal seas, (12) inland seas.

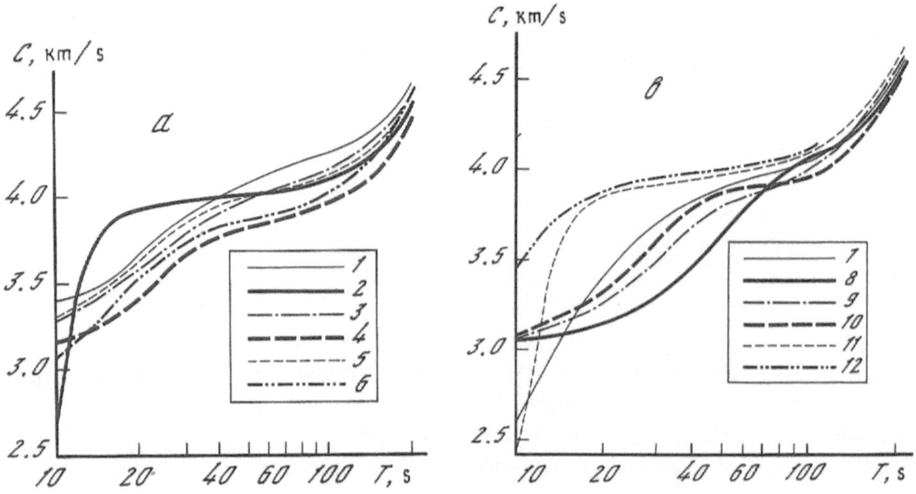

Fig. 6.3. Theoretical dispersion curves of phase velocity for the fundamental mode of Rayleigh waves in regions 1—6 (a) and 7—12 (b). Legend as for Figure 6.2.

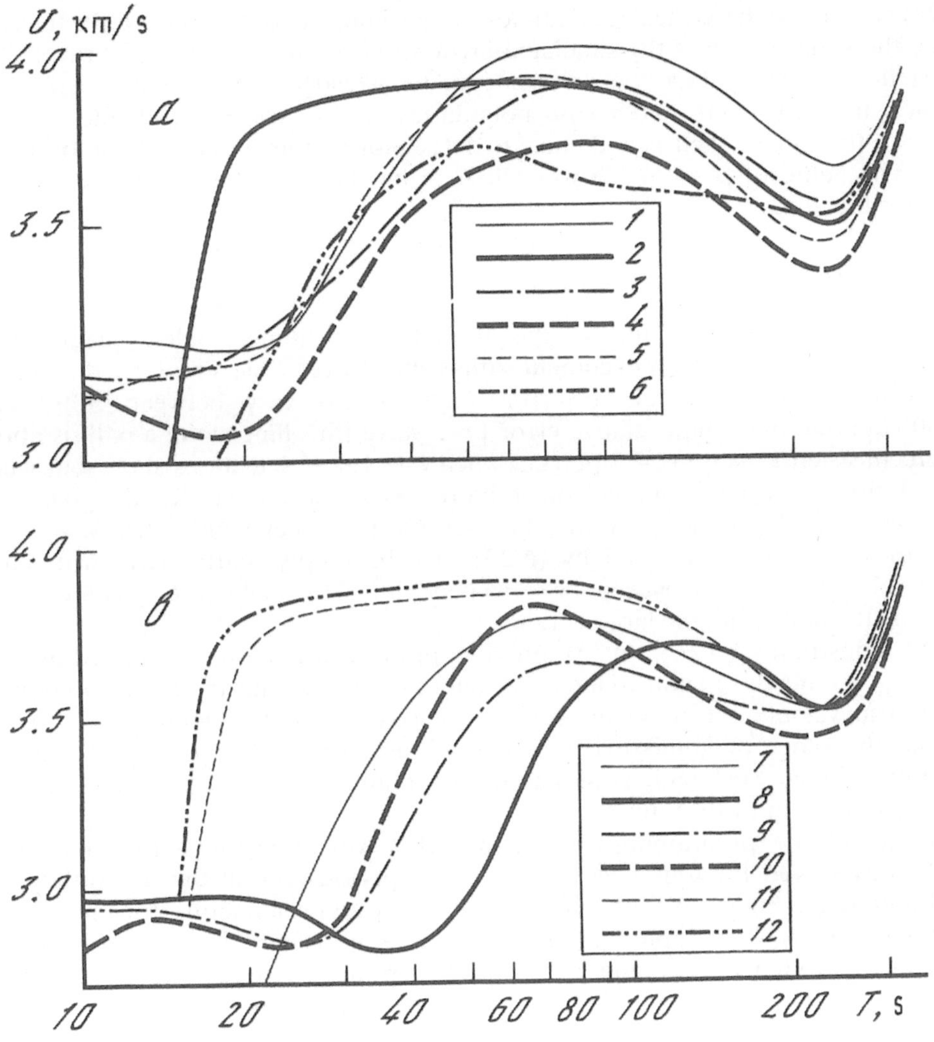

Fig. 6.4. Theoretical dispersion curves of group velocity for the fundamental mode of Rayleigh waves in regions 1—6 (a) and 7—12 (b). Legend as for Figure 6.2.

Comparison of quadratic programming and least squares. The method of quadratic programming was used to determine regional dispersion curves for the same ten variants of modelled data that were employed in the LS method; (6.12) was minimized under the constraints

$$\left| \sum_{n=1}^{12} \Delta_{nj} x_n - t_j \right| \leqslant k\sigma_j, \tag{6.23}$$

$$v_n = x_n^{-1} \leqslant v_{\max} = v'_{\max} + 1.5(v'_{\max} - v'_{\min}), \tag{6.24}$$

$$v_n = x_n^{-1} \geqslant v_{\min} = v'_{\min} - 1.5(v'_{\max} - v'_{\min}), \tag{6.25}$$

(6.23) constrains theoretical travel times to be within $k\sigma_j$ of the observed values, σ_j being the square root of the simulated error variance in (6.21) for the travel time along the jth path. Since we did not plan to investigate the origin of errors in relation to path length and region boundaries a wave traverses, all the σ_j were assumed the same for all js and equal to 10 sec for all the ten variants of modelled data. The value of p in (6.22) was taken to be six; this means that the t_js vary between the limits

$$t_j^0 - \frac{p\sigma_j}{2} \leqslant t_j \leqslant t_j^0 + \frac{p\sigma_j}{2}.$$

Here t_j^0 is the true travel time for the jth path; k is a dimensionless quantity that characterizes the ratio of maximum admissible travel time error to the mean-square deviation of simulated error. Path lengths vary between 3000 and 12 000 km, the maximum relative error for a wave travelling along a path is about 3 percent when $k = 2$ and 2.2 percent when $k = 1.5$. Calculations show that when $k > 3$, no gain results from the constraint (6.23), because the polyhedral set is too large and the LS solution is within that set. On the other hand, when k is small (one or less), the set defined by (6.23) may be empty, particularly with large observational errors. This can be obviated either by making k greater or by rejecting the paths with the largest errors.

The constraints (6.24), (6.25) provide prior information on the admissible velocities within the regions. Here, $v'_{\max}$ and $v'_{\min}$ are the maximum and minimum theoretical velocities (for all the regions) at some period T, while $v_{\max}$ and $v_{\min}$ expand the range of admissible velocities by a factor of 1.5 compared with the theoretical range. Such wide ranges allow one to investigate solutions that are fairly removed from our prior notions.

The quadratic programming problem (6.12) with constraints (6.23) through (6.25) was solved for four fixed values of the period and all the ten variants of modelled data. Figure 6.5 shows phase velocity determinations for the twelve Eurasian regions, the adopted parameters being $\sigma = 10$ sec, $k = 2$, $p = 6$. Phase velocities bounded by $v_{\min}$ and $v_{\max}$ at each period (horizontal bars) are plotted along the vertical axis, and the region number (N) along the horizontal. The results are not markedly different from the LS ones for sufficiently extended regions. The maximum range of velocities for regions 6, 11 is now twice as small, 5 to 7 percent. This effect is due to (6.24), (6.25); the constraint (6.23) does not affect the result much when $k = 2$. The results for the three remaining periods are similar, except that the scatter is diminished for regions 4 and 7. This is related to the fact that the admissible velocity range $v_{\min}$, $v_{\max}$ becomes narrower with the period T increasing.

We shall vary k, σ and the velocity bounds for individual regions using the same ten variants of modelled data. We define an estimate for the mean-square error of velocity determination in a region in order to provide an objective assessment of the accuracy in regional velocity determinations for the different methods, as follows:

$$\vartheta_n(v) = \sqrt{\frac{\sum_{m=1}^{10} (v_n^{(m)} v_n^0)}{10}} \bigg/ v_n \times 100\% \qquad (6.26)$$

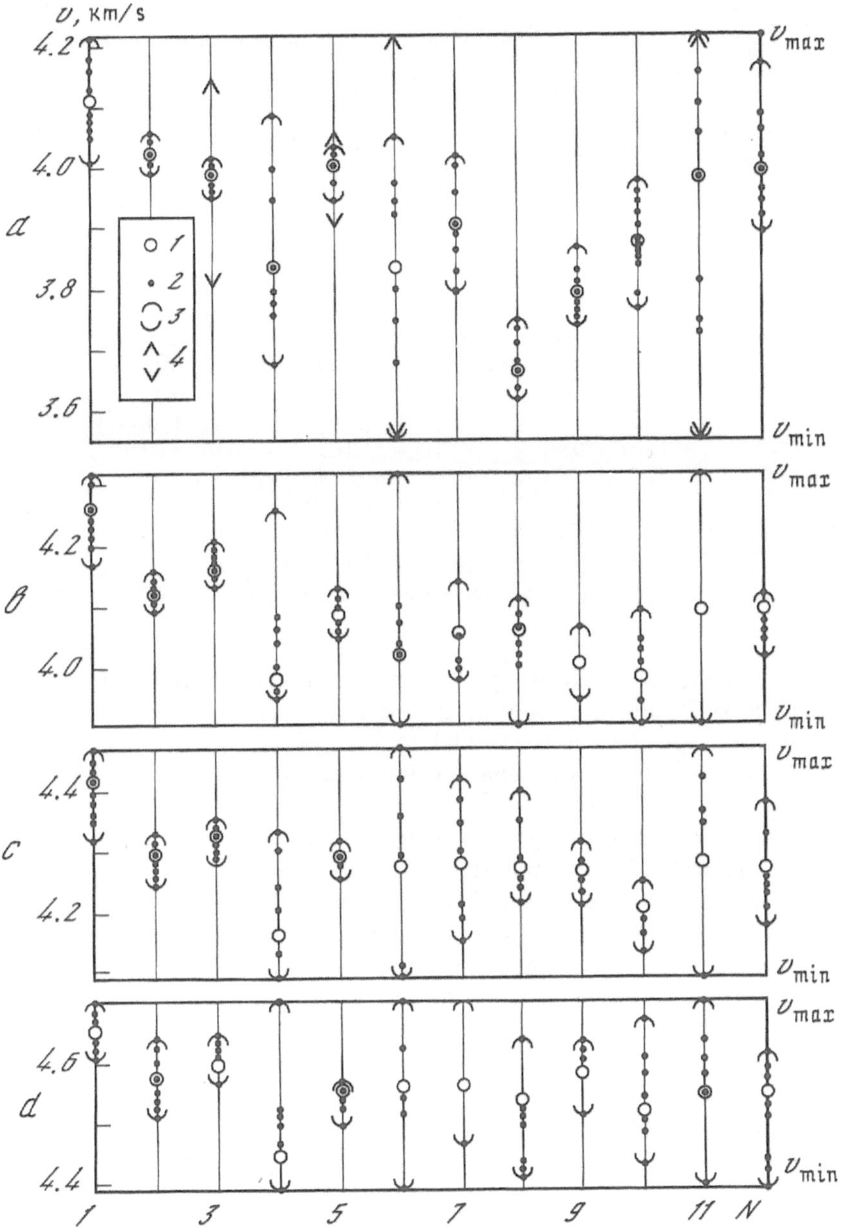

Fig. 6.5. Numerical experiments for phase velocity at periods (a) 50 sec, (b) 100 sec, (c) 150 sec, (d) 200 sec. (1) true velocity, (2) individual calculation variants, (3, 4) greatest and least velocity found by linear and quadratic programming, respectively.

where $m = 1, \ldots, 10$; $v_n^{(m)}$ is the velocity estimated for the nth region in the mth model; v_n^0 is the true velocity.

We have calculated (6.26) for LS and quadratic programming ($\sigma = 10$, $k = 2$, $p = 6$) as shown in Table 6.II (third and fourth column, respectively).

TABLE 6.II.
Relative rms uncertainty of regional phase velocities (in percent).

region number	Least squares uncertainty	Quadratic programming uncertainty		
		$k = 2$	$k = 2^*$	$k = 1,5^*$
1	1.55	1.57	1.2	0.91
2	0.52	0.53	0.43	0.33
3	0.67	0.51	0.42	0.30
4	3.12	2.97	0.26	0.26
5	0.36	0.38	0.46	0.34
6	5.32	4.47	0.26	0.26
7	1.87	1.90	0.26	0.26
8	1.17	1.90	1.38	0.57
9	0.90	0.82	0.87	0.69
10	2.01	1.74	1.37	1.35
11	7.59	5.5	0.25	0.25
12	2.06	2.08	0.25	0.20

* Additional constraints are introduced for regions 4, 6, 7, 11, 12 (see text).

One can see that quadratic programming improves the estimates for the following regions: island arcs from 3.12 to 2.97 percent, intracontinental rifts from 5.32 to 4.47 percent, 'China' from 2.01 to 1.74 percent, marginal seas from 7.59 to 5.5 percent. The changes in phase velocity accuracy for the other regions are very slight.

We now introduce additional information on velocity boundaries for regions 4, 6, 7, 11, 12. These are the smallest regions, their total area being 10 percent at most of the study area. We fix the following upper bound for them: $v_{n\,\max} = v_n^0 + 0.0025\, v_n^0$, with the lower as $v_{n\,\min} = v_n^0 - 0.0025\, v_n^0$, corresponding to the largest admissible range of velocities being a mere 0.25 percent. These are extremely rigid constraints, virtually equivalent to fixing exact values of velocity for every region. For all that (see Table 6.II, fourth column), the accuracy of velocity determination for the most extensive regions, such as aseismic platforms or mountain areas, remains unaffected. Velocities in the larger regions are thus well determined by any of the methods. Moderately sized regions typically have diminished errors: 1.55 to 1.2 percent for the older shields, 0.67 to 0.42 for the Alpine belt, 2.01 to 1.37 for 'China'.

Let us tighten the constraint (6.23) by fixing $k = 1.5$ (see Table 6.II, fifth column). The overall picture remains what it was before: the large regions yield the same error, a further reduction occurring in the error for moderately sized regions: 1.55 to 0.91 percent for the older shields (region 1), 0.52 to 0.23 for the oceanic regions (2), 0.67 to 0.30 for the Alpine belt (3), 1.17 to 0.57 for 'Tibet' (8).

One can conclude from the above that quadratic programming enhances the accuracy for moderately sized regions when extra constraints are imposed on smaller regions. An advantage of this method is that it enables the investigator to introduce extra information on the regions, while preserving the statistical validity of the solution.

Comparison between the results derived by linear and quadratic programming. The results from linear programming (period $T = 50$ sec with $k = 2$) for regions 3, 5, 6 and 11 are compared in Figure 6.5 with the results from quadratic programming. Linear programming yields somewhat worse results than quadratic programming, because in this case the statistical properties of the noise are not used in determining the travel times. In fact, the most unfavorable situation is being explored: we minimize (maximize) the velocity in a fixed region, the other variables being free, but remaining within the bounds of the polyhedral set defined by the system of constraints. A distinct pattern emerges when we analyse the results: the greater is the portion of the paths for a region (the greater is its size), the closer are the results of linear programming to those of quadratic programming (region 6). With decreasing region size the scatter in the velocities increases (twice as large for region 3), until $x_{n\,\min}$ and $x_{n\,\max}$ reach their bounds. This occurs for the small regions 6 and 11. Although the results of linear programming are much worse than those of quadratic programming, they can be quite helpful in evaluating the dimensions of the polyhedral set of solutions and can be used in choosing the values of σ and k.

6.5. BACKUS–GILBERT METHOD

The method based on the Backus—Gilbert formalism in application to estimation of the horizontal distribution of surface wave velocities was developed for the case of a plane surface, that is, for wave propagation in a half-space of x, y, z. No essential difficulty arises in extending the method to deal with a spherical model, but the computational scheme is much simpler for x, y than for θ, φ. For this reason it is convenient to change from the spherical set of coordinates to the rectangular and work with the inverse problem in the latter. The transformation from the spherical to the rectangular system is made using the following simple transformation, which keeps travel times between the corresponding points invariant:

$$x = R_0 \ln \tan (\theta/2), \tag{6.27}$$

$$y = R_0\, \varphi, \tag{6.28}$$

$$V(x, y) = v(\theta, \varphi)/\sin \theta, \tag{6.29}$$

where R_0 is the Earth's radius. The velocity distribution over the area is not strongly distorted by this transformation, provided $\sin \theta$ does not vary by too much within the surface area under consideration; this is the case when the area is close to the equator, where $\mathrm{d}/\mathrm{d}\theta\,(\sin \theta) \simeq 0$. Obviously, in considering a function defined on the surface of a spherical Earth, one must not necessarily make the pole of the spherical system coincide with the geographic pole, it is always possible to place it at a point such that the spherical area in question would be near the equator of the coordinate system chosen. The preliminary transformations of input data thus consist in a suitable choice of the pole of the spherical system and in performing the transformation (6.27) through (6.29). Travel times along rays on a spherical surface will, as has been mentioned, remain the same in the transformed system too. When solution has been found in the rectangular system, one can transform to the spherical system again using (6.27) through (6.29). For this reason subsequent treatment will be for a rectangular coordinate system, the surface area under consideration being assumed to be a rectangle: $0 < x < X$, $0 < y < Y$.

The travel time between (x_{0j}, y_{0j}) and (x_{1j}, y_{1j}) can be

$$t_j = \int_{(x_{0j}, y_{0j})}^{(x_{1j}, y_{1j})} V^{-1}(x, y)\, \mathrm{d}s \tag{6.30}$$

where the integration is in general performed along rays whose configuration is controlled by the distribution of phase velocity (see Chapter 2, 2.2). $V(x, y)$ may stand for group or phase velocity depending on what the observations are. The essential hypothesis underlying the method (as well as those discussed in the last section) is that the horizontal variations in structure are assumed to be small, so that the corresponding variations in phase velocity would not make straight lines strongly curved. The variations in group velocity may be rather large. We thus assume the integration in (6.30) to be performed along the straight line between the points (x_{0j}, y_{0j}), (x_{1j}, y_{1j}).

Take some middle value of slowness V^{-1} for the study area (V_0^{-1}, say) and compute the travel times along fixed paths, t_{0j}, with this value of velocity. The difference between the desired distribution $V^{-1}(x, y)$ and V_0^{-1} is denoted $\delta V^{-1}(x, y)$; then obviously, the difference between the observed and theoretical travel time under this assumption, that is, $\delta t_j = t_j - t_{0j}$, is expressed as

$$\delta t_j = \int_{(x_{0j}, y_{0j})}^{(x_{1j}, y_{1j})} \delta V^{-1}(x, y)\, \mathrm{d}s \tag{6.31}$$

the integration being along a straight line.

Define a dimensionless relative correction to the slowness $m(x, y) = V_0\, \delta\, V^{-1}(x, y)$. Then (6.31) can be rewritten in the form

$$\delta t = \int_{(x_{0j}, y_{0j})}^{(x_{1j}, y_{1j})} m(x, y) \frac{\mathrm{d}s}{V_0} \,. \tag{6.32}$$

(6.32) can also be written as

$$\iint_S G_j(x, y)\, m(x, y)\, \mathrm{d}x\, \mathrm{d}y = \delta t_j \tag{6.33}$$

where $G_j(x, y)$ is a data kernel which is singular on the jth path, equal to zero elsewhere and satisfies the relationship

$$\iint_S G_j(x, y)\, \mathrm{d}x\, \mathrm{d}y = t_{0j}. \tag{6.34}$$

S in (6.33) and (6.34) is an area of study whose shape is not restricted, apart from the requirement that it must contain all the paths.

If $m(x, y)$ is to be determined from a finite set of functionals (6.33), an additional constraint should be imposed on it. Since the data do not provide information on the behaviour of velocity at points not on any path the most natural assumption is a smoothness of the unknown function $m(x, y)$ (this is analogous to the assumptions used in spline interpolation of a one-dimensional-function). The simplest condition of smoothness can be written as

$$\iint_S (\nabla m)^2 \, dx \, dy = \min. \tag{6.35}$$

This is however insufficient to make the solution unique: there must be some requirement at the boundary of S. If the requirement in question is adopted in the form

$$\left.\frac{\partial m}{\partial n}\right|_{C_s} = 0 \tag{6.36}$$

the variational problem of finding a function $m(x, y)$ that would minimize (6.35) and satisfy (6.33), (6.36) reduces to Poisson's equation

$$\Delta m(\mathbf{r}) = -2\pi \sum_j \lambda_j G_j(\mathbf{r}) \tag{6.37}$$

under the boundary condition (6.36) (von Neumann's problem). Here Δ is the two-dimensional Laplace operator, $\mathbf{r} = (x, y)$, the λ_j are indeterminate Lagrange multipliers to be found from (6.33). The factor -2π is inserted into the right-hand side of (6.37) for convenience of subsequent calculations.

We choose Green's function of the Laplace operator as

$$g(\mathbf{r}, \mathbf{r}_0) = -\frac{1}{2\pi} \ln |\mathbf{r} - \mathbf{r}_0| + \Psi(\mathbf{r}, \mathbf{r}_0)$$

where $\Psi(\mathbf{r}, \mathbf{r}_0)$ is a function that is harmonic within S. Using Green's formula and (6.36), we get

$$m(\mathbf{r}) = 2\pi \sum_j \lambda_j \iint_S G_j(\mathbf{r}') \, g(\mathbf{r}', \mathbf{r}) \, d\mathbf{r}' - \int_{C_s} m(\mathbf{r}') \frac{\partial g(\mathbf{r}', \mathbf{r})}{\partial n} \, dl' \tag{6.38}$$

where C_s is the contour of S, $d\mathbf{r}' = dx' \, dy'$.

Choosing $\Psi(\mathbf{r}_1 \mathbf{r}_0)$ so as to make $\partial g/\partial n = 0$ at the contour, we obtain

$$m(\mathbf{r}) = 2\pi \sum_j \lambda_j \iint g(\mathbf{r}', \mathbf{r}) \, d\mathbf{r}'. \tag{6.39}$$

For simplicity of calculation we remove the contour of S away to infinity (it has been mentioned that the shape of S is under to restriction, apart from the requirement that it contain all the paths). We write (6.38) as

$$m(\mathbf{r}) = -\sum_j \lambda_j \int_S\int G_j(\mathbf{r}') \ln|\mathbf{r}' - \mathbf{r}|\, \mathrm{d}\mathbf{r}' + 2\pi \sum_j \lambda_j \int_S\int G_j(\mathbf{r}')\, \Psi(\mathbf{r}, \mathbf{r}')\, \mathrm{d}\mathbf{r}' -$$

$$- \int_{C_S} m(\mathbf{r}') \frac{\partial g(\mathbf{r}', \mathbf{r})}{\partial n'}\, \mathrm{d}l'. \tag{6.40}$$

For the last term on the right-hand side of (6.40) to be bounded at $\mathbf{r}$ finite it is necessary, firstly, that $\Psi(\mathbf{r}_1\, \mathbf{r}_0)$ should be a constant (for any harmonic function not a constant increases at infinity at least as rapidly as $|\mathbf{r}|$) and, secondly, that $m(\mathbf{r})$ should be bounded at infinity. In that case for large $\rho = |\mathbf{r} - \mathbf{r}_0|$, where $\mathbf{r}_0$ is a point on a path, $m(\mathbf{r})$ behaves as $-\ln \rho\, \Sigma_j\, \lambda_j t_{0j} + A(\mathbf{r})$, where $A(\mathbf{r})$ is bounded. The requirement of $m(\mathbf{r})$ to be bounded at infinity yields

$$\sum_j \lambda_j t_{0j} = 0. \tag{6.41}$$

Since $m(\mathbf{r})$ is a harmonic function at infinity, $A(\mathbf{r})$ must be a constant.

We can thus write $m(\mathbf{r})$ as

$$m(\mathbf{r}) = -\sum_j \lambda_j \int_{L_{0j}} \ln|\mathbf{r} - \mathbf{r}_j| \frac{\mathrm{d}l_j}{V_0} + C \tag{6.42}$$

the λ_j obeying (6.41).

The λ_j and C are found by using (6.33). For brevity we denote $\Lambda = (\lambda_1, \lambda_2, \ldots, \lambda_n)$, $t_0 = (t_{01}, t_{02}, \ldots, t_{0n})$, $\delta\mathbf{t} = (\delta t_1, \delta t_2, \ldots, \delta t_n)$. Substituting (6.42) into (6.33), we obtain

$$\mathbf{S}\Lambda + C\mathbf{t}_0 = \delta\mathbf{t}_0 \tag{6.43}$$

where

$$S_{ij} = -\int_{L_{0i}}\int_{L_{0j}} \ln|\mathbf{r}_i - \mathbf{r}_j| \frac{\mathrm{d}l_i}{V_0}\frac{\mathrm{d}l_j}{V_0}. \tag{6.44}$$

Condition (6.41) can be rewritten as

$$\Lambda^T\mathbf{t}_0 = 0. \tag{6.45}$$

We then find from (6.44), (6.45):

$$C = \frac{\mathbf{t}_0^T\mathbf{S}^{-1}\,\delta\mathbf{t}}{\mathbf{t}_0^T\mathbf{S}^{-1}\,\mathbf{t}_0}, \qquad \Lambda = \mathbf{S}^{-1}\,\delta\mathbf{t} - \frac{\mathbf{t}_0^T\mathbf{S}\,\delta\mathbf{t}}{\mathbf{t}_0^T\mathbf{S}^{-1}\,\mathbf{t}_0}\mathbf{S}^{-1}\,\mathbf{t}_0\,. \tag{6.46}$$

Denoting

$$K_j(\mathbf{r}) = -\int_{L_{0j}} \ln|\mathbf{r} - \mathbf{r}_j| \frac{\mathrm{d}l_j}{V_0} \tag{6.47}$$

and treating the $K_j(\mathbf{r})$ as the components of $\mathbf{K}(\mathbf{r})$, we can write the desired function $m(\mathbf{r})$ in the form

$$m(\mathbf{r}) = \Lambda^T \mathbf{K} + C = \mathbf{K}^T \mathbf{S}^{-1}\, \delta\mathbf{t} + \frac{1 - \mathbf{t}_0^T \mathbf{S}^{-1} \mathbf{K}}{\mathbf{t}_0^T \mathbf{S}^{-1} \mathbf{t}_0} \mathbf{t}_0^T \mathbf{S}^{-1}\, \delta\mathbf{t} \tag{6.48}$$

or

$$m(\mathbf{r}) = \mathbf{a}^T\, \delta\mathbf{t} \tag{6.49}$$

where

$$\mathbf{a}(\mathbf{r}) = \mathbf{S}^{-1}\, \mathbf{K} + \frac{1 - \mathbf{t}_0^T \mathbf{S}^{-1} \mathbf{K}}{\mathbf{t}_0^T \mathbf{S}^{-1} \mathbf{t}_0} \mathbf{S}^{-1}\, \mathbf{t}_0. \tag{6.50}$$

It is readily shown that an exactly identical solution can be derived from entirely different considerations, namely, by starting to construct a solution for $m(\mathbf{r})$ as a smooth one

$$\langle m(\mathbf{r}) \rangle = \int\!\!\int_s A(\mathbf{r}', \mathbf{r})\, m(\mathbf{r}')\, \mathrm{d}\mathbf{r}' \tag{6.51}$$

where $m(\mathbf{r}')$ is any of the solutions that satisfy (6.33), and by making the averaging kernel obey the δ-ness criterion

$$s(\mathbf{r}) = \int\!\!\int_s |\mathbf{E}(\mathbf{r}', \mathbf{r}) - \boldsymbol{\varepsilon}(\mathbf{r}', \mathbf{r})|^2\, \mathrm{d}\mathbf{r}' = \min \tag{6.52}$$

where

$$\operatorname{div} \mathbf{E}(\mathbf{r}', \mathbf{r}) = A(\mathbf{r}', \mathbf{r}),\ \boldsymbol{\varepsilon} = \frac{1}{2\pi} \nabla' \ln |\mathbf{r}' - \mathbf{r}| \tag{6.53}$$

as well as the normalization requirement

$$\int\!\!\int_s A(\mathbf{r}', \mathbf{r})\, \mathrm{d}\mathbf{r}' = 1. \tag{6.54}$$

The criterion (6.52), (6.54) is a generalization of Backus and Gilbert's (1969) δ-ness criterion for the averaging kernel in the one-dimensional case, namely,

$$\begin{aligned} s(r_0) &= \int_0^R \left[\varepsilon(r - r_0) - \int_0^r A(r', r_0)\, \mathrm{d}r' \right]^2 \mathrm{d}r = \min \\ &\int_0^R A(r', r)\, \mathrm{d}r' = 1 \end{aligned} \tag{6.55}$$

expressing the closeness of $A(r', r_0)$ to the δ-function in the sense of the integral of the kernel being close to the Heaviside function $\varepsilon(r - r_0)$. We just wish to note that, in contrast to (6.55), the value of $s(\mathbf{r})$ as given by (6.52) becomes unbounded at infinity. As we shall see below, however, this singularity is caused by the contribution due to the constant term, and this is of no importance in finding a minimum.

To demonstrate that the solution (6.48) and the smooth solution (6.51) under the requirements (6.52), (6.54) are equivalent, we use the relations

$$A(\mathbf{r}', \mathbf{r}) = \sum_j a_j(\mathbf{r})\, G_j(\mathbf{r}'), \qquad \mathbf{E}(\mathbf{r}', \mathbf{r}) = -\frac{1}{2\pi} \sum_j a_j(\mathbf{r}) \nabla K_j(\mathbf{r}) \tag{6.56}$$

The expression for $s(\mathbf{r})$ can then be written as follows:

$$s(\mathbf{r}) = \sum_{ij} a_i a_j P_{ij} - 2 \sum_j a_j Q_j(\mathbf{r}) + \int_s\int |\varepsilon(\mathbf{r}', \mathbf{r})|^2 \, d\mathbf{r}' \tag{6.57}$$

where

$$P_{ij} = \frac{1}{4\pi^2} \int_s\int (\nabla K_i, \nabla K_j)\, d\mathbf{r}$$

$$Q_j(\mathbf{r}) = -\frac{1}{4\pi^2} \iint (\nabla K_j, \nabla \ln |\mathbf{r}' - \mathbf{r}|)\, d\mathbf{r}'. \tag{6.58}$$

Condition (6.54) is combined with (6.56) to give

$$\mathbf{a}^T \mathbf{t}_0 = 1 \tag{6.59}$$

where $\mathbf{a} = (a_1, a_2, \ldots, a_n)$.

Minimization of (6.57) under the condition (6.59) yields the set of equations

$$2\mathbf{Pa} - 2\mathbf{Q} + \lambda \mathbf{t}_0 = 0 \tag{6.60}$$

where the elements of $\mathbf{P}$ and $\mathbf{Q}(\mathbf{r})$ are given by (6.58), λ being a Lagrange multiplier. We transform (6.58) by using the Green formula

$$P_{ij} = \frac{1}{2\pi} \int_s\int K_j G_i \, d\mathbf{r} + \frac{1}{4\pi^2} \int_{C_s} \frac{\partial K_i}{\partial n} K_j \, dl$$

$$Q_j(\mathbf{r}) = \frac{1}{2\pi} K_j(\mathbf{r}) - \frac{1}{4\pi^2} \int_{C_s} \frac{\partial \ln |\mathbf{r}' - \mathbf{r}|}{\partial n'} K_j(\mathbf{r}')\, dl'. \tag{6.61}$$

The first of these expressions can be rewritten using (6.47):

$$P_{ij} = \frac{1}{2\pi} S_{ij} + D_{ij} \tag{6.62}$$

where

$$D_{ij} = -\frac{1}{4\pi^2} \int_{C_s} \frac{\partial K_i}{\partial n} K_j \, dl.$$

Denoting

$$F_j = -\frac{1}{4\pi^2} \int_{C_s} \frac{\partial \ln |\mathbf{r}' - \mathbf{r}|}{\partial n} K_j(\mathbf{r}')\, dl'$$

and treating the F_j as the components of $\mathbf{F}$, we can write (6.60) as

$$\frac{1}{2\pi} \mathbf{Sa} + \mathbf{Da} + \frac{\lambda}{2} \mathbf{t}_0 = \frac{1}{2\pi} \mathbf{K} + \mathbf{F}. \tag{6.63}$$

If the contour of S is made to tend to infinity, then it can be shown (Ditmar and Yanovskaya, 1987) by using (6.59) that, apart from quantities of order $1/\rho$, where ρ is the distance from the paths to points on the contour C_s, that the following relation is true

$$\mathbf{Da} = \mathbf{F}. \tag{6.64}$$

Thus, using (6.59) we obtain from (6.63):

$$\mathbf{a} = \mathbf{S}^{-1}\mathbf{K} + \frac{1 - \mathbf{K}^T \mathbf{S}^{-1} \mathbf{t}_0}{\mathbf{t}_0^T \mathbf{S}^{-1} \mathbf{t}_0} \mathbf{S}^{-1} \mathbf{t}_0. \tag{6.65}$$

If the expression (6.56) for the averaging kernel $A(\mathbf{r}', \mathbf{r})$ is substituted in (6.51), the solution $\langle m(\mathbf{r}) \rangle$ is expressed in terms of $\mathbf{a}$ in the form of a linear combination (6.48), the $\mathbf{a}$ being exactly identical with (6.50). We thus see that the solution based on the δ-ness criterion for the averaging kernel in the form (6.52), (6.54) is indeed exactly identical with that derived by making the desired function obey (6.35), (6.36).

We note that the solution may be based on different prior constraints. These are chosen from a specific idea of the solution, and also from the type of the data.

If we know, for example, that $m = 0$ at the boundary of a finite region, this condition will naturally replace (6.36), so that the variational problem (6.35), (6.33) will again reduce to Poisson's equation (6.37). Its solution will have the form (6.39), but the Green function $g(\mathbf{r}', \mathbf{r})$ must then be chosen so as to vanish at the region boundary.

One could demand greater smoothness for $m(\mathbf{r})$, for example, assuming it to satisfy

$$\int_s\!\!\int (\nabla^2 m)^2 = \min \tag{6.66}$$

or

$$\int_s\!\!\int \| \nabla\nabla\, m \|^2 = \min \tag{6.67}$$

where $\nabla\nabla\, m$ is the tensor of the second derivatives of $m(\mathbf{r})$.

The variational problem (6.66), (6.33) under the boundary conditions

$$\left.\frac{\partial m}{\partial n}\right|_{C_s} = 0, \qquad \left.\frac{\partial \Delta m}{\partial n}\right|_{C_s} = 0$$

is readily shown to reduce to the equation

$$\Delta^2 m = -2\pi \sum_j \lambda_j G_j(\mathbf{r}) \tag{6.68}$$

the problem (6.67), (6.33) being reduced to the same equation, if the boundary condition is taken in the form

$$\left.\frac{\partial \nabla m}{\partial n}\right|_{C_s} = 0.$$

The solution is constructed as in the preceding, except that $g(\mathbf{r}', \mathbf{r})$ is the Green function of the biharmonic equation. The smoothness conditions (6.66) or (6.67) are convenient to use for constructing a solution when the data set includes velocities at individual points, in addition to travel times over paths.

We turn again to (6.48). Treating this solution as a smoothed one in the form (6.61), we can evaluate the resolution of the data by defining the effective radius of a smoothing region. To evaluate this, we consider an averaging kernel $A(\mathbf{r}', \mathbf{r})$ that would be a constant equal to $1/\pi R^2$ within a circle of radius R centered at $\mathbf{r}$ and vanish outside the circle. The radius R can be regarded as an estimation of linear size of the smoothing region, provided the value of $s(\mathbf{r})$ corresponding to the kernel given by (6.52) would equal that corresponding to the solution obtained. For such a kernel

$$2\pi \mathbf{E}(\mathbf{r}', \mathbf{r}) = \begin{cases} (\mathbf{r}' - \mathbf{r})/\rho^2 & \rho = |\mathbf{r}' - \mathbf{r}| \geqslant R \\ (\mathbf{r}' - \mathbf{r})/R^2 & \rho < R \end{cases} \tag{6.69}$$

and, as

$$2\pi \boldsymbol{\varepsilon}(\mathbf{r}', \mathbf{r}) = (\mathbf{r}' - \mathbf{r})/\rho^2$$

we have

$$s(\mathbf{r}) = \frac{1}{4\pi^2} \int_0^{2\pi} \mathrm{d}\varphi \int_0^R \left(\frac{1}{R^2} - \frac{1}{\rho^2} \right) \rho^3 \, \mathrm{d}\rho. \tag{6.70}$$

As has been mentioned, the value of $s(\mathbf{r})$ given by (6.52) contains a singularity, so that, if (6.70) and (6.57) are to be compared, the singularity should be isolated in both expressions. This can be done if, in integrating over S, we isolate a small vicinity of $\mathbf{r}$ of radius ε. (6.70) then becomes

$$2\pi s(\mathbf{r}) = -3/4 + \ln R - \ln \varepsilon \tag{6.71}$$

while (6.57) combined with (6.61), (6.62), (6.64) is expressed as

$$2\pi s(\mathbf{r}) = -\mathbf{a}^T \mathbf{S} \mathbf{a} + 2\mathbf{K}^T \mathbf{a} - \ln \varepsilon + O(\varepsilon). \tag{6.72}$$

Comparing (6.71) and (6.72), we evaluate the effective radius of the smoothing region

$$R = \exp(3/4 - \mathbf{a}^T \mathbf{S} \mathbf{a} + 2\mathbf{K}^T \mathbf{a}) \tag{6.73}$$

which extends the Backus and Gilbert's (1970) one-dimensional 'spread' to the case of two dimensions.

So far, we constructed our solution on the assumption of error-free data. We are going to show that, if the data are subject to errors, the solution can be constructed in the same way (i.e., by (6.48)), but **S** must be replaced by $\mathbf{S} + \alpha\mathbf{I}$, where α is a stabilizing parameter determined by the ratio of the error variance to the prior variance of the desired function $m(\mathbf{r})$. We restrict ourselves to the case of independent observations with equal variances.

We must then solve, not the variational problem (6.35) under conditions (6.34), but minimize the functional

$$\alpha \int_s\int |\nabla m|^2 \,\mathrm{d}x\,\mathrm{d}y + \sum_j \left[\int_s\int G_j m \,\mathrm{d}x\,\mathrm{d}y - \delta t_j\right]^2 . \tag{6.74}$$

Minimization of (6.74), $m(\mathbf{r})$ being constrained to be bounded at infinity, yields (6.37) in which the λ_j are now functionals of the unknown function sought:

$$\lambda_j = \alpha^{-1}\left[\int\int G_j m \,\mathrm{d}x\,\mathrm{d}y - \delta t_j\right]^2 . \tag{6.75}$$

Strictly speaking, this is no longer Poisson's equation, so the λ_j cannot be determined using the approach outlined above. If, however, we still try to determine the coefficients by treating them to be independent of m and expressing $m(\mathbf{r})$ in terms of these through (6.42), the resulting coefficients will satisfy (6.75). We are going to show this.

If the error variance is σ^2, the residuals must satisfy the requirement

$$\sum_j \left[\int\int G_j m \,\mathrm{d}x\,\mathrm{d}y - \delta t_j\right]^2 = n\sigma^2 = \delta^2 \tag{6.76}$$

or

$$\Lambda^T\Lambda = \alpha^{-2}\,\delta . \tag{6.77}$$

We assume the solution to be of the form (6.42), as before. Substituting it into (6.76), we obtain

$$(\mathbf{S}\Lambda + C\mathbf{t}_0 - \delta\mathbf{t})^T(\mathbf{S}\Lambda + C\mathbf{t}_0 - \delta\mathbf{t}) = \delta^2 . \tag{6.78}$$

If we now put $\tilde{\mathbf{S}} = \mathbf{S} + \alpha\mathbf{I}$ and replace **S** by $\tilde{\mathbf{S}}$ in (6.78), then with (6.77) taken into account, some simple manipulations yield the following equation:

$$(\tilde{\mathbf{S}}\Lambda + C\mathbf{t}_0 - \delta\mathbf{t})^T(\tilde{\mathbf{S}}\Lambda - 2\alpha\Lambda + C\mathbf{t}_0 - \delta\mathbf{t}) = 0 . \tag{6.79}$$

It turns out that the solution corresponding to

$$\tilde{\mathbf{S}}\Lambda + C\mathbf{t}_0 - \delta\mathbf{t} = 0 \tag{6.80}$$

leads to values of Λ that satisfy relations (6.75). Now since (6.80) is identical with (6.43) in form, except that $\tilde{\mathbf{S}} = \mathbf{S} + \alpha\mathbf{I}$ stands for **S**, it is obvious that the desired solution for $m(\mathbf{r})$ must be of the form (6.48) with **S** replaced by $\mathbf{S} + \lambda\mathbf{I}$.

Model example. The possibilities offerred by the method may be illustrated on the following model examples (Ditmar and Yanovskaya, 1987). Two velocity distributions $v(x, y)$ were taken having different correlation lengths (Figure 6.7 a,

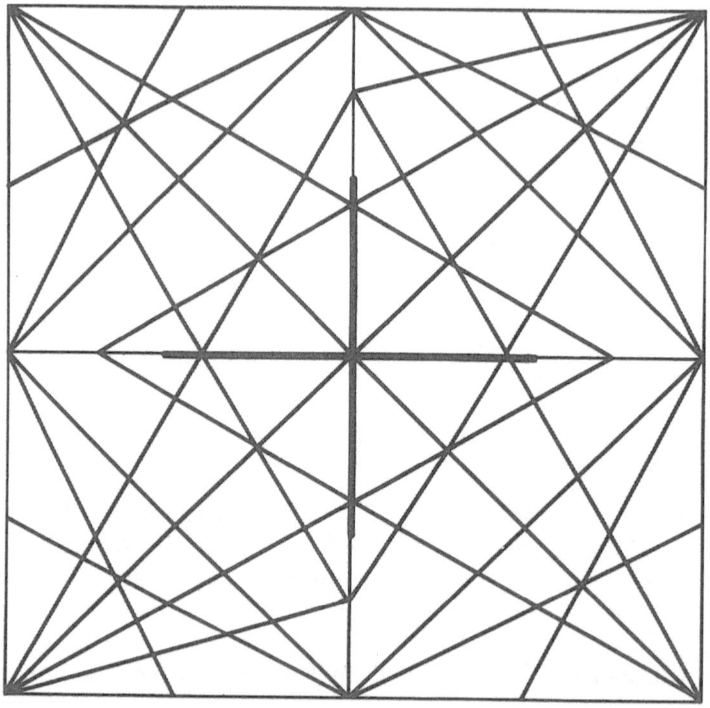

Fig. 6.6. Paths the travel times for which have been used in a test inversion by the Backus—Gilbert technique.

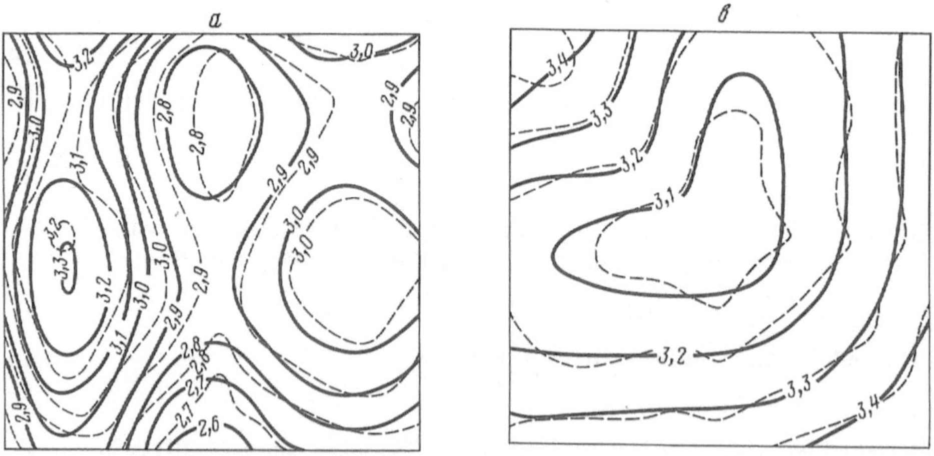

Fig. 6.7. Velocity distributions obtained from travel times along the paths shown in Figure 6.6, for different correlation lengths (*a*, *b*). Solid lines show the initial velocity distribution, dashed lines-the result of inversion.

b). The correlation length in Figure 6.7a is about 350, while in Figure 6.7b it is comparable with the size of the entire area (2000 × 2000). Travel times were calculated for paths shown in Figure 6.6 for each velocity distribution. The results of velocity inversion based on these data are shown in Figure 6.7 a, b by a dashed line. Figure 6.8 shows the distribution of effective smoothing radius given by (6.73) which summarizes the resolution of the data in hand.

As was to be expected, the large-scale inhomogeneity is found with greater accuracy: the misfit between the true velocities and the inverted values does not exceed 0.01 (Figure 6.7a), while smaller inhomogeneities are recovered from the same data set (the paths) with greater errors, reaching 0.1 at some points; even so, all the anomalous zones are identified reliably enough, as can be seen from Figure 6.7b.

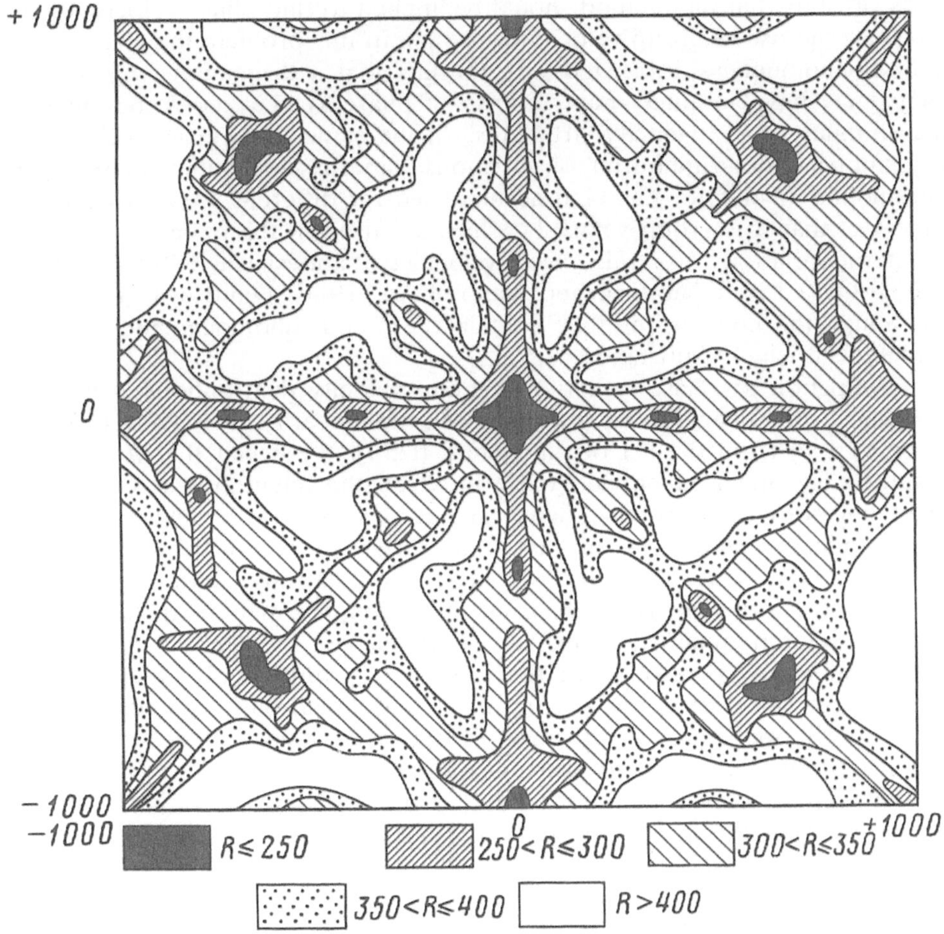

Fig. 6.8. A pattern of effective smoothing radius.

6.6. SURFACE WAVE TOMOGRAPHY BASED ON A COMBINED USE OF PHASE AND GROUP VELOCITIES FOR DIFFERENT PERIODS

The method described in the last section assumes knowledge of either phase or group velocities of surface waves along different paths for a fixed period. The solution will represent the velocity distribution as a function of the lateral coordinates for the period. This approach is not however always practicable for the following reasons.

First, observations do not always provide enough data for the particular period under consideration, while data may be available for near periods along a fairly large number of paths, and they contain certain information on the velocity for that period which the above approach ignores.

Secondly, if we remember that the resulting velocity distributions $V(x, y)$ ($V = C$ or U) are to be used for determining crustal and upper mantle structure, then it is fairly obvious that the method should be applied to find phase velocity distributions, since the use of group velocity dispersion in this problem may lead to much greater nonuniqueness associated with the fact that a dispersion curve of group velocity may have infinitely many curves of phase velocity corresponding to it, hence infinitely many velocity structures. At the same time, group velocities are much simpler to determine from observations than phase velocities; there are quite large amounts of data of this kind accumulated for different areas. This accounts for the fact that it is group velocities which are used in the many papers dealing with lateral variations of surface wave velocities (Yanovskaya, 1982; Yanovskaya and Nikolova, 1984; Sabitova and Yanovskaya, 1986; Dmitrieva *et al.*, 1986), even though the conclusions they contain on lateral inhomogeneities cannot be considered sufficiently substantiated.

Determination of phase rather than group velocity distributions is important, also because this enables one to incorporate nonlinearity of inverse problem in finding $V(x, y)$ from the set of functionals (6.30). We recall that (6.30) can be reduced to the linear functional (6.32) in which the integration is done along a straight line, provided the ray shape can be assumed to be determined by the starting model, which is V_0 = constant in our case. However, if the lateral variation of velocity is not small enough, one should accomodate ray curvature; this can be done by successive approximations, the initial approximation at each iteration being the solution obtained at the last iteration, i.e., the path of integration in (6.32) is taken to correspond to the velocity distribution obtained at the last iteration. Now since ray shape is controlled by the distribution of phase velocity, then obviously, successive approximations cannot be constructed, unless it is the phase velocity that one obtains at the last iteration. However, these drawbacks of the above approach can be obviated, if one recalls the well-known relation between phase and group velocity

$$U^{-1}(\omega) = C^{-1}(\omega) + \omega \frac{\mathrm{d}C^{-1}(\omega)}{\mathrm{d}\omega} \tag{6.81}$$

so that the dispersion curve of phase velocity can be completely determined from that of group velocity, provided the constant of integration is known. In the case of a single dimension it is sufficient to know phase velocity at a single value of frequency. This fact can form the basis of a technique for determining phase velocity distributions for different frequencies in laterally varying structures from the values of group velocity along different paths, provided we know a few values of phase velocity along some paths in the study area. The technique in question was proposed by Yanovskaya *et al.* (1987).

The raw data are phase $C_i^{(k)}$ $(i = 1, 2, \ldots, m_k)$ and/or group velocities $U_j^{(k)}$ $(j = 1, 2, \ldots, n_k)$ for periods $T_i^{(k)}$) along paths $L_k (k = 1, 2, \ldots, K)$ traversing the study area. A path is as before defined by giving the coordinates of the initial and end point; the problem will be solved in the linearized formulation, i.e., paths are assumed to be straight lines L_{0k}. The data can be converted into travel time residuals to be related to the unknown distributions of phase and group velocities $C(x, y, T)$ and $U(x, y, T)$ through

$$\int_{L_{0k}} \delta C^{-1}(x, y, T_i^{(k)})\, \mathrm{d}l = \delta t_{ik}^{ph}$$
$$\int_{L_{0k}} \delta U^{-1}(x, y, T_j^{(k)})\, \mathrm{d}l = \delta t_{jk}^{gr} \tag{6.82}$$

where $\delta C^{-1} = C^{-1}(x, y, T) - C_0^{-1}(T)$, $\delta U^{-1} = U^{-1}(x, y, T) - U_0^{-1}(T)$, $C_0(T)$ and $U_0(T)$ being dispersion curves of phase and group velocity, respectively, averaged over the area. It should also be remembered that C and U are related by (6.81).

The idea of the method is that the phase slowness as a function of frequency $\omega = 2\pi/T$ is fitted by a polynomial whose coefficients are unknown functions of the lateral coordinates

$$C^{-1}(x, y, \omega) = \sum_{q=0}^{N} a_q(x, y)\, \omega^q . \tag{6.83}$$

Accordingly, group slowness is also fitted by a polynomial of the same degree:

$$U^{-1}(x, y, \omega) = \sum_{q=0}^{N} (1 + q)\, a_q(x, y)\, \omega^q . \tag{6.84}$$

To find $C(x, y, \omega)$, we must determine the coefficients a_q which are now functions of the spatial coordinates x, y only. This problem can in turn be reduced to that discussed in the last section.

We represent the 'mean' dispersion curve $C_0(\omega)$ in the form (6.83) again:

$$C_0^{-1}(\omega) = \sum_{q=0}^{N} a_{0q}\, \omega^q \tag{6.85}$$

where the a_{0q} are constants to be determined from $C_0(\omega)$. Accordingly, $U_0(\omega)$ is also expressible in terms of the same corefficients:

$$U_0^{-1}(\omega) = \sum_{q=0}^{N} a_{0q}(1 + q)\, \omega^q . \tag{6.86}$$

Equations (6.82) for the kth path can now be written as

$$\sum_{q=0}^{N} (\omega_i^{(k)})^q a_{0q} \int_{L_{0k}} m_q(x, y)\, dl = \delta t_{ik}^{ph} (i = 1, 2, \ldots, m_k)$$

$$\sum_{q=0}^{N} (\omega_j^{(k)})^q a_{0q}(1 + q) \int_{L_{0k}} m_q(x, y)\, dl = \delta t_{jk}^{gr} (j = 1, 2, \ldots, n_k) \tag{6.87}$$

where $m_q(x, y) = (a_q(x, y) - a_{0q})/a_{0q}$. These equations for all paths $k = 1, 2, \ldots, K$ can be combined to form the set of equations

$$\sum_{q=0}^{N} b_{rq} \int_{L_{0k}} m_q(x, y)\, dl = \gamma_r, \qquad r = 1, 2, \ldots, M \tag{6.88}$$

where $\gamma_r = \delta t_{ik}^{ph}$ or δt_{jk}^{gr}, $k = k(r)$, $b_{rq} = a_{0q}(\omega_{i(j)}^{(k)})^q(1 + pq)$, $p = 0(1)$ corresponding to phase (group) velocity; $M = \Sigma_k(m_k + n_k)$.

One can find $m_q(x, y)$ from (6.88) using a method which is an extension of that described in the last section.

We assume the $m_q(x, y)$ to be components of a vector fuction $\mathbf{m}(x, y)$ and demand this function to obey a smoothness requirement similar to (6.35):

$$\int_S\int \|\nabla \mathbf{m}\|^2\, dx\, dy = \min \tag{6.89}$$

where $\nabla \mathbf{m}$ is a matrix of the derivatives $\partial m_q/\partial x_k$. We also assume $\mathbf{m}(x, y)$ to be bounded at infinity. It then turns out that each of the $m_q(x, y)$ must satisfy Poisson's equation (6.37); its solution has the form

$$m_q(\mathbf{r}) = \sum_{i=1}^{M} \lambda_i b_{iq} \int_{L_{0i}} \ln|\mathbf{r} - \mathbf{r}_i|\, dl_i + C_q. \tag{6.90}$$

We have taken path number to correspond to equation number in (6.88), so that different indices may correspond to one and the same path.

The boundedness of $\mathbf{m}(\mathbf{r})$ at infinity yields the requirement that the λ_i must satisfy

$$\sum_{i=1}^{M} \lambda_i b_{iq} l_i = 0 \tag{6.91}$$

where l_i is the length of the ith path.

We define matrices $\mathbf{S}$ and $\mathbf{D}$ as follows:

$$S_{ij} = \sum_{q=0}^{M} b_{iq} b_{jq} \int_{L_{0i}}\int_{L_{0j}} \ln|\mathbf{r}_i - \mathbf{r}_j|\, dl_i\, dl_j, \qquad 1 \leqslant i, j \leqslant M \tag{6.92}$$

$$D_{ij} = b_{ij} l_i, \quad 1 \leqslant i \leqslant M, \quad 0 \leqslant j \leqslant N. \tag{6.93}$$

The coefficients λ_i (the vector Λ) and the constant terms C_q (the vector $\mathbf{C}$) satisfy the following set of equations:

$$\mathbf{S\Lambda} + \mathbf{DC} = \boldsymbol{\gamma} \qquad \mathbf{D}^T\mathbf{\Lambda} = 0. \tag{6.94}$$

The set (6.94) is similar to (6.43), (6.45). If we now define a matrix $\mathbf{K(r)}$:

$$K_{iq}(\mathbf{r}) = b_{iq} \int_{L_{0i}} \ln|\mathbf{r} - \mathbf{r}_i| \, \mathrm{d}l \qquad 1 \leqslant i \leqslant M, \quad 0 \leqslant q \leqslant N \tag{6.95}$$

we obtain a solution for $\mathbf{m(r)}$ that is an extension of (6.48):

$$\mathbf{m(r)} = \mathbf{K}^T\mathbf{S}^{-1}\boldsymbol{\gamma} + (\mathbf{I} - \mathbf{K}^T\mathbf{S}^{-1}\mathbf{D})(\mathbf{D}^T\mathbf{S}^{-1}\mathbf{D})^{-1}\mathbf{D}^T\mathbf{S}^{-1}\boldsymbol{\gamma}. \tag{6.96}$$

Note that $\mathbf{S}$ is degenerate, if there are more data points for a path than the coefficients of the fitting polynomial. This means that equations (6.88) must be satified approximately rather than exactly. If we assume them to be satisfied in the sense of the minimum squared residuals, the appropriate solution is (6.96) again, but the inverse matrix $\mathbf{S}^{-1}$ is now to be the generalized inverse $\mathbf{S}^+$ based on the singular value decomposition of $\mathbf{S}$.

Model example. This method has been tested on the following model example. The distribution of phase velocity in the region $-1000 \leqslant x \leqslant 1000, -1000 \leqslant y \leqslant 1000$ is given as

$$C(x, y, \omega) = 1.5(3.0 - \tan^{-1}(2\omega^{3/2} H(x, y)) \tag{6.97}$$

where $H(x, y) = 1 + \exp[-(x/1000)^2 - 3(y/1000)^2]$. Phase and group travel times were calculated along the 28 paths shown in Figure 6.9 for $0.2 \leqslant \omega \leqslant 0.5$, and a sample of 100 travel times was used to determine phase velocity in the study area.

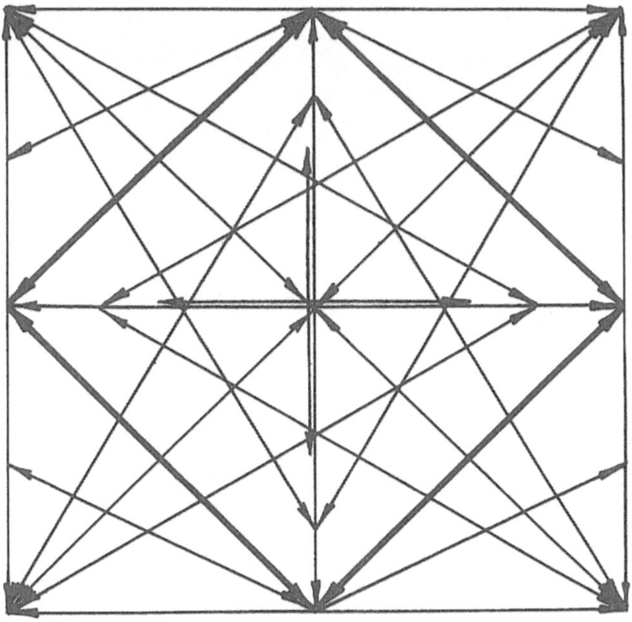

Fig. 6.9. A pattern of paths the travel times for which have been used in a test inversion. Solid lines are paths for which phase velocities were known.

The number of data points for a path was 3 to 5, the phase slowness was fitted by a straight line, hence $N = 1$. The phase travel times were taken only for the 4 paths indicated by heavy lines in Figure 6.9.

The inversion gave distributions of phase velocity for four frequencies shown in Figure 6.10 a—d together with the original velocity distributions. One can see that, even though few data points of phase velocity were used, the resulting distributions of phase velocity reflect the principal features of the original distributions. One can thus expect that, also in the interpretation of real seismological data, a few observations of phase velocity are sufficient, when combined with a large sample of group velocity measurements, to derive fine detail in the distribution of phase velocity.

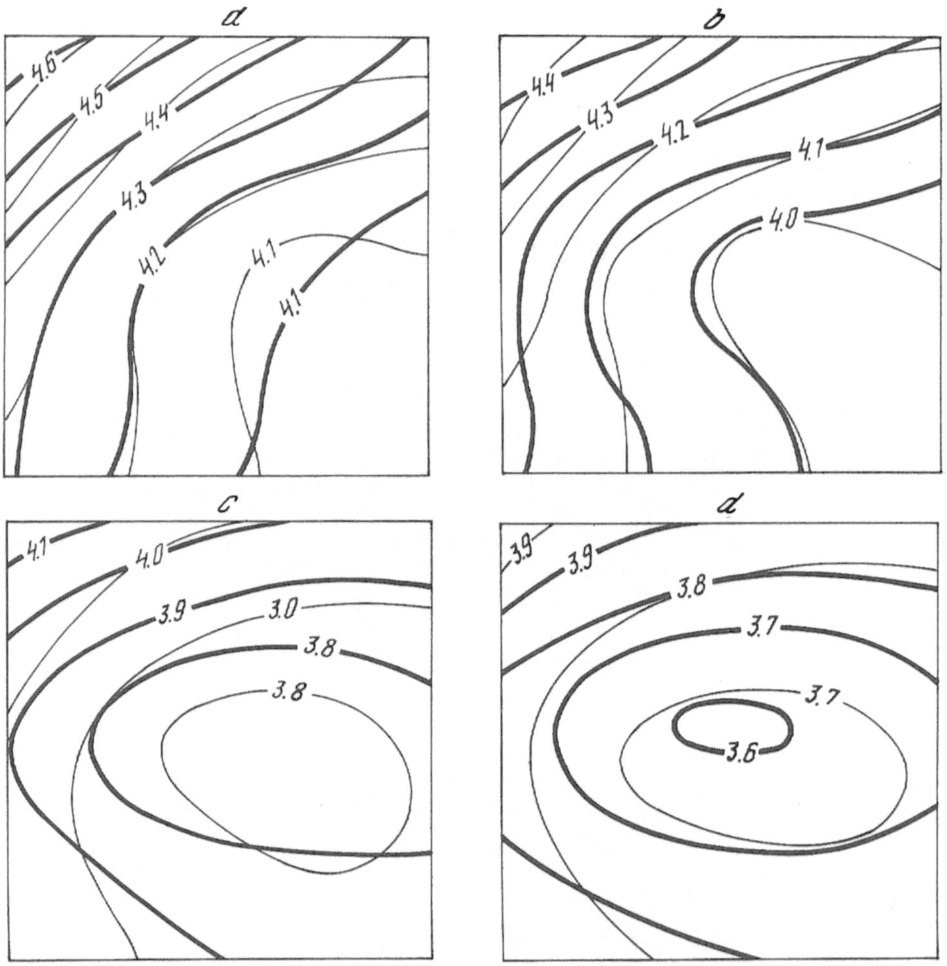

Fig. 6.10. Results of joint inversion of phase and group velocities. Solid lines show the initial phase velocity distributions, thin lines the results of inversion, for the frequencies: (a) $\omega = 0.21$ rad/sec, (b) $\omega = 0.28$ rad/sec, (c) $\omega = 0.42$ rad/sec, (d) $\omega = 0.50$ rad/sec.

6.7. THE INFLUENCE OF LATERAL INHOMOGENEITY ON THE MEASUREMENT OF SURFACE WAVE ATTENUATION

Surface wave observations are widely used to study dissipation within the Earth. With this end in view, observations are utilized to determine the surface wave attenuation coefficient $\alpha_{k\mathrm{D}}(\omega)$ and the related apparent surface wave quality factor, $Q_{k\mathrm{D}}(\omega)$ (see (1.55)). These data provide a basis for solving the inverse problem of true quality factor as a function of depth, $Q(z)$. Recalling (1.53), we can attack this inverse problem in a linear formulation by making use of the least squares, singular value decomposition, and the Backus–Gilbert technique (Backus and Gilbert, 1968, 1970), as well as of linear (Dorofeev and Zharkov, 1987) or quadratic programming.

We shall not discuss inverse problems of this class, but examine effects caused in measurements of $\alpha_{k\mathrm{D}}$ and $Q_{k\mathrm{D}}$ by lateral inhomogeneity. To do this, we shall point out the methods for determining $\alpha_{k\mathrm{D}}(\omega)$ from surface wave observations. Since we speak of measurements specific to a particular mode and wave type, the suffixes k and D will be omitted in what follows.

The data used for measurements include amplitude spectra that are either found by direct spectral transformation of seismograms or calculated from the results of frequency–time analysis with the correction factor (5.96) incorporated. The effects of factors other than absorption on the amplitude wave spectra are eliminated by using a number of techniques: source mechanism, depth, and spectral shape are determined independently (Mendiguren, 1977; Nakanishi and Kanamori, 1982; Weidner and Aki, 1973) and appropriate corrections are made in the observed spectrum; one may use observations at two stations on the same great-circle path with the epicenter and determine the ration between amplitude spectra of the same wave; the ratio of amplitude spectra is determined for waves that have travelled around the Earth in any one direction and were recorded at a single station (Burton, 1973; Kijko and Mitchell, 1983; Mills, 1978; Patton, 1965, 1980; Tsai and Aki, 1969).

All these techniques assume, implicitly or explicitly, that the medium is laterally homogeneous. The theoretical analysis of Chapters 2 and 3 can be used to estimate distortion that measurements of this sort involve due to a smooth or sharp lateral inhomogeneity (Levshin, 1984).

The effect of a smooth inhomogeneity. Suppose a hypocenter is at the point $M_0'(0, 0, h)$ of a half-space, while we observe at the point $M(r, \varphi, 0)$ at epicentral distance r. Denote the observed amplitude spectrum of the pth displacement component in a surface wave by $|\hat{u}^p|(\omega)$. In a laterally homogeneous medium this is

$$|\hat{u}^p|(\omega) = \frac{1}{\sqrt{8\pi}}\,\frac{\exp[-\alpha^*(\omega)r]}{\sqrt{\xi r}}\left[\frac{|\varepsilon_p|}{\sqrt{CUI^{(0)}}}\right]_{M_0}\left[\frac{|W(\omega, h, \varphi)|}{\sqrt{CUI^{(0)}}}\right]_{M_0} \tag{6.98}$$

(we assume that W is a source-controlled factor and that the medium has the same structure as that beneath the epicenter, at the point $M_0(0, 0, 0)$). Because of lateral inhomogeneity, the attenuation coefficient α^* as given by (6.98) is an apparent one. Its departure from the local $\alpha(\omega, x, y)$ or the mean $\bar{\alpha}$ averaged over the epicenter–station path is determined by the lateral inhomogeneity.

Suppose the medium varies from M_0 to M in accordance with the postulates laid down in Section 2.1. Then (6.98) is replaced by the following expression for observed amplitude spectrum (where (2.9) has been taken into account)

$$|\hat{u}^p|(\omega) = \frac{1}{\sqrt{8\pi}} \frac{\exp[-\bar{\alpha}(\omega)L]}{\sqrt{[\xi J]_M}} \left[\frac{|\varepsilon_p|}{\sqrt{CUI^{(0)}}}\right]_M \left[\frac{|W(\omega, h, \varphi)|}{\sqrt{CUI^{(0)}}}\right]_{M_0} \tag{6.99}$$

where L is the path length from M_0 to M, $L = \int_{M_0}^{M} \mathrm{d}s$; $\bar{\alpha}(\omega)$ is the mean attenuation coefficient along the path:

$$\bar{\alpha} = \int_{M_0}^{M} \alpha(\omega, s)\, \mathrm{d}s/L. \tag{6.100}$$

From (6.99) and (6.100) it follows that by ignoring lateral inhomogeneity we determine α^* from (6.98) instead of $\bar{\alpha}$:

$$\alpha^* = \bar{\alpha}\, \frac{L}{r} + \frac{1}{r} \ln \frac{[J]_M}{r} + \frac{1}{r} \ln R^p_{M_0 M}(\omega) \tag{6.101}$$

where

$$R^p_{M_0 M}(\omega) = \sqrt{\frac{[UI^{(0)}]_{M_0}}{[UI^{(0)}]_M}} \left|\frac{[\varepsilon_p]_M}{[\varepsilon_p]_{M_0}}\right|. \tag{6.102}$$

To estimate the distorting effect on $\bar{\alpha}$ due to the R^p term in (6.101), we discuss calculation for two lithosphere models from Patton (1980). Model 1 is Gutenberg's continental platform model, while model 2 is for the Pamir mountain region, being essentially different from model 1 in the structure of the uppermost 70-km lithosphere (Table 6.III). Figure 6.11a shows group velocity $U(T)$ and the quantity $I^{(0)}\,(T)$ plotted for the fundamental mode Rayleigh wave in both models. The largest misfit between the models is in the range 30 to 50 sec (as much as 50 percent in $I^{(0)}$ and 25 percent in U). This produces a noticeable frequency dependence of $R^z\,(T)$ (Figure 6.11b), the fluctuations reaching 30 percent. Figure 6.12 shows the behavior of the R^z term in (6.101) as a function of period for four values of epicentral distance r (marked by numbers above the curves). Also shown are plots of $\alpha(T)$ for the laterally homogeneous model 1 with two different Q distributions (Q is assumed to be independent of frequency, being associated with shear dissipation alone). Most of the experimental values of α found by Burton (1973) and Patton (1980) lie between the two curves. One can see that the bias caused by neglecting the R^z term is comparable with the observed and theoretical values of that quantity. It cannot be eliminated by averaging within an azimuth window designed to suppress geometrical effects of a lateral inhomogeneity, if the stations used lie within the same lithosphere block or structurally similar ones. In the case considered, α as determined from observation is overestimated, because group velocity and depth of wave penetration (hence the factor $I^{(0)}$) increase from model 2 to model 1; consequently, the spectral amplitude decreases. This effect is erroneously attributed to attenuation. With a wave propagating in the opposite direction the effect is reversed, so that attenuation may be grossly underestimated.

Similar errors may be incurred in estimating α from spectra recorded at two

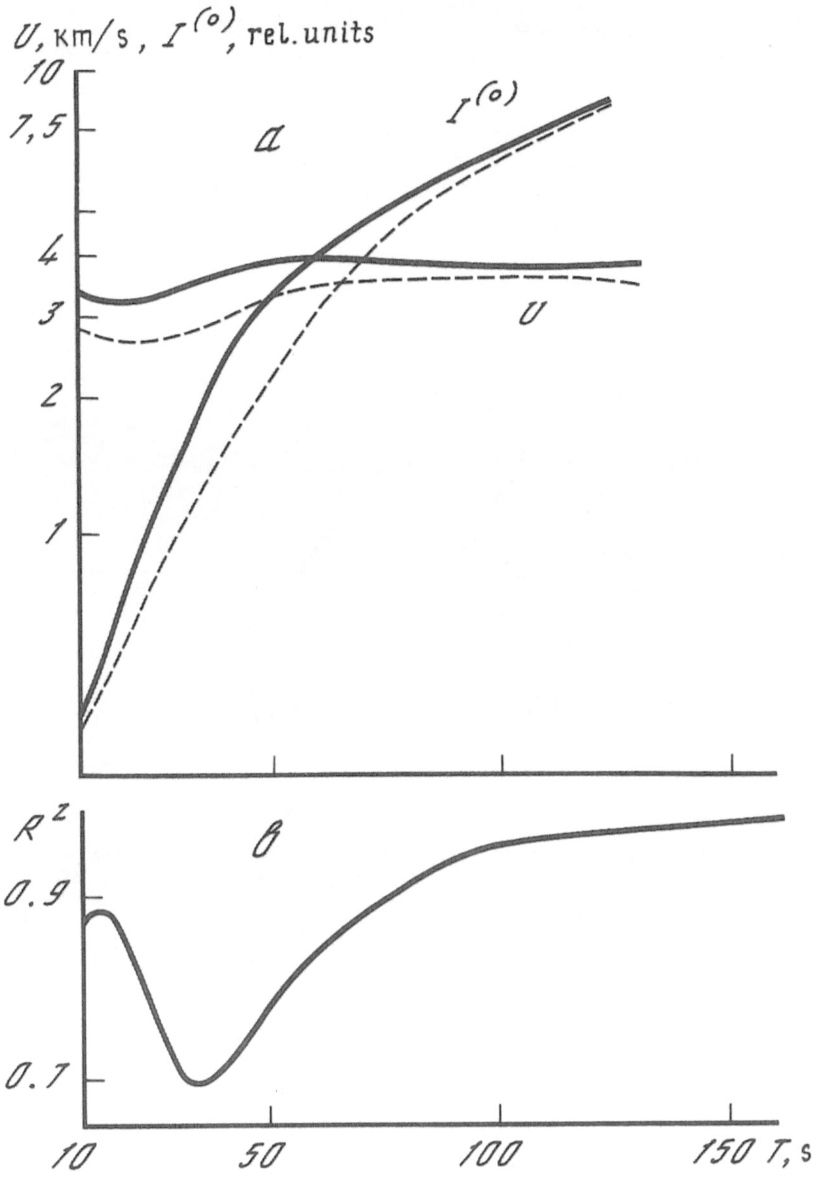

Fig. 6.11. Absolute and relative spectral characteristics for fundamental Rayleigh mode in model 1 (solid line) and 2 (dashed line).

stations that lie within a narrow azimuth sector relative to the source. In that case M_0 in (6.101) corresponds to the station closer to the source, M to the farther one, r is the great circle interstation distance, the second term being replaced by $1/2r$ $(\ln [J]_M - \ln [J]_{M_0})$.

The effect can be compensated if, in two-station observations, one uses surface waves excited by two sources that lie on the same great circle but on opposite

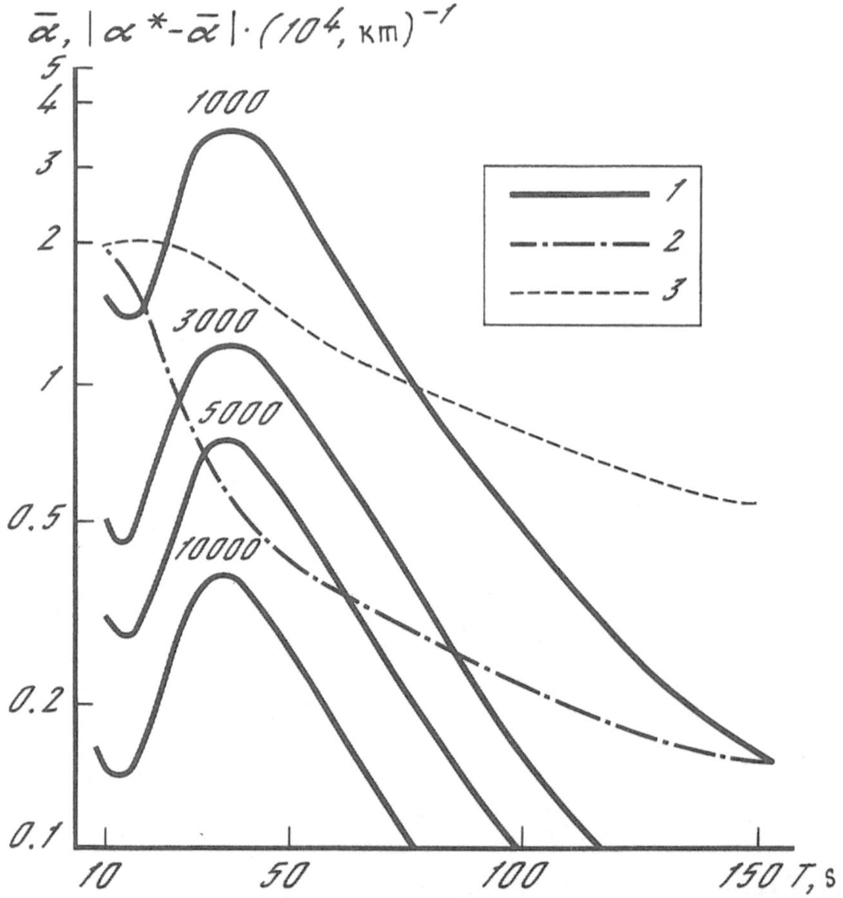

Fig. 6.12. Bias in the determination of mean attenuation coefficient $\bar{\alpha}$ when lateral inhomogeneity is disregarded. (1) $(\alpha^* - \bar{\alpha})$, numerals at the curves indicating path length in km; (2, 3) $\bar{\alpha}$ for several distributions of Q over depth: (2) $Q = 400$ in the crust and mantle, (3) $Q = 400$ in the crust and $Q = 100$ in the mantle.

sides of the interstation path. Summing (6.101) for two measurements, we arrive at an expression for α^* that does not involve the R^p term:

$$\alpha^* = \bar{\alpha}\,\frac{L}{r} + \frac{1}{4r}\left(\ln\frac{[J]^{\mathrm{I}}_{M}}{[J]^{\mathrm{I}}_{M_0}} + \frac{1}{r}\ln\frac{[J]^{\mathrm{II}}_{M_0}}{[J]^{\mathrm{II}}_{M}}\right) \tag{6.103}$$

in which the values of geometrical spreadings J^{I}, J^{II} correspond to the two respective sources. A similar method of eliminating R^p can be used for single-station observations of waves that have travelled around the Earth different numbers of times; the measured α then corresponds to the station-epicenter great circle.

The apparent Q of surface waves along the $M_0 M$ path is usually found from

$$\bar{Q} = \omega/2\bar{\alpha}\,\bar{U}$$

where $\bar{\alpha}$, $\bar{U}$ are means for the path:

$$\bar{\alpha} = \frac{1}{L} \int_{M_0}^{M} \alpha(\omega, s)\, \mathrm{d}s = \frac{\omega}{2L} \int_{M_0}^{M} \frac{\mathrm{d}s}{U(s)\, Q(s)},$$

$$\bar{U} = L \bigg/ \int_{M_0}^{M} \frac{\mathrm{d}s}{U(s)}.$$

Then the mean $\bar{Q}$ for the path is

$$\bar{Q} = \int_{M_0}^{M} \frac{\mathrm{d}s}{U(s)} \bigg/ \int_{M_0}^{M} \frac{\mathrm{d}s}{U(s)\, Q(s)}$$

that is, being influenced by local variations in Q, it also depends on the variation of group velocity along the path, which is usually ignored in the interpretation of Q data. Also, when $\bar{Q}$ is based on an α^* that is distorted by lateral inhomogeneity, the result is biased

$$Q^* = \frac{\bar{\alpha}}{\alpha^*}\, \bar{Q}.$$

Still greater bias in $\bar{\alpha}$ and Q can be caused by geometrical effects, in case the path involves regions of increased or lower velocity, while the wave periods are not large (see examples in Section 4.3). Additional bias is caused by later arrivals like reflections from vertical contacts, in case the time delays involved are such as not to allow complete separation of these signals from the principal one.

To sum up, measurements of α in the presence of lateral inhomogeneity are subject to significant bias, unless the conditions along the path and around the station are incorporated. For this reason distributions of Q in the Earth derived from such measurements must be revised by taking account of the effects we have found.

Based on the above, we can outline the following approach to interpretation of observations in the presence of weak lateral inhomogeneity. With many paths traversing an area under study and the data on earthquake mechanisms available, one determines dispersion curves along the paths. Data for some fixed periods are then used to construct the scalar fields $C(x, y)$, $U(x, y)$ following the procedure of Section 6.4. The scalar fields yield surface wave paths for given source-station pairs, and geometrical spreading estimated for the paths. Models are chosen for the structures around the source and the stations; all these data are used to determine α along a wave path from formula (6.99), and then to find the scalar field $\alpha(x, y)$ or $Q(x, y)$ by the technique described in Section 6.5.

Although the above procedure is rather more complicated than those used for simple laterally homogeneous models, it also has advantages in that it eliminates bias in the estimation of dissipation within the Earth.

Effect of a vertical contact. We now discuss the question of how an estimate of α made under the assumption of lateral homogeneity can be affected by a vertical contact between two laterally homogeneous media, when the source and the receiver are on opposite sides of the contact and far from it. With the assumptions made in Section 3.3, we use the results of Section 1.3 for the spectrum of a wave recorded at point M_1 in medium 1 and excited at point M_2 in medium 2 (Figure 6.13) to obtain

$$|\hat{u}_p|(\omega) = \frac{1}{\sqrt{8\pi}} \frac{\exp[-\alpha_1 l_1 - \alpha_2(\omega) l_2]}{[\sqrt{\xi J}]_{M_1}} \left[\frac{|\varepsilon_p V^{(i)}|}{\sqrt{CUI^{(0)}}}\right]_{M_1} \left[\frac{|W(\omega, h, \varphi)|}{\sqrt{CUI^{(0)}}}\right]_{M_2} \times$$

$$\times K_{21}(\theta_2, \omega) \sqrt{\frac{\cos\theta_1}{\cos\theta_2}}. \tag{6.104}$$

The subscripts 1 and 2 indicate the relevant medium; l_1 and l_2 are the distances that the wave travels in medium 1 and 2, and $l_1 = h_1/\cos\theta_1$, $l_2 = h_2/\cos\theta_2$; h_1 and h_2 are the distances from M_1 and M_2 to the discontinuity; θ_1 and θ_2 are the angles of incidence and refraction for the ray connecting M_1 and M_2: $\sin\theta_1/C_1 = \sin\theta_2/C_2$; $K_{21}(\theta_2, \omega)$ is the ratio of displacement amplitude at the free surface due to a wave propagating from medium 2 toward medium 1 as determined according to what was said in Section 3.3. The geometrical spreading then is

$$[J]_M = \frac{\cos\theta_1 \cos\theta_2}{C_1} \left(\frac{h_1 C_1}{\cos^3\theta_1} + \frac{h_2 C_2}{\cos^3\theta_2}\right).$$

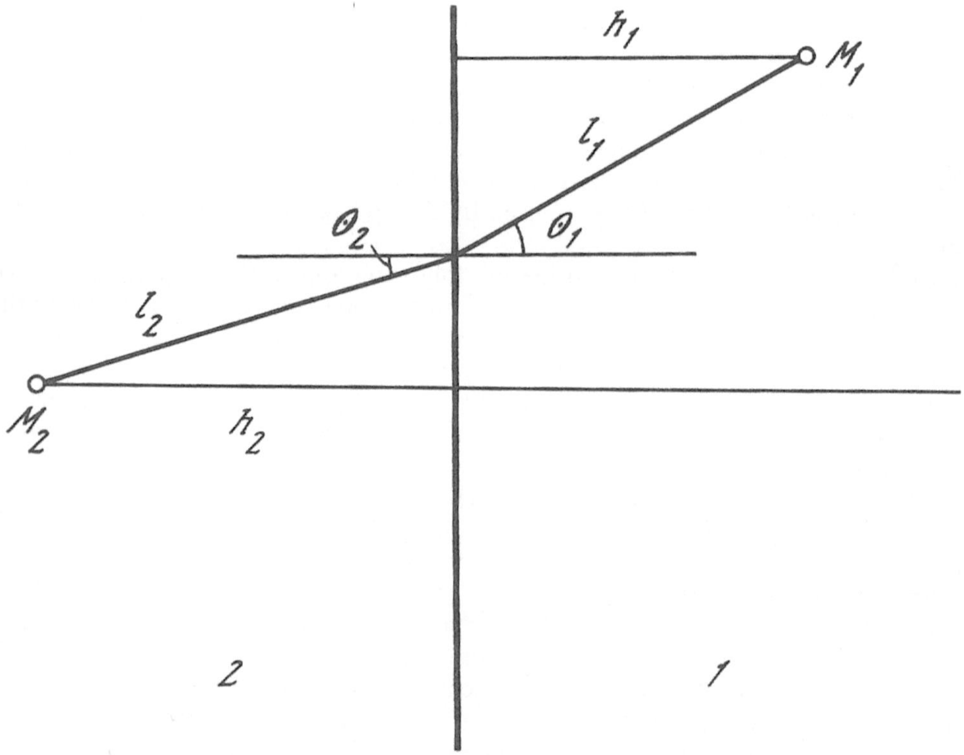

Fig. 6.13. Surface waves propagating across a vertical contact.

Comparing (6.104) to (6.98), we see that the assumption of lateral homogeneity (the medium is everywhere that around the source) leads to the following formula for the apparent attenuation coefficient

$$\alpha^* = \bar{\alpha}\, \frac{l_1 + l_2}{r} + \frac{1}{2r} \ln \frac{[J]_M}{r} + \frac{1}{2r} \ln \Re^{p}_{M_1 M_2}(\omega, \theta_2) \tag{6.105}$$

where

$$\bar{\alpha} = (\alpha_1 l_1 + \alpha_2 l_2)/(l_1 + l_2)$$

$$\Re^{p}_{M_2 M_1} = K_{21}(\theta_2, \omega) \sqrt{\frac{\cos \theta_1}{\cos \theta_2} \left| \frac{[\varepsilon_p]_{M_1}}{[\varepsilon_p]_{M_2}} \right|}.$$

Calculations of $\Re^{\rightleftarrows}_{M_2 M_1}$ for a fundamental Rayleigh mode incident at an angle close to the normal ($\theta_2 \approx 0$), when the wave is transmitted from the medium of lithosphere model 2 into that of model 1 (Table 6.III), demonstrate a practically identical behavior of $R^{\rightleftarrows}$ and $\Re^{\rightleftarrows}$ (Figure 6.14a). This means that the change in displacement amplitude for a transmitted wave is largely controlled by changes in its energy profile over depth and in group velocity (speed of energy transport) rather than by losses due to reflected and converted waves. This is in agreement with the fact that reflection and conversion coefficients for angles of incidence below 30° reach a few hundredths at most (Figure 6.14b). This situation is similar for a wave propagating in the opposite direction, from medium 1 toward medium 2: the overall behavior of $(R^{\rightleftarrows})^{-1}$ and $\Re^{\rightleftarrows}_{M_1 M_2}$ is similar, but the misfit is greater than

TABLE 6.III.
Models of lithosphere.

h, km	a, km/sec	b, km/sec	ρ, g/cm^3	h, km	a, km/sec	b, km/sec	ρ, g/cm^3
Model 1				Model 2			
19	6.14	3.55	2.76	4	4.41	2.55	2.41
19	6.58	3.80	3.00	26	5.58	3.18	2.66
12	8.20	4.65	3.32	30	6.50	3.76	2.90
10	8.17	4.62	3.34	20	8.10	4.51	3.36
10	8.14	4.57	3.35	10	8.07	4.46	3.37
10	8.10	4.51	3.36	10	8.02	4.41	3.38
10	8.07	4.46	3.37	25	7.93	4.37	3.39
10	8.02	4.41	3.38	25	7.85	4.35	3.41
25	7.93	4.37	3.39	25	7.89	4.36	3.43
25	7.85	4.35	3.41	25	7.98	4.38	3.46
25	7.89	4.36	3.43	25	8.10	4.42	3.48
25	7.98	4.38	3.46	25	8.21	4.46	3.50
25	8.10	4.42	3.48	50	8.38	4.54	3.53
25	8.21	4.46	3.50	50	8.62	4.68	3.58
50	8.38	4.54	3.53	50	8.87	4.85	3.62
50	8.62	4.68	3.58	—	9.15	5.04	3.69
50	8.87	4.85	3.62				
—	9.15	5.04	3.69				

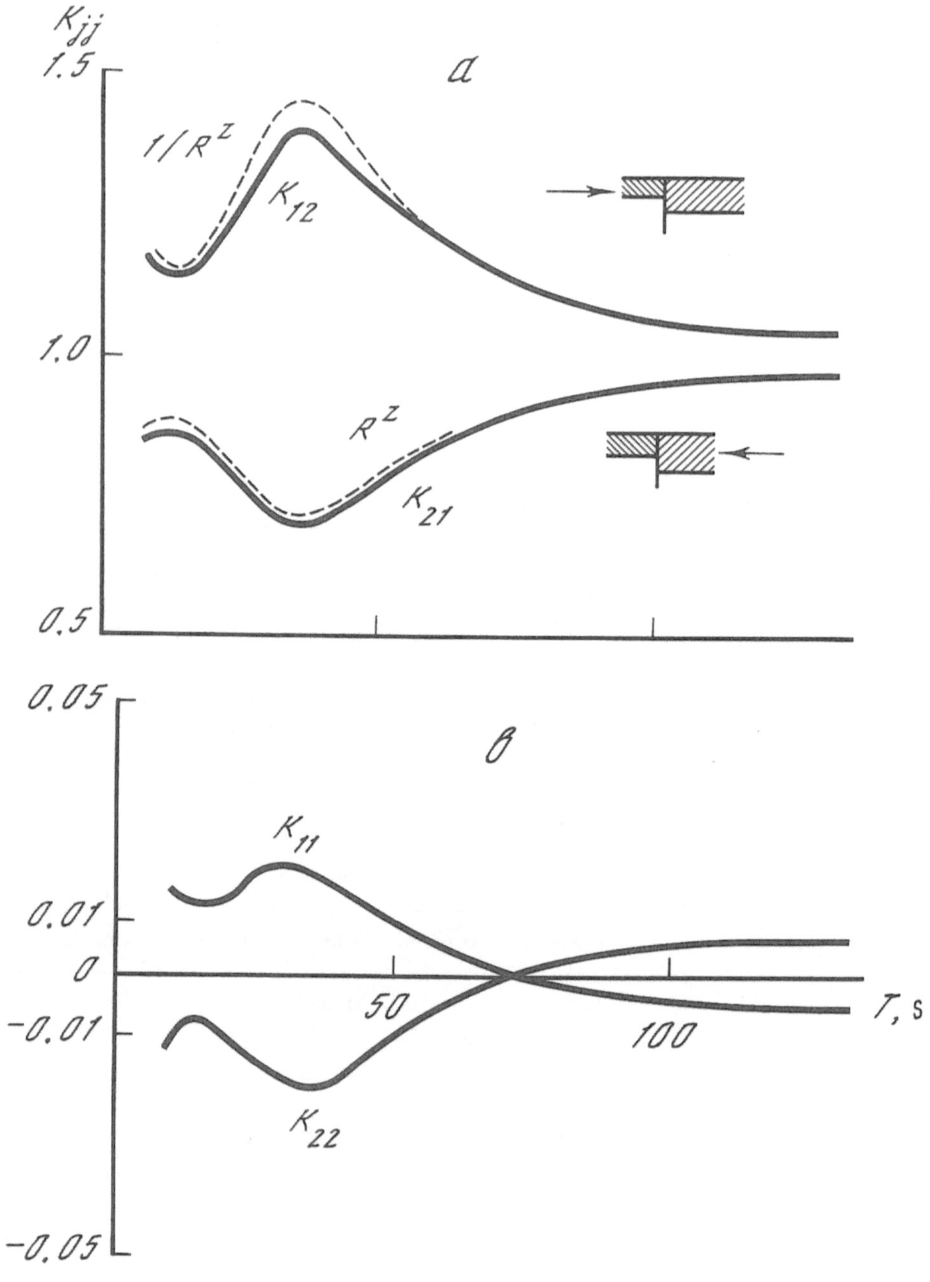

Fig. 6.14. Variation of displacement amplitude in Rayleigh wave at a vertical contact (fundamental mode, normal incidence). (a) transmission coefficients K_{21} and K_{12}; (b) reflection coefficients K_{22} and K_{11}.

in the preceding case (Figure 6.14b). This seems to be due to the fact that the contribution into body wave generation is somewhat greater when a wave passes from a structure with a thinner crust into one with a thicker one than in with the opposite situation.

These calculations show that the error caused by a neglected vertical contact is close to that found for a smooth lateral inhomogeneity. It may be particularly large for mixed ocean—continent paths.

6.8. ANISOTROPY OR INHOMOGENEITY?

The existence of seismic anisotropy has been established by numerous laboratory and field experiments. The anisotropy properties of rocks are closely associated with their geologic evolution and contain significant information on earth dynamics and structure (Chesnokov, 1977; Crampin, 1981). In this context the great interest shown by geophysicists in all seismic phenomena that can be interpreted within the framework of anisotropy is understandable. However, incomplete analyses may treat as anisotropy those phenomena which are of totally different origin, for example, ones due to large-scale inhomogeneities (Bukchin, Levshin and Ratnikova, 1984). We use numerical modelling to demonstrate how large-scale inhomogeneities can produce such effects as (1) departure of the polarization plane of a surface wave from the theoretical position for an isotropic medium; (2) discrepancies between the dispersion curves of Love and Rayleigh waves when working with an isotropic, laterally homogeneous model. Phenomena of this sort are commonly thought to indicate anisotropy (Crampin, 1981; Leveque, 1980; Schlue and Knopoff, 1978).

Polarization in higher modes of surface waves. It has been repeatedly pointed out that the polarization in higher modes of surface waves is extremely sensitive even to a very slight anisotropy (Crampin, 1981). Anisotropy was also invoked to explain observed polarization in higher modes for long continental paths: analysis of three-component records revealed a persistent elliptical motion at arrival times of higher modes occurring in a vertical plane not identical with that containing the theoretical epicenter-station path.

To verify if such things must necessarily be associated with anisotropy, we computed spectral characteristics and theoretical seismograms for the first higher modes of Love and Rayleigh waves (denoted $2L$ and $2R$ in what follows) in a medium (Table 6.IV) whose parameters are close to those of an anisotropic model in Crampin (1981).

The spectral characteristics are shown in Figure 6.15, where one can see that the phase velocity curves of $2R$ and $2L$ modes are quite close to one another when $T < 11$ sec and intersect twice at $T \approx 6$ and 10 sec. Now suppose some layer in the model to be weakly anisotropic. Then, according to Crampin (1981), the degenerate $2L$ and $2R$ modes are replaced by the so-called generalized $3G$ and $4G$ modes. What is the relation between their dispersion curves and the $2L$ and $2R$ curves? It seems that they must be extremely close to one another and coalesce in the limit (no anisotropy). However, modes in the anisotropic model are numbered in a different way from the isotropic one, where Love and Rayleigh modes are numbered independently, namely, $C_{4G} > C_{3G}$ for any period (see Figure 6.15). As a result, when anisotropy vanishes, adjacent spectral portions for a mode ($3G$ or $4G$) become spectral portions for different modes in an isotropic medium ($2L$ or $2R$, as the case may be). Naturally enough, such portions are dramatically different in polarization when there is weak anisotropy. This creates the impression of higher mode polarization being particularly sensitive to anisotropy.

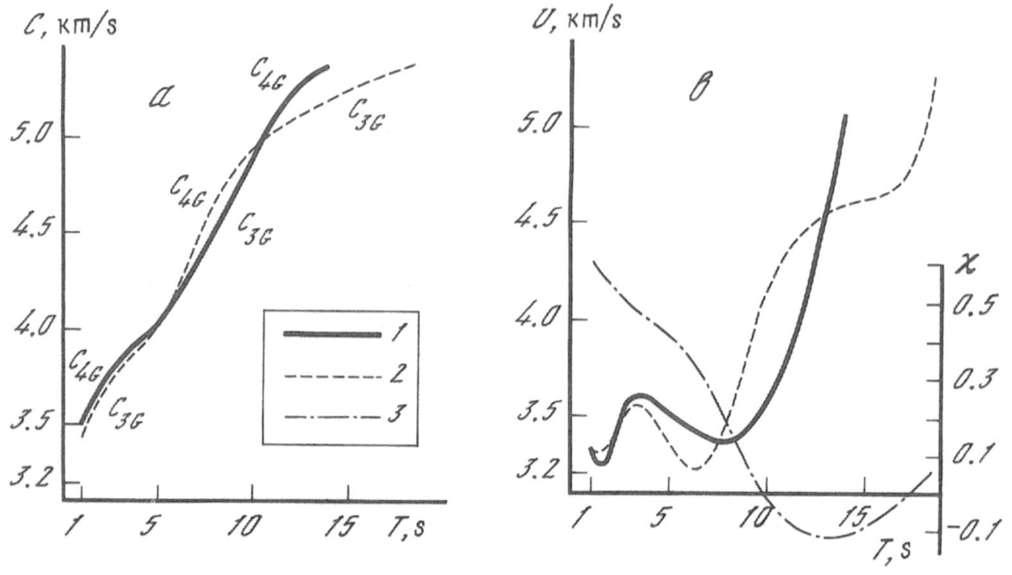

Fig. 6.15. Spectral characteristics of higher modes $2R$ and $2L$. (a) phase velocity, (b) group velocity and ellipticity. (1) $2L$ wave, (2) $2R$ wave; (3) ellipticity χ for $2R$ wave.

As a matter of fact, it is only a sum of $3G$ and $4G$ modes that has a physical meaning; the combined oscillation produces a displacement whose vertical and radial component are quite similar to that in the $2R$ wave, and whose tangential component is similar to the $2L$ wave. As anisotropy gets stronger, C_{3G} and C_{4G} depart further from C_{2L} and C_{2R} producing azimuth-dependent polarization effects. The inadequacy in the description of weakly coupled interfering fields (quasi-SH waves and quasi-P, -SV waves in our case) by means of a single set of eigenfunctions which one encounters here is also well-known in other wave-motion situations.

As the group velocities of higher $2R$ and $2L$ modes are close to one another (Figure 6.15b), analyses of three-component records within the relevant ranges of time and period always deal with combined oscillations whose polarization depends on the relation between the intensities of these modes. The relation is rather variable, being a function of source mechanism and depth, as well as of the source-station geometry. There is an important quantity, however, that depends on these factors but slightly, namely, the amplitude ratio of vertical to radial component in the time domain, because this is largely controlled by the spectral ratio between displacement amplitudes at the surface for the $2R$ mode, $\chi_2(T)$, which is a function of the model only. Figure 6.15 shows this ratio as a function of period for our model. One can see that the radial displacement at the surface in the range 8 to 15 sec (i.e., in the group velocity range 4.6 to 3.5 km/sec) does not exceed 0.1 of the vertical in absolute value; consequently, in contrast to the fundamental $1R$ mode, the $2R$ mainly carries vertical displacements.

Theoretical seismograms of higher modes were computed from the asymptotic formulas (1.31) for a point source at $h = 3$ km depth that modelled pure shear slip on a vertical fault plane. The time history in the interval ($0 \leqslant t \leqslant \tau$) followed the $0.5\,(1 - \cos(\pi t/\tau))$ law, the parameter τ being taken equal to 3 sec to counteract the cut-off effect in high frequecy spectrum. The radiation toward the receiver was assumed to occur with the same intensity for Love and Rayleigh waves, corresponding to an angle of $\pi/8$ between the direction of slip and the line toward the station. The resulting seismograms and particle motion due to the combined surface wave in the $z\varphi$-plane (a vertical plane perpendicular to the direction of wave propagation) are presented in Figure 6.16. Practically the whole motion occurs in that plane and is decidedly elliptic (because the phase difference between the $2L$ wave and the vertical $2R$ component is close either to $\pi/2$ or $3\pi/2$). Crampin (1981) notes a similar character of motion for his analyses of experimental higher mode seismograms. The ellipses may strongly vary in shape, depending on the relation of Love and Rayleigh wave intensities radiated in a given direction, but the radial displacement is persistently small. Thus, the observed effects of higher mode polarization can be accounted for in the frame of vertically varying models.

Transmission of surface waves through a laterally varying medium. Consider an Earth consisting of two vertically varying quarter-spaces in contact that have different distributions of elastic wave velocities or density over depth. It follows from the results of Chapter 3 that, when both source and receiver are far from the contact compared with the wavelengths of interest, the observed dispersion is close to that obtained by averaging the dispersions in each medium:

$$\bar{C} = \frac{l_1 + l_2}{l_1/C_1 + l_2/C_2}, \qquad \bar{U} = \frac{l_1 + l_2}{l_1/U_1 + l_2/U_2}.$$

The subscripts 1 and 2 denote the left and right paths, l_1 and l_2 being the respective path lengths. Suppose observations have provided such averaged dispersion curves of Love and Rayleigh waves, and we seek a laterally homogeneous model that would fit the observations. If the desired model has too few adjustable parameters, the problem may prove to be intractable. In such a case the use of anisotropic laterally homogeneous models, which are intrinsically defined to have more parameters, may yield a solution that would fit the observations, but poorly represent the real geophysical situation. We take as an example the results of model computations and their interpretation for a fairly simple layered model (Figure 6.17a). Love and Rayleigh wave dispersion in each medium is illustrated by the plots in Figure 6.17 b–d; also shown are fictitious 'observed' curves for a long path divided by the contact into equal paths. The parameters of the crustal model with a vertical contact are: thickness in medium 1 is 35 km, that in medium 2 is 50 km; velocity $a = 6.0$ km/sec, $b = 3.46$ km/sec, density $\rho = 2.8$ g/cm^3; velocities in the underlying half-space are $a = 7.72$ km/sec, $b = 4.60$ km/sec, $\rho =$ 3.3 g/cm^3.

The 'hedgehog' method (Valyus, 1968) was used to search for three-layered models fitting the 'observed' dispersion curves of phase and group velocities in the range 15 to 90 sec. We varied shear velocities and thicknesses in all the layers. Models have been found that fit either Love or Rayleigh wave dispersion to within 0.01 km/sec; no model has been found, however, that would fit both types of data to within 0.02 km/sec rms, the maximum misfit for some periods being as large as 0.05 km/sec or even larger. Under these conditions, the appeal to anisotropy

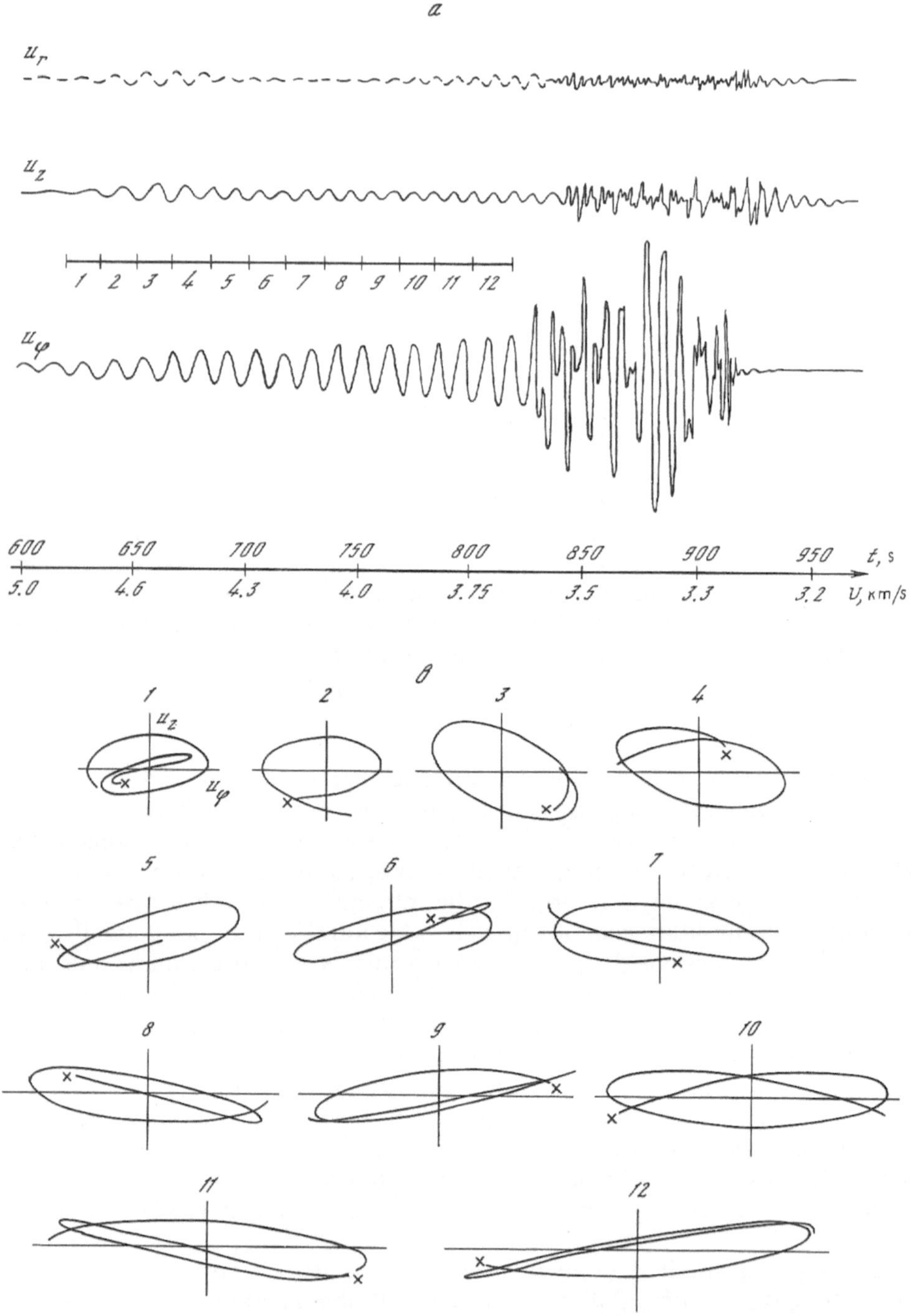

Fig. 6.16. Polarization of higher modes in an isotropic model. (a) theoretical seismograms: u_z, u_r components correspond to $2R$ mode, u_φ component corresponds to $2L$ mode. The dashed portion of u_r is magnified by a factor of 10; (b) particle motion in the total surface wave (projected onto the $z\varphi$-plane); the snapshots are for the intervals shown below the u_z trace.

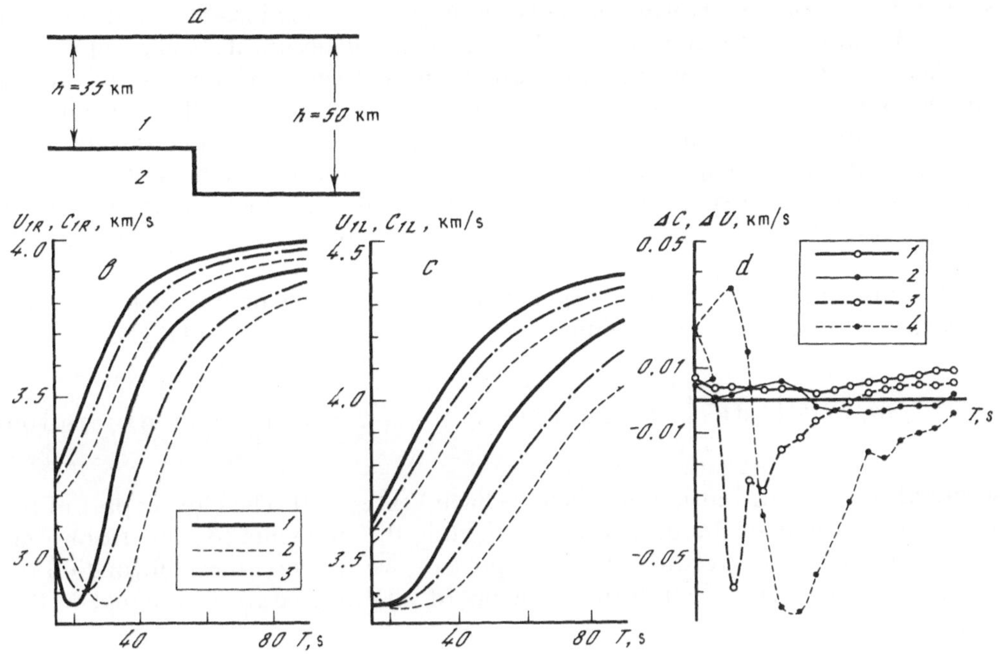

Fig. 6.17. Influence of lateral inhomogeneity on the interpretation of 'observed' dispersion. (a) the structure, 1 and 2 being layer numbers; (b) phase and group velocity curves of Rayleigh waves in laterally homogeneous models with the layer thickness 35 km (1), 50 km (2), and the 'observed' curves (3); (c) the same as in (b), Love waves; (d) misfit between 'observed' and theoretical curve for the best model: phase (1) and group (2) velocities for Love waves, phase (3) and group (4) velocities for Rayleigh waves.

could seem natural, but the resulting "well adjusted" model would not have provided a correct geophysical interpretation of the data.

In our opinion, such effects can have been operative in deriving the current Earth model: discrepancies between types of observations for the upper mantle (obtained by inevitable averaging of data over regions of significantly different deep structure) were considered by Anderson and Dziewonski (1982), Dziewonski and Anderson (1981) as indicating a transversely anisotropic upper-mantle model.

The above examples should by no means be considered as refuting the existence of anisotropy: they merely alert one to be wary of introducing new, ever more complicated models on the basis of observations that do not provide sufficient resolution and do not admit of a unique interpretation.

6.9. ESTIMATION OF EARTHQUAKE SOURCE PARAMETERS

In 1.2 an earthquake source region was defined as a region at each point of which the tensor $\dot{\Gamma}_{ij}$ is not identically zero. The source duration is the time during which anelastic motion occurs at various points within the source region, i.e., $\dot{\Gamma}_{ij}$ is different from zero. Along with these concepts, we shall deal with an instantaneous source, i.e., a volume involved in anelastic motion at some instant t, and the local source duration, i.e., the time during which such motion occurs at some point $\mathbf{x}$.

This section contains formulas for estimating the geometry of a source region and the time-averaged geometry of an instantaneous source, as well as formulas for estimating the source duration and the space-averaged local source duration. All these parameters are expressed in terms of spatio-temporal moments of $\dot{\Gamma}_{ij}(\mathbf{x}, t)$ of total degree (both in space and time) 0, 1, and 2. In turn, these moments can, as shown below, be estimated by using long-period records of the displacement at a number of sites at the free surface.

The moment $\dot{\Gamma}^{(m,n)}(\mathbf{q}, \tau)$ of spatial degree m and temporal degree n with respect to point $\mathbf{q}$ and instant of time τ is a tensor of order $m+2$ and is given by

$$\dot{\Gamma}^{(m,n)}_{ij;k_1\ldots k_m}(\mathbf{q}, \tau) = \int_\Omega \mathrm{d}V_x \int_0^\infty \dot{\Gamma}_{ij}(\mathbf{x}, t)\,(x_{k_1} - q_{k_1}) \ldots (x_{k_m} - q_{k_m})\,(t-\tau)^n\,\mathrm{d}t \qquad (6.106)$$

where Ω is a volume outside of which we have $\dot{\Gamma}(\mathbf{x}, t) \equiv 0$. The closing part of this section is devoted to the question of extracting the moments (6.106) from long-period displacement records: for the present we discuss the information on earthquake source involved in the moments of total degree $m+n$, equal to 0, 1 and 2.

When i and j are fixed, the moments (6.106) characterize the spatio-temporal configuration of the scalar field $\dot{\Gamma}_{ij}(\mathbf{x}, t)$. A seismic source is described by six (because $\dot{\Gamma}_{ij}$ is symmetric) different scalar functions, making an interpretation of its moments difficult. Following Backus (1977), we set up a correspondence between an earthquake source and the scalar field

$$c(\mathbf{x}, t) = \dot{\Gamma}^{(0,0)}_{ij}\,\dot{\Gamma}_{ij}(\mathbf{x}, t) \qquad (6.107)$$

regarding the geometrical and temporal parameters of this field as estimates of the respective source parameters.

Note that $\dot{\Gamma}^{(0,0)}_{ij}$ is a limiting value (as $t \to \infty$) of the seismic moment tensor $\int_\Omega \Gamma_{ij}(\mathbf{x}, t)\,\mathrm{d}V_x$ (see Kostrov, 1970). The moments of $c(\mathbf{x}, t)$ are given by a formula similar to (6.106):

$$c^{(m,n)}_{k_1\ldots k_m}(\mathbf{q}, \tau) = \int_\Omega \mathrm{d}V_x \int_0^\infty c(\mathbf{x}, t)\,(x_{k_1} - q_{k_1}) \ldots (x_{k_m} - q_{k_m})\,(t-\tau)^n\,\mathrm{d}t. \qquad (6.108)$$

If the moments of $\dot{\Gamma}_{ij}(\mathbf{x}, t)$ are known, the moments of $c(\mathbf{x}, t)$ can be got from

$$c^{(m,n)}_{k_1\ldots k_m}(\mathbf{q}, \tau) = \dot{\Gamma}^{(0,0)}_{ij}\,\dot{\Gamma}^{(m,n)}_{ij;k_1\ldots k_m}. \qquad (6.109)$$

Estimation of temporal and geometrical parameters of an earthquake source. Suppose we know the moments (6.108) of $c(\mathbf{x}, t)$ as given by (6.107). The inequality $c(\mathbf{x}, t) \geqslant 0$ is assumed to be true. According to Backus (1977), many types of seismic sources are consistent with it. Thus, it is readily shown that, when the source is a plain ideal fault (the displacement discontinuity vector lies in the rupture plane) the inequality is equivalent to the requirement of the slip velocity vector being within $\pi/2$ of some fixed direction. Backus (1977) puts forward some estimates of the temporal and geometrical parameters of an earthquake source based on the assumption of $c(\mathbf{x}, t)$ being nonnegative. Source location is estimated by the spatial centroid $\mathbf{q}_c$ of the field $c(\mathbf{x}, t)$ defined as

$$\mathbf{q}_c = \int_\Omega \mathrm{d}V_x \int_0^\infty c(\mathbf{x}, t)\, \mathbf{x}\mathrm{d}t \Bigg/ \int_\Omega \mathrm{d}V_x \int_0^\infty c(\mathbf{x}, t)\, \mathrm{d}t, \tag{6.110}$$

which can be written in the form

$$c^{(0,0)}\mathbf{q}_c = \mathbf{c}^{(1,0)}(0). \tag{6.111}$$

If $m(\mathbf{x}) = \int_0^\infty c(\mathbf{x}, t)\, \mathrm{d}t$ is regarded as a distribution of mass in space, its center of mass is identical with the centroid $\mathbf{q}_c$. In a similar fashion, the temporal centroid τ_c is estimated by

$$c^{(0,0)}\tau_c = c^{(1,0)}(0). \tag{6.112}$$

The values of $\mathbf{q}_c$ and τ_c combined define the spatio-temporal centroid of a source. The source duration is estimated by $2\Delta\tau$, where

$$(\Delta\tau)^2 = \int_\Omega \mathrm{d}V_x \int_0^\infty c(\mathbf{x}, t)\,(t - \tau_c)^2\, \mathrm{d}t \Bigg/ \int_\Omega \mathrm{d}V_x \int_0^\infty c(\mathbf{x}, t)\, \mathrm{d}t \tag{6.113}$$

or, in our notation,

$$(\Delta\tau)^2 = c^{(0,2)}(\tau_c)/c^{(0,0)}. \tag{6.114}$$

Let $\mathbf{r}$ be a unit vector. The mean source size along $\mathbf{r}$ is estimated by $2l_r$, where

$$l_r^2 = \int_\Omega \mathrm{d}V_x \int_0^\infty c(\mathbf{x}, t)\,[(\mathbf{x} - \mathbf{q}_c)\cdot \mathbf{r}]^2\, \mathrm{d}t \Bigg/ \int_\Omega \mathrm{d}V_x \int_0^\infty c(\mathbf{x}, t)\, \mathrm{d}t. \tag{6.115}$$

We define the matrix

$$\mathbf{W} = \mathbf{c}^{(2,0)}(\mathbf{q}_c)/c^{(0,0)}. \tag{6.116}$$

Then l_r^2 can be written as a quadratic form:

$$l_r^2 = \mathbf{r}^T \mathbf{W}\, \mathbf{r} \tag{6.117}$$

where $\mathbf{r}^T$ is the transpose of $\mathbf{r}$.

From (6.117) it follows that a source region has the least size along that eigenvector of $\mathbf{W}$ corresponding to the least eigenvalue and the greatest size along that eigenvector of the same matrix corresponding to the greatest eigenvalue.

To interpret the moment $\mathbf{c}^{(1,1)}$, the last of the moments of $c(\mathbf{x}, t)$ of total degree 0, 1, and 2, we consider the following problem. Among points $\mathbf{x} = \mathbf{x}_0 + \mathbf{v}t$ that move at uniform velocity, find the one around which $c(\mathbf{x}, t)$ is concentrated in the best manner in the sense of making the function

$$\Phi(\mathbf{x}_0, \mathbf{v}) = \int_\Omega \mathrm{d}V_x \int_0^\infty c(\mathbf{x}, t)\,(\mathbf{x} - \mathbf{x}_0 - \mathbf{v}t)^T(\mathbf{x} - \mathbf{x}_0 - \mathbf{v}t)\, \mathrm{d}t \tag{6.118}$$

a minimum. As shown by Backus (1977), the solution is

$$\mathbf{x}_0 = \mathbf{q}_c - \mathbf{v}\tau_c, \qquad \mathbf{v} = \mathbf{w}/(\Delta\tau)^2 \tag{6.119}$$

where

$$\mathbf{w} = \mathbf{c}^{(1,1)}(\mathbf{q}_c, \tau_c)/c^{(0,0)}.$$

Instantaneous source and estimation of its mean geometrical parameters. Anelastic motion involves various portions of the space as a result of source action. The source region discussed above is the union of all these portions. An instantaneous source is understood to be a time-variable volume which at each instant of time includes points where anelastic motion occurs, i.e., the glut stresses are functions of time. In our opinion, even approximate estimates of time-averaged geometrical parameters for such a region may be useful in studying the processes taking place at the earthquake source.

Thus, a point $\mathbf{x}$ at time t_0 belongs to an instantaneous source, if $\dot{\Gamma}_{ij}(\mathbf{x}, t)$ is not identically zero within any vicinity of t_0. In a similar fashion to the above, we consider a scalar field $c(\mathbf{x}, t)$ given by (6.117) instead of a tensor field $\dot{\Gamma}_{ij}(\mathbf{x}, t)$, at each instant of time the instantaneous source being a region at each point of which $c(\mathbf{x}, t)$ is not identically zero within any vicinity of t_0.

An instantaneous centroid is understood to be $\mathbf{q}(t)$ as given by

$$\mathbf{q}(t) = \int_\Omega c(\mathbf{x}, t)\mathbf{x}\, \mathrm{d}V_x \bigg/ \int_\Omega c(\mathbf{x}, t)\, \mathrm{d}V_x. \tag{6.120}$$

As $c(\mathbf{x}, t) \geqslant 0$, (6.120) defines the function for which the following functional attains its minimum

$$F[\mathbf{z(t)}] = \int_0^\infty \mathrm{d}t \int_\Omega c(\mathbf{x}, t)\,[\mathbf{x} - \mathbf{z}(t)]^T\,[\mathbf{x} - \mathbf{z}(t)]\, \mathrm{d}V_x. \tag{6.121}$$

As to the times t at which $c(\mathbf{x}, t) \equiv 0$ for all $\mathbf{x}$, $\mathbf{q}(t)$ may be defined in any manner for these.

Similarly to (6.115), the mean size of an instantaneous source along $\mathbf{r}$ is estimated by $2d_r$, where

$$d_r^2 = \int_\Omega \mathrm{d}V_x \int_0^\infty c(\mathbf{x}, t)\,\{[\mathbf{x} - \mathbf{q}(t)]\cdot\mathbf{r}\}^2\, \mathrm{d}t \bigg/\!\!\bigg/ \int_\Omega \mathrm{d}V_x \int_0^\infty c(\mathbf{x}, t)\, \mathrm{d}t. \tag{6.122}$$

An estimate of d_r is thus obtained from knowledge of $\mathbf{q}(t)$ that minimizes (6.121). The solution (6.120) is expressed in terms of the spatial moments of $c(\mathbf{x}, t)$ which we do not know. We wish to express d_r in terms of the spatio-temporal moments (6.108) of $c(\mathbf{x}, t)$. In the general case this can only be done approximately. Namely, the $\mathbf{q}(t)$ in (6.122) will be fitted by a function $\mathbf{y}(t)$ that minimizes (6.121) among polynomials of degree n.

We are going to show that such a minimum exists for any n and is unique.

$\mathbf{y}(t)$ is a polynomial of degree n:

$$\mathbf{y}(t) = \sum_{j=0}^{n} \mathbf{a}_j(t - \tau_c)^j. \tag{6.123}$$

Let the ith component of the vector $\mathbf{a}_j$ be α_{ij}. Then

$$y_i(t) = \sum_{j=0}^{n} \alpha_{ij}(t-\tau_c)^j, \qquad i = 1, 2, 3 \tag{6.124}$$

and

$$F[\mathbf{y}(t)] = \int_{\Omega} \mathrm{d}V_x \int_0^{\infty} c(\mathbf{x}, t) \sum_{i=1}^{3} \left[x_i - \sum_{j=0}^{n} \alpha_{ij}(t-\tau_c)^j \right]^2 \mathrm{d}t. \tag{6.125}$$

Differentiation of (6.125) with respect to α_{ij} gives

$$\frac{\partial F}{\partial \alpha_{ij}} = -2 \int_{\Omega} \mathrm{d}V_x \int_0^{\infty} c(\mathbf{x}, t) \left[x_i - \sum_{k=0}^{n} \alpha_{ik}(t-\tau_c)^k \right] (t-\tau_c)^j \, \mathrm{d}t. \tag{6.126}$$

From the condition of stationarity $\partial F/\partial \alpha_{ij} = 0$ in the notation of (6.108), we get a set of $3(n+1)$ equations for α_{ik}:

$$c_i^{(1,j)}(0, \tau_c) - \sum_{k=0}^{n} \alpha_{ij} c^{(0,k+j)}(\tau_c) = 0,$$
$$i = 1, 2, 3; \qquad j = 0, 1, \ldots, n. \tag{6.127}$$

We impose an ordering upon the α_{ij} and $c_i^{(1,j)}$, setting up a correspondence between these and $3(n+1)$-dimensional vectors $\mathbf{b}$ and $\mathbf{p}$ in such a manner that

$$\alpha_{ij} = b_k, \quad c_i^{(1,j)}(0, \tau_c) = p_k, \quad k = (i-1)(n+1) + j. \tag{6.128}$$

(6.127) then yields the following equation for $\mathbf{b}$:

$$\mathbf{p} = \mathbf{B}\mathbf{b}, \tag{6.129}$$

where $\mathbf{B}$ is a $3(n+1) \times 3(n+1)$ block matrix:

$$\mathbf{B} = \begin{Vmatrix} \mathbf{A} & 0 & 0 \\ 0 & \mathbf{A} & 0 \\ 0 & 0 & \mathbf{A} \end{Vmatrix}$$

the elements of $\mathbf{A}$ being given by

$$A_{ij} = c^{(0, i+j-2)}(\tau_c), \qquad i, j = 1, \ldots, n+1. \tag{6.130}$$

From (6.126) one can see that the second derivatives of F with respect to b_k are

$$\frac{\partial^2 F}{\partial b_i \, \partial b_j} = 2B_{ij}, \tag{6.131}$$

where the B_{ij} are elements of $\mathbf{B}$.

The matrix **A** can be shown to be positive definite. For,

$$A_{ij} = (f_i, f_j), \tag{6.132}$$

where $f_i = (t - \tau_c)^{i-1}$, while the scalar product (f_i, f_j) is given by:

$$(f_i, f_j) = \int_0^\infty \nu(t) f_i(t) f_j(t) \, \mathrm{d}t, \tag{6.133}$$

where $\nu(t) = \int_\Omega c(\mathbf{x}, t) \, \mathrm{d}V_x \geqslant 0$. From (6.132) it follows that **A** is Gramm's matrix and is positive definite, because the f_i are linearly independent. Note that det $\mathbf{B} = (\det \mathbf{A})^3$ and, since **A** is positive definite, so is **B**. Now in such a case (6.129) has a unique solution that minimizes F.

To sum up, the solution of (6.129) defines coefficients to fit $\mathbf{q}(t)$ by a polynomial of degree n. Substituting the resulting polynomial $\mathbf{y}(t)$ for $\mathbf{q}(t)$ in (6.122), one can express an approximate value of the mean instantaneous source size along any direction in terms of the spatio-temporal moments of $c(\mathbf{x}, t)$.

We are going to derive explicit formulas for a linear approximation of $\mathbf{q}(t)$, i.e., when $n = 1$. The minimizing function (6.125) is then identical with (6.118) discussed in Backus (1977). Dividing equation (6.129) by $c^{(0,0)}$, we can write its solution in the form

$$\mathbf{b}^T = (q_{c_1}, w_1/(\Delta\tau)^2, q_{c_2}, w_2/(\Delta\tau)^2, q_{c_3}, w_3/(\Delta\tau)^2) \tag{6.134}$$

whence

$$\mathbf{y}(t) = \mathbf{q}_c + (t - \tau_c)\mathbf{w}/(\Delta\tau)^2, \tag{6.135}$$

i.e., $\mathbf{y}(t)$ is the radius-vector of a point moving at a constant velocity $\mathbf{w}/(\Delta\tau)^2$, and which is at $\mathbf{q}_c$ at time τ_c. Note that $\mathbf{y}(t)$ is identical with the solution (6.119) minimizing (6.118). In the derivation of (6.134) and (6.135) we utilized the relations

$$\mathbf{c}^{(1,1)}(0, \tau_c) = \mathbf{c}^{(1,1)}(\mathbf{q}_c, \tau_c), \qquad c^{(0,1)}(\tau_c) = 0.$$

Substituting $\mathbf{y}(t)$ from (6.135) into (6.122) and denoting the approximate value of d_r by $\hat{d}_r$, we get

$$\hat{d}_r^2 = \mathbf{r}^T \left(\mathbf{W} - \frac{\mathbf{w}\mathbf{w}^T}{(\Delta\tau)^2} \right) \mathbf{r}. \tag{6.136}$$

Note that the estimate $\hat{d}_r$ is an upper bound on d_r, i.e., the inequality $d_r \leqslant \hat{d}_r$ holds.

Within the approximation considered, an instantaneous source has the least mean size along the eigenvector of $\mathbf{W} - [\mathbf{w}\mathbf{w}^T/(\Delta\tau)^2]$ corresponding to the least eigenvalue, while the greatest size is along the eigenvector corresponding to the greatest eigenvalue of the same matrix.

Estimation of averaged local source duration. Anelastic motion is excited at different points of the source region during different time intervals that make up the source duration whose estimation was discussed above. The local source duration at point $\mathbf{x}_0$ is understood to be the time interval during which anelastic motion occurs at the point, i.e., $c(\mathbf{x}_0, t)$ is different from zero.

The local time centroid $\tau(\mathbf{x})$ is defined to be

$$\tau(\mathbf{x}) = \int_0^\infty c(\mathbf{x}, t)\, t\mathrm{d}t \bigg/ \int_0^\infty c(\mathbf{x}, t)\, \mathrm{d}t. \tag{6.137}$$

Here $\mathbf{x}$ belongs to the source region, so that $\int_0^\infty c(\mathbf{x}, t)\, \mathrm{d}t \neq 0$.

Since $c(\mathbf{x}, t) \geqslant 0$, (6.137) minimizes the functional

$$Q[\varphi(\mathbf{x})] = \int_\Omega \mathrm{d}V_x \int_0^\infty c(\mathbf{x}, t)\, [t - \varphi(\mathbf{x})]^2\, \mathrm{d}t, \tag{6.138}$$

i.e., $\varphi(\mathbf{x}) = \tau(\mathbf{x})$ is the point where Q is a minimum.

The space-averaged local source duration will be estimated by $2\Delta\theta$ where, similarly to (6.113),

$$(\Delta\theta)^2 = \int_\Omega \mathrm{d}V_x \int_0^\infty c(\mathbf{x}, t)\, [t - \tau(\mathbf{x})]^2\, \mathrm{d}t \bigg/ \int_\Omega \mathrm{d}V_x \int_0^\infty c(\mathbf{x}, t)\, \mathrm{d}t. \tag{6.139}$$

From (6.138), (6.139) it follows that

$$(\Delta\theta)^2 = Q[\tau(\mathbf{x})]/c^{((0,0))}. \tag{6.140}$$

The estimate $\Delta\theta$ is this obtained by minimizing Q. In a similar fashion to the estimation of the mean instantaneous source sizes, $\Delta\theta$ can only approximately be expressed in terms of the spatio-temporal moments of $c(\mathbf{x}, t)$. $\tau(\mathbf{x})$ in (6.140) will be fitted by a function $\theta(\mathbf{x})$ that minimizes Q among polynomials of degree n in $\mathbf{x}$. We are going to prove the existence and uniqueness of this minimum. We begin by considering a general case in which $c(\mathbf{x}, t)$ is distributed within some three-dimensional volume. Let $\theta(\mathbf{x})$ be a polynomial of degree n:

$$\theta(\mathbf{x}) = \sum_{|\alpha| \leqslant n} a_\alpha (\mathbf{x} - \mathbf{q}_c)^\alpha, \tag{6.141}$$

where $\alpha = (k, l, m)$ is a multiindex; $|\alpha| = k + l + m$; $\mathbf{x}^\alpha = x_1^k x_2^l x_3^m$.

If N is the number of all possible values of α such that $|\alpha| \leqslant n$, i.e., the number of terms in (6.141), then

$$N = (n + 1)(n + 2)(n + 3)/6. \tag{6.142}$$

Let the set of possible values of α be ordered in some manner and $\alpha_i = (k_i, l_i, m_i)$, its ith value. Then we have

$$\begin{aligned}\theta(\mathbf{x}) &= \sum_{i=1}^N a_i (\mathbf{x} - \mathbf{q}_c)^{\alpha_i} \\ &= \sum_{i=1}^N a_i (x_1 - q_{c_1})^{k_i} (x_2 - q_{c_2})^{l_i} (x_3 - q_{c_3})^{m_i},\end{aligned} \tag{6.143}$$

where $a_i = a_{\alpha_i}$ is the ith component of the N-dimensional vector $\mathbf{a}$ (the coefficients in (6.143)), and

$$Q[\theta(\mathbf{x})] = \int_\Omega \mathrm{d}V_x \int_0^\infty c(\mathbf{x}, t)\,[t - $$

$$- \sum_{i=1}^{N} a_i (x_1 - q_{c_1})^{k_i} (x_2 - q_{c_2})^{l_i} (x_3 - q_{c_3})^{m_i}]^2 \,\mathrm{d}t. \tag{6.144}$$

Differentiation of (6.144) with respect to a_i gives

$$\frac{\partial Q}{\partial a_i} = -2 \int_\Omega \mathrm{d}V_x \int_0^\infty c(\mathbf{x}, t) \times$$

$$\times (x_1 - q_{c_1})^{k_i} (x_2 - q_{c_2})^{l_i} (x_3 - q_{c_3})^{m_i} \times$$

$$\times \left[t - \sum_{j=1}^{N} a_j (x_1 - q_{c_1})^{k_j} (x_2 - q_{c_2})^{l_j} (x_3 - q_{c_3})^{m_j} \right] \mathrm{d}t. \tag{6.145}$$

The conditions of stationarity $\partial Q/\partial a_i = 0$ yield N equations for the a_i:

$$b_i = D_{ij}\, a_j, \qquad i, j = 1, \ldots, N \tag{6.146}$$

where

$$b_i = \int_\Omega \mathrm{d}V_x \int_0^\infty c(\mathbf{x}, t)\, (x_1 - q_{c_1})^{k_i} (x_2 - q_{c_2})^{l_i} (x_3 - q_{c_3})^{m_i}\, t \,\mathrm{d}t \tag{6.147}$$

are elements of $\mathbf{c}^{(p_i, 1)}(\mathbf{q}_c, 0)$, $\qquad p_i = k_i + l_i + m_i$;

$$D_{ij} = \int_\Omega \mathrm{d}V_x \int_0^\infty c(\mathbf{x}, t)\, (x_1 - q_{c_1})^{k_i + k_j} (x_2 - q_{c_2})^{l_i + l_j} \times \tag{6.148}$$

$\times (x_3 - q_{c_3})^{m_i + m_j}\, \mathrm{d}t$ are elements of $\mathbf{c}^{(p_i + p_j, 0)}(\mathbf{q}_c)$.

Differentiation of (6.145) with respect to a_j gives the matrix of second derivatives of Q:

$$\frac{\partial^2 Q}{\partial a_i\, \partial a_j} = 2 D_{ij}. \tag{6.149}$$

Let us show that the $\mathbf{D}$ whose elements are given by (6.148) is positive definite. To do this, consider the set of functions

$$f_i = (x_1 - q_{c_1})^{k_i} (x_2 - q_{c_2})^{l_i} (x_3 - q_{c_3})^{m_i}$$

and define a scalar product (f_i, f_j) as

$$(f_i, f_j) = \int_\Omega \nu(\mathbf{x})\, f_i(\mathbf{x})\, f_j(\mathbf{x})\, \mathrm{d}V_x, \tag{6.150}$$

where $\nu(\mathbf{x}) = \int_0^\infty c(\mathbf{x}, t)\, dt \geqslant 0$. Then

$$D_{ij} = (f_i, f_j) \tag{6.151}$$

whence it follows that **D** is Gramm's matrix and, since the f_i are linearly independent, it is positive definite. This fact ensures the existence and uniqueness of the solution to (6.146) and the minimum of Q for this solution.

Let the polynomial $\theta(\mathbf{x})$ be a solution of (6.146) and $Q[\theta(\mathbf{x})]$ be the minimum of Q. Then $(\Delta\theta)^2$ will be approximated by

$$(\Delta\hat{\theta})^2 = Q[\theta(\mathbf{x})]/c^{(0,0)}. \tag{6.152}$$

One can easily see that $\Delta\theta \leqslant \Delta\hat{\theta}$ must be true. Since $Q[\theta(\mathbf{x})]$ is of the form (6.144), formula (6.152) expresses $(\Delta\hat{\theta})^2$ in terms of the spatio-temporal moments of $c(\mathbf{x}, t)$.

We consider the linear approximation of $\tau(\mathbf{x})$, i.e., $n = 1$, in some detail. The expression (6.143) for $\theta(\mathbf{x})$ can then be written as

$$\theta(\mathbf{x}) = \tau_0 + \mathbf{u}(\mathbf{x} - \mathbf{q}_c), \tag{6.153}$$

where $\tau_0 = a_1$; $u_1 = a_2$; $u_2 = a_3$; $u_3 = a_4$. From (6.153) one can see that $\theta(\mathbf{x})$ is the arrival time at point $\mathbf{x}$ for a plane with the normal $\mathbf{u}/|\mathbf{u}|$ propagating at velocity $u/(\mathbf{u}\cdot\mathbf{u})$, so that $\theta(\mathbf{q}_c) = \tau_0$. We denote this plane by Σ_θ. Here $\mathbf{u}$ is the slowness vector.

Using the relations

$$\mathbf{c}^{(1,0)}(\mathbf{q}_c) = 0 \quad \text{and} \quad \mathbf{c}^{(1,1)}(\mathbf{q}_c, 0) = \mathbf{c}^{(1,1)}(\mathbf{q}_c, \tau_c),$$

one can reduce (6.146) to the set of equations

$$\tau_c = \tau_0, \qquad \mathbf{w} = \mathbf{W}\,\mathbf{u}, \tag{6.154}$$

whence we obtain

$$\theta(\mathbf{x}) = \tau_c + \mathbf{w}^T\mathbf{W}^{-1}(\mathbf{x} - \mathbf{q}_c), \tag{6.155}$$

$$(\Delta\hat{\theta})^2 = (\Delta\tau)^2 - \mathbf{w}^T\mathbf{W}^{-1}\,\mathbf{w}. \tag{6.156}$$

All these relations are also valid for a $c(\mathbf{x}, t)$ that is concentrated on a plane or a straight line. In these cases $\mathbf{x}$ entering the expressions for $\theta(\mathbf{x})$ and for the moments of $c(\mathbf{x}, t)$ is to be treated as a two-dimensional vector or scalar.

Formulas (6.114) and (6.117) provide estimates for total source duration and the mean source sizes, formulas (6.156) and (6.136) estimate space-averaged local source duration and the time-averaged sizes of an instantaneous source. We now derive a relation connecting these quantities. Let $2l_u$ estimate the source size along the normal to Σ_θ, and let $2\hat{d}_u$ be the time-averaged size of an instantaneous source in the same direction. Then we put $\mathbf{r} = \mathbf{u}/\sqrt{\mathbf{u}\cdot\mathbf{u}}$ in (6.117) and (6.136) and, remembering that $\mathbf{u} = \mathbf{W}^{-1}\,\mathbf{w}$, derive from (6.117), (6.136) and (6.156):

$$\hat{d}_u/l_u = \Delta\hat{\theta}/\Delta\tau. \tag{6.157}$$

Provided the local source duration is small compared with the total source duration, i.e., $\Delta\theta/\Delta\tau \ll 1$, it follows from (6.117), (6.136) and (6.156) that the approximate relation $\Delta\tau \approx |\mathbf{u}|\, l_u$ holds, so from (6.157) we get

$$\hat{d}_u/\Delta\hat{\theta} \approx 1/|\mathbf{u}| \tag{6.158}$$

(one should remember that $1/|\mathbf{u}|$ is the velocity of the plane Σ_θ). Relations (6.157) and (6.158) demonstrate consistency between the estimates of spatial (6.117), (6.136) and temporal (6.113), (6.156) parameters of an earthquake source.

Consider an example, a source which is a particular case of the plane ideal fault. Suppose the displacement discontinuity has a constant direction **b** and the rupture propagates in a rectangular plane region Σ at a constant velocity **v** along a side of the rectangle.

The displacement discontinuity $[\mathbf{u}(\mathbf{x}, t)]$ is thus given on Σ in the form

$$[\mathbf{u}(\mathbf{x}, t)] = \mathbf{b}\Delta u(\mathbf{x}, t), \tag{6.159}$$

where **b** is a constant unit vector; $\Delta u(\mathbf{x}, t) = |[\mathbf{u}(\mathbf{x}, t)]|$. If **n** is the unit normal to the plane of rupture Σ, the stress glut tensor for such a source has the form (1.15). Substituting $[\mathbf{u}(\mathbf{x}, t)]$ as given by (6.159) into these formulas, we get

$$\dot{\Gamma}_{jk}(\mathbf{x}, t) = \mu(\mathbf{x})\, \Delta \dot{u}(\mathbf{x}, t)\, (b_j n_k + b_k n_j)\, \delta_\Sigma,$$

whence

$$\dot{\Gamma}_{jk}^{(0,0)} = (b_j n_k + b_k n_j) \int_\Sigma \mu(\mathbf{x})\, \Delta u(\mathbf{x}, \infty)\, d\Sigma.$$

Since the movement is an ideal one, the displacement discontinuity $[\mathbf{u}(\mathbf{x}, t)]$ lies on Σ. Consequently, **b** lies on the same plane and is orthogonal to **n**, i.e., $b_j n_j = 0$ is true, yielding

$$\sum_{j,k=1}^{3} (b_j n_k + b_k n_j)^2 = 2.$$

Using this relation and setting the shear modulus $\mu(\mathbf{x})$ egual to a constant value μ on the rupture plane, we get for the scalar field $c(\mathbf{x}, t)$ defined by (6.107):

$$c(\mathbf{x}, t) = \alpha \Delta \dot{u}(\mathbf{x}, t)\, \delta_\Sigma, \tag{6.160}$$

where $\alpha = 2\mu^2 \int_\Sigma \Delta u(\mathbf{x}, \infty)\, d\Sigma$.

Now we choose a set of coordinates in which the rupture plane is identical with the (x_1, x_2)-plane, and the sides of the rectangle where the rupture propagates are parallel to the axes: the velocity of rupture propagation is parallel to the x_1-axis and the rupture front to the x_2-axis. The law that governs rupture propagation can then be given in the form

$$\Delta u(\mathbf{x}, t) = \begin{cases} f(t - x_1/v), & \mathbf{x} \in \Sigma \\ 0, & \mathbf{x} \notin \Sigma, \end{cases} \tag{6.161}$$

where Σ is the rectangular region in question

$$(0 < x_1 < L_1, 0 < x_2 < L_2, x_3 = 0);$$

$f(\tau)$ is a function which steadily increases from zero for $\tau \leqslant 0$ to some limiting value D for $T \leqslant \tau$. The derivative $\varphi(\tau) = \dot{f}(\tau)$ is a nonnegative function vanishing for $\tau \leqslant 0$ and $T \leqslant \tau$. From formulas (6.160) and (6.161) for $c(\mathbf{x}, t)$ we have

$$c(\mathbf{x}, t) = \alpha\varphi(t - x_1/v)\,\delta(x_3),$$

where $\alpha = 2\mu^2 L_1 L_2 D$. The spatio-temporal moments of $c(\mathbf{x}, t)$ are expressed in terms of the model parameters as follows:

$$c^{(0,0)} = 2(\mu L_1 L_2 D)^2, \tag{6.162}$$

$$\mathbf{q}_c^T = (L_1/2, L_2/2, 0), \tag{6.163}$$

$$\tau_c = \tau_\varphi + L_1/(2v), \tag{6.164}$$

$$W_{11} = L_1^2/12, \quad W_{22} = L_2^2/12, \; W_{33} = W_{12} = W_{13} = W_{23} = 0, \tag{6.165}$$

$$\mathbf{w}^T = (L_1^2/(12v), 0, 0), \tag{6.166}$$

$$(\Delta\tau)^2 = (\Delta\tau_\varphi)^2 + L_1^2/(12v^2), \tag{6.167}$$

where $\tau_\varphi = \int_0^\infty \varphi(\tau)\,\tau\,d\tau / \int_0^\infty \varphi(\tau)\,d\tau$ is the time centroid of $\varphi(\tau) = \dot{f}(\tau)$; $(\Delta\tau_\varphi)^2 = \int_0^\infty \varphi(\tau)\,(\tau - \tau_\varphi)^2\,d\tau / \int_0^\infty \varphi(\tau)\,d\tau$; $2\Delta\tau_\varphi$ is an estimate of the time of $f(\tau)$ growth.

Denoting the estimates of the mean source sizes by $2l_{x_1}$, $2l_{x_2}$, $2l_{x_3}$, we obtain from (6.165) and (6.117):

$$l_{x_1} = L_1/(2\sqrt{3}), \quad l_{x_2} = L_2/(2\sqrt{3}), \quad l_{x_3} = 0.$$

In analogous notation, the estimates of the mean instantaneous source sizes along the respective directions are given by

$$\hat{d}_{x_1} = (\Delta\tau_\varphi/\Delta\tau)L_1/(2\sqrt{3}), \quad \hat{d}_{x_2} = l_{x_2}, \quad \hat{d}_{x_3} = 0.$$

Formula (6.156) combined with (6.165), (6.166) and (6.167) provides an estimate for the mean local source duration:

$$\Delta\hat{\theta} = \Delta\tau_\varphi. \tag{6.168}$$

Since, for the model considered, (6.154) together with (6.166) gives

$$\mathbf{u}^T = (1/v, 0, 0), \tag{6.169}$$

relations (6.157) and (6.158) become

$$\hat{d}_{x_1}/l_{x_1} = \Delta\hat{\theta}/\Delta\tau, \qquad \hat{d}_{x_1}/\Delta\hat{\theta} \approx v.$$

The above formulas can be used to describe earthquake sources by means of this model. They estimate the model parameters using the spatio-temporal moments of $\dot{\Gamma}_{ij}$ of total degree 0, 1 and 2. Suppose we know those moments and they are consistent with the hypothesis of a plane ideal fault. The moments of $c(\mathbf{x}, t)$ are obtained from (6.109). The axes of the set of coordinates defined above are parallel to the eigenvectors of $\mathbf{W}$, the x_3-axis being parallel to the eigenvector for the zero eigenvalue and the x_1-axis parallel to $\mathbf{w}$ ($\mathbf{w}$ must be parallel to an eigenvector of $\mathbf{W}$, otherwise our model would not describe the source). Transforming $\mathbf{W}$ to the diagonal form, we estimate the sizes of the rectangle Σ from (6.165). Substitution of the resulting values of L_1, L_2 and $\mathbf{q}_c$ into (6.163) gives the location of the coordinate origin at the corner of Σ. Solving (6.154) for $\mathbf{u}$ and substituting this into (6.169), we obtain rupture velocity v. Fixing a value of the shear modulus around the rupture, we can estimate the maximum slip D from (6.162). Lastly, (6.164) and (6.168) yield the time centroid of $\varphi(\tau)$ and the time of slip growth $2\Delta\tau_\varphi$. The above relations thus provide estimates of the quantities characterizing source processes based on the moments of the stress glut tensor as determined from long-period records of seismic waves. These same relations can be used to estimate parameters in some models of the earthquake source.

Relation between the moments of a stress glut tensor and the moments of an equivalent force. The displacement field excited by a source with the stress glut tensor $\Gamma_{ij}(x, t)$ is the solution to the boundary value problem (1.1) through (1.6) (the force f_i should be replaced by the equivalent force g_i given by (1.12)). Since the problem is correctly formulated, the equivalent force $\mathbf{g}(\mathbf{x}, t)$ is uniquely determined by the displacement field $\mathbf{u}(\mathbf{x}, t)$ due to it, while the stress glut tensor $\Gamma_{ij}(\mathbf{x}, t)$, as can be deduced from (1.12), is determined by the displacements (through $\mathbf{g}(\mathbf{x}, t)$), apart from the tensor $\tilde{\Gamma}_{ij}(\mathbf{x}, t)$ for which the following equality is true

$$\tilde{\Gamma}_{ij,j} \equiv 0.$$

Determination of $g_i(\mathbf{x}, t)$ based on (1.1) through (1.3) requires knowledge of the displacement $\mathbf{u}(\mathbf{x}, t)$ at any point within the medium. If we are interested in the moments of the equivalent force $g^{(m,n)}_{i;k_1\ldots k_m}$ or in those of its time derivative $\dot{g}^{(m,n)}_{i;k_1\ldots k_m}$, these can, as we show below, be expressed in terms of the displacement or its spectrum at some finite number of points at the free surface (the moments $\mathbf{g}^{(m,n)}$ and $\dot{\mathbf{g}}^{(m,n)}$ are given by (6.106) with Γ_{ij} replaced by g_i or $\dot{g}_i$, respectively).

We wish to find out whether the moments of $\dot{\Gamma}_{ij}$ can be expressed in terms of those of the equivalent force. From the definition of moments of $\dot{g}_i$ and $\dot{\Gamma}_{ij}$ (6.106), formula (1.12), and the Gauss—Ostrogradsky theorem we have

$$g_i^{(0,n)}(\tau) = \dot{g}_i^{(0,n)}(\tau) \equiv 0, \tag{6.170}$$

$$\dot{g}^{(m,n)}_{i;k_1\ldots k_m}(\mathbf{q}, \tau) = \dot{\Gamma}^{(m-1,n)}_{ik_1;k_2\ldots k_m}(\mathbf{q}, \tau) + \dot{\Gamma}^{(m-1,n)}_{ik_2;k_3\ldots k_m k_1}(\mathbf{q}, \tau) + \ldots + \dot{\Gamma}^{(m-1,n)}_{ik_m;k_1\ldots k_{m-1}}(\mathbf{q}, \tau), \tag{6.171}$$

$$n \geqslant 0, \quad m \geqslant 1; \quad i, k_1, \ldots, k_m = 1, 2, 3.$$

Relation (6.170) is natural, since the zero (over the space) moment of $\mathbf{g}$ is the resultant of internal forces. The summation on the right-hand side of (6.171) involves all elements of the moment $\dot{\Gamma}^{(m-1,n)}$ obtained by cyclic permutation of $k_1, \ldots, k_m$.

Fixing m and n, and assuming some values of $i, k_1, \ldots, k_m$, one can obtain from (6.171) a set of equations for the elements of $\dot{\Gamma}^{(m-1,n)}_{ij;k_1\ldots k_{m-1}}(\mathbf{q}, \tau)$. When $m = 1$, (6.171) becomes

$$\dot{g}^{(1,n)}_{i;k}(\mathbf{q}, \tau) = \dot{\Gamma}^{(0,n)}_{ik}(\tau). \tag{6.172}$$

From (6.106) and the symmetry of $\dot{\Gamma}_{ij}$ one deduces the symmetry of $\dot{\Gamma}^{(m,n)}_{ij;k_1\ldots k_m}$ with respect to i, j and $k_1, \ldots, k_m$, the moment $\dot{g}^{(m,n)}_{i;k_1\ldots k_m}$ being symmetric with respect to $k_1, \ldots, k_m$. Formula (6.172) gives the result that $\dot{g}^{(1,n)}_{i;k}$ is symmetric with respect to i and k. In the general case the number of different elements of $\dot{\Gamma}^{(m-1,n)}_{ij;k_1\ldots k_{m-1}}$

which has the above symmetries is given by

$$N_\Gamma = 3m(m+1).$$

To sum up, the number of unknowns in (6.171) is N_Γ. The number of equations is determined by the number of different elements of $\dot{g}^{(m,n)}_{i;k_1\ldots k_m}$; taking its symmetry for $m \geqslant 2$ into account, this number is

$$N_g = 3(m+1)(m+2)/2.$$

The discussion in Backus and Mulcahy (1976a) shows that the equations are linearly independent. Thus, when $m \geqslant 2$, formula (6.171) defines a set of N_g equations in N_Γ unknowns, and

$$N = N_\Gamma - N_g = 3(m+1)(m-2)/2. \tag{6.173}$$

It is easy to see that we have $N = 0$ when $m = 2$, i.e., the number of equations equals the number of unknowns. Formula (6.171) then becomes

$$\dot{g}^{(2,n)}_{i;k_1k_2}(\mathbf{q},\tau) = \dot{\Gamma}^{(1,n)}_{ik_1;k_2}(\mathbf{q},\tau) + \dot{\Gamma}^{(1,n)}_{ik_2;k_1}(\mathbf{q},\tau). \tag{6.174}$$

A cyclic permutation of i, k_1, k_2 on the left-hand side of (6.174) yields

$$\dot{g}^{(2,n)}_{k_1;k_2i}(\mathbf{q},\tau) = \dot{\Gamma}^{(1,n)}_{k_1k_2;i}(\mathbf{q},\tau) + \dot{\Gamma}^{(1,n)}_{k_1i;k_2}(\mathbf{q},\tau), \tag{6.175}$$

$$\dot{g}^{(2,n)}_{k_2;ik_1}(\mathbf{q},\tau) = \dot{\Gamma}^{(1,n)}_{k_2i;k_1}(\mathbf{q},\tau) + \dot{\Gamma}^{(1,n)}_{k_2k_1;i}(\mathbf{q},\tau). \tag{6.176}$$

The unique solution of (6.174) through (6.176) is

$$\dot{\Gamma}^{(1,n)}_{ik_1;k_2}(\mathbf{q},\tau) = 0.5[\dot{g}^{(2,n)}_{i;k_1k_2}(\mathbf{q},\tau) + \dot{g}^{(2,n)}_{k_1;k_2i}(\mathbf{q},\tau)_i - \dot{g}^{(2,n)}_{k_2;ik_1}(\mathbf{q},\tau)]. \tag{6.177}$$

To sum up, the moments of $\dot{\Gamma}^{(m-1,n)}$ for $m \leqslant 2$ are uniquely expressed in terms of $\dot{\mathbf{g}}^{(m,n)}$ by formulas (6.172) and (6.177). When $m \geqslant 2$, the number of unknowns in (6.171) exceeds the number of equations (see (6.173)) and moments of the stress glut tensor cannot be uniquely expressed in terms of those of the equivalent force. This can however be done by using some prior information on the source (Backus, 1977).

We shall demonstrate, by way of example, how the moments of $\dot{\Gamma}^{(m-1,n)}$ can be uniquely expressed in terms of those of $\dot{\mathbf{g}}^{(m,n)}$, assuming that the source is a plane ideal fault. Let Σ be the plane of displacement discontinuity with the unit normal $\mathbf{n}$. Then the stress glut tensor is given by (1.15). For the moment $\dot{\Gamma}^{(0,0)}_{jk}$ we have

$$\dot{\Gamma}^{(0,0)}_{jk} = n_j a_k + n_k a_j \tag{6.178}$$

where

$$a_k = \int_0^\infty dt \int_\Sigma \mu(\mathbf{x})\,[\dot{u}_k(\mathbf{x},t)]\,d\Sigma = \int_\Sigma \mu(\mathbf{x})\,[u_k(\mathbf{x},\infty)]\,d\Sigma.$$

Since $[u_k(\mathbf{x},t)]$ lies in Σ, the vector $\mathbf{a}$ lies in the same plane and is orthogonal to $\mathbf{n}$, i.e., $a_k n_k = 0$ is true. Let $\mathbf{b} = \mathbf{a} \times \mathbf{n}$ be a vector that is orthogonal both to $\mathbf{a}$ and $\mathbf{n}$. We then have

$$n_k b_k = a_k b_k = 0.$$

Multiplying both sides of (6.178) by b_k and summing over k, we get $\dot{\Gamma}^{(0,0)}_{jk} b_k = 0$, i.e., the matrix $\dot{\Gamma}^{(0,0)}$ has the eigenvector **b** corresponding to the zero eigenvalue. It is also easy to see that $\dot{\Gamma}^{(0,0)}$ has a zero trace, for from (6.178) we have

$$\operatorname{Tr} \dot{\Gamma}^{(0,0)} = 2n_k a_k = 0.$$

However, since the sum of the eigenvalues of $\dot{\Gamma}^{(0,0)}$ equals the trace, while one of the eigenvalues is zero, it follows that the sum of the other two must be zero. Hence, denoting the eigenvalues of $\dot{\Gamma}^{(0,0)}$ as $\lambda_1, \lambda_2, \lambda_3$, we have $\lambda_1 = \lambda$, $\lambda_2 = -\lambda$, $\lambda_3 = 0$. Since $\dot{\Gamma}^{(0,0)}$ is symmetric, the eigenvectors are mutually orthogonal and form a basis on which $\dot{\Gamma}^{(0,0)}_{jk}$ is

$$\dot{\Gamma}^{(0,0)}_{jk} = \lambda \delta_{j1}\, \delta_{k1} - \lambda \delta_{j2}\, \delta_{k2}, \tag{6.179}$$

i.e., all elements of $\dot{\Gamma}^{(0,0)}_{jk}$ are zero, except $\dot{\Gamma}^{(0,0)}_{11} = \lambda$ and $\dot{\Gamma}^{(0,0)}_{22} = -\lambda$. Multiplying both sides of (6.178) by n_k and summing over k, we get

$$\dot{\Gamma}^{(0,0)}_{jk}\, n_k = a_j. \tag{6.180}$$

On the other hand, from (6.179) it follows that the following relation must hold on the basis composed of the eigenvectors of $\dot{\Gamma}^{(0,0)}$:

$$\dot{\Gamma}^{(0,0)}_{jk}\, n_k = \lambda (n_1\, \delta_{j1} - n_2\, \delta_{j2}). \tag{6.181}$$

On the same basis, from (6.180) and (6.181) we have

$$a_j = \lambda (n_1\, \delta_{j1} - n_2\, \delta_{j2}).$$

Multiplying this by n_j, we get $\lambda(n_1^2 - n_2^2) = 0$ or $n_1^2 = n_2^2$ ($a_3 = n_3 = 0$, $\mathbf{a} \perp \mathbf{b}$, $\mathbf{n} \perp \mathbf{b}$). Hence **n** bisects the vertical angles made by those eigenvectors of $\dot{\Gamma}^{(0,0)}$ corresponding to nonzero eigenvalues. The vector **a** bisects the other pair of angles made by the same vectors. One cannot distinguish between the two vectors from knowledge of the moment $\dot{\Gamma}^{(0,0)}$ alone. Only two elements of $\dot{\Gamma}^{(0,0)}$ are different from zero on the basis $(\mathbf{n}, \mathbf{a}/|\mathbf{a}|, \mathbf{b}/|\mathbf{b}|)$, namely, $\dot{\Gamma}^{(0,0)}_{21} = \dot{\Gamma}^{(0,0)}_{21} = |\mathbf{a}|$, as can be deduced from (6.178). On the same basis, the only elements of $\dot{\Gamma}_{ij}(\mathbf{x}, t)$ that are different from zero in the general case are $\dot{\Gamma}_{12} = \dot{\Gamma}_{21}$, $\dot{\Gamma}_{13} = \dot{\Gamma}_{31}$.

Suppose the moments $\dot{\mathbf{g}}^{(m,n)}(\mathbf{q}_0, \tau_0)$ are knwon from observed displacements. From (6.106) one can see that, knowing the moments of a function with respect to the point $\mathbf{q}_0$ and time τ_0, one can obtain the moments with respect to any other point **q** and any other instant τ. The moments $\dot{\Gamma}^{(0,n)}$ and $\dot{\Gamma}^{(1,n)}$ for any n can be found from (6.172) and (6.177). The values of $\mathbf{q}_c$ and τ_c are obtained from (6.109), (6.111), and (6.112). Suppose $\dot{\Gamma}^{(0,0)}$ has the eigenvalues $\lambda_1 = -\lambda_2$ and $\lambda_3 = 0$. The basis is chosen to be the unit vectors $\mathbf{r}_1$, $\mathbf{r}_2$, and $\mathbf{r}_3$, where $\mathbf{r}_1$ and $\mathbf{r}_2$ are parallel to the bisectors of the angles made by those eigenvectors of $\dot{\Gamma}^{(0,0)}$ corresponding to λ_1 and λ_2, while $\mathbf{r}_3$ is the eigenvector corresponding to $\lambda_3 = 0$. Only two elements of $\dot{\Gamma}^{(0,0)}$ are different from zero on this basis: $\dot{\Gamma}^{(0,0)}_{12} = \dot{\Gamma}^{(0,0)}_{21}$. Assuming the source to be a plane ideal fault, we can assert that we must have either $\mathbf{n}^T = (1, 0, 0)$ or $\mathbf{n}^T = (0, 1, 0)$ on the basis chosen, where **n** is the vector normal to Σ, the plane of displacement discontinuity. Besides, it follows from (1.15) that the diagonal elements $\dot{\Gamma}_{ii}(\mathbf{x}, t)$ of $\dot{\Gamma}(\mathbf{x}, t)$ are zero. This allows one to express the elements $\dot{\Gamma}^{(2,0)}_{12;11}(\mathbf{q}_c) = \dot{\Gamma}^{(2,0)}_{21;11}(\mathbf{q}_c)$ and $\dot{\Gamma}^{(2,0)}_{12;22}(\mathbf{q}_c) = \dot{\Gamma}^{(2,0)}_{21;22}(\mathbf{q}_c)$ of the tensor $\dot{\Gamma}^{(2,0)}(\mathbf{q}_c)$ uniquely in terms of the respective elements of the tensor $\dot{\mathbf{g}}^{(3,0)}(\mathbf{q}_c)$. For, since $\dot{\Gamma}^{(2,0)}_{11;21} = \dot{\Gamma}^{(2,0)}_{22;12} = 0$, from (1.171) we have

$$\dot{\Gamma}^{(2,0)}_{12;11}(\mathbf{q}_c) = \dot{g}^{(3,0)}_{1;211}(\mathbf{q}_c), \qquad \dot{\Gamma}^{(2,0)}_{12;22}(\mathbf{q}_c) = \dot{g}^{(3,0)}_{2;122}(\mathbf{q}_c). \tag{6.182}$$

Further, from (6.109), (6.172), and (6.182) we get

$$c_{11}^{(2,0)}(\mathbf{q}_c) = 2\dot{g}_{1;2}^{(1,0)}\, \dot{g}_{1;211}^{(3,0)}(\mathbf{q}_c)$$
$$c_{22}^{(2,0)}(\mathbf{q}_c) = 2\dot{g}_{1;2}^{(1,0)}\, \dot{g}_{2;122}^{(3,0)}(\mathbf{q}_c). \tag{6.183}$$

Let $2l_1$ be the mean source size along $\mathbf{r}_1$, and $2l_2$ that along $\mathbf{r}_2$. Then from (6.116) and (6.117) we have

$$l_1^2 = c_{11}^{(2,0)}(\mathbf{q}_c)/c^{(0,0)}, \qquad l_2^2 = c_{22}^{(2,0)}(\mathbf{q}_c)/c^{(0,0)}$$

and, if our assumption about the source is true, than at least one of the quantities in (6.183) is zero, because in that case either $\mathbf{r}_1$ or $\mathbf{r}_2$ coincides with the normal to the rupture plane.

Let $c_{11}^{(2,0)}(\mathbf{q}_c) = 0$, and $c_{22}^{(2,0)}(\mathbf{q}_c) \neq 0$. In that case $\mathbf{n} = \mathbf{r}_1$ and the rupture plane Σ coincides with the $x_1 = q_{c_1}$ plane, whence (see (6.106)) all those elements $\dot{\Gamma}_{ij;k_1\ldots k_m}^{(m,n)}(\mathbf{q}_c, \tau)$ of the moment $\dot{\Gamma}^{(m,n)}(\mathbf{q}_c, \tau)$ are zero for which at least one of the k_i equals 1. If no one of the k_i equals 1, then the desired relation can be derived from (6.171) using the symmetry of $\dot{\Gamma}_{ij}(\mathbf{x}, t)$ and its moments with respect to the suffixes i and j, and remembering that only the elements $\dot{\Gamma}_{12} = \dot{\Gamma}_{21}$, $\dot{\Gamma}_{13} = \dot{\Gamma}_{31}$ are different from zero:

$$\dot{\Gamma}_{ij;k_1\ldots k_m}^{(m,n)}(\mathbf{q}_c, \tau) = (\delta_{i1}\,\delta_{j2} + \delta_{i2}\,\delta_{j1})\,\dot{g}_{2;1k_1\ldots k_m}^{(m+1,n)}(\mathbf{q}_c, \tau) + $$
$$+ (\delta_{i1}\,\delta_{j3} + \delta_{i3}\,\delta_{j1})\,\dot{g}_{3;1k_1\ldots k_m}^{(m+1,n)}(\mathbf{q}_c, \tau). \tag{6.184}$$

When $c_{22}^{(2,0)}(\mathbf{q}_c) = 0$, while $c_{11}^{(2,0)}(\mathbf{q}_c) \neq 0$, then $\mathbf{n} = \mathbf{r}_2$, and the elements $\dot{\Gamma}_{ij;k_1\ldots k_m}^{(m,n)}(\mathbf{q}_c, \tau)$ are zero, provided at least one of the k_i equals 2. The relation derived from (6.184) by replacing the superscripts 1 by 2 and 2 by 1 holds for the nonzero elements of $\dot{\Gamma}^{(m,n)}$.

If, simultaneously, $c_{11}^{(2,0)}(\mathbf{q}_c) = 0$ and $c_{22}^{(2,0)}(\mathbf{q}_c) = 0$, then the source region is a straight segment along $\mathbf{r}_3$:

$$x_1 = q_{c_1}, \quad x_2 = q_{c_2}.$$

In that case only those elements $\dot{\Gamma}_{ij;k_1\ldots k_m}^{(m,n)}(\mathbf{q}_c, \tau)$ are different from zero which satisfy the following conditions: $k_1 = k_2 = \ldots = k_m = 3$ and i, j do not equal 3 simultaneously. Let $i \neq 3$. From (6.171) we then have

$$\dot{\Gamma}_{ij;k_1\ldots k_m}^{(m,n)}(\mathbf{q}_c, \tau) = \dot{g}_{j;ik_1\ldots k_m}^{(m+1,n)}(\mathbf{q}_c, \tau)\,\delta_{k_1 3}\ldots\delta_{k_m 3}. \tag{6.185}$$

We have thus shown the moments $\dot{\Gamma}^{(m,n)}$ to be uniquely representable in terms of moments $\dot{\mathbf{g}}^{(m+1,n)}$ for any m and n on the assumption of an ideal plane slip source.

Relation between the displacement field and the moments of the equivalent force. We are going to discuss relations that connect observed displacements with the moments of the equivalent force and can be used to estimate the moments.

Replacing f_i by $\dot{g}_j$ in (1.8), we get a formula that expresses the displacement u_i in terms of the derivative of the equivalent force $\dot{g}_j$:

$$u_i(\mathbf{x}, t) = \int_0^{t_c} \mathrm{d}\tau \int_\Omega H_{ij}(\mathbf{x}, \mathbf{y}, t - \tau)\,\dot{g}_j(\mathbf{y}, \tau)\,\mathrm{d}V_y.$$

Replacing in this expression the function $H_{ij}(\mathbf{x}, \mathbf{y}, t-\tau)$ by its Taylor series in powers of $\mathbf{y}$ and in powers of τ, we get

$$u_i(\mathbf{x}, t) = \sum_{m=1}^{\infty} \sum_{n=0}^{\infty} \frac{(-1)^n}{m!\, n!} \dot{g}^{(m,n)}_{j;k_1 \ldots k_m}(0, 0) \times$$

$$\times \frac{\partial^n}{\partial t^n} \frac{\partial}{\partial y_{k_1}} \cdots \frac{\partial}{\partial y_{k_m}} H_{ij}(\mathbf{x}, \mathbf{y}, t)|_{y=0}. \tag{6.186}$$

Expanding H_{ij} in powers of $\mathbf{y}$, we assume the elastic parameters to be sufficiently smooth. We have for the Fourier transforms $\hat{u}_i(\mathbf{x}, \omega)$ and $\hat{H}_{ij}(\mathbf{x}, \mathbf{y}, \omega)$ from (6.136):

$$\hat{u}_i(\mathbf{x}, \omega) = \sum_{m=1}^{\infty} \sum_{n=0}^{\infty} \frac{(-1)^n}{m!\, n!} \dot{g}^{(m,n)}_{j;k_1 \ldots k_m}(0, 0) \times$$

$$\times (i\omega)^n \frac{\partial}{\partial y_{k_1}} \cdots \frac{\partial}{\partial y_{k_m}} \hat{H}_{ij}(\mathbf{x}, \mathbf{y}, \omega)|_{y=0}. \tag{6.187}$$

Since (6.186) and (6.187) involve infinite series, these relations cannot be used to compute the moments $\dot{g}^{(m,n)}$. However, when the displacement function $u_i(\mathbf{x}, t)$ and the integral of Green function $H_{ij}(\mathbf{x}, \mathbf{y}, t)$ have been lowpass filtered, the terms in (6.186) and (6.187) start to decrease with m and n increasing and one might then restrict oneself to considering finite sums only.

Most of the low-frequency displacement energy is usually transported in surface waves. Chapter 5 is concerned with methods for separating individual surface wave and measuring the relevant spectral parameters. The long-period displacements or their spectrum in a surface wave can be represented by (6.186) and (6.187), respectively. H_{ij} is to be replaced by the integral of the appropriate long-period surface wave Green function for the model and wave type in hand, and H_{ij} by the spectrum $(i\omega)^{-1} \hat{G}_{ij}$.

The above discussion was based on the most general assumptions about the structure outside the source region. The surface wave Green function can therefore be described both by formula (1.31) for a laterally homogeneous medium and by formulas like (2.29) and (6.104) for various laterally varying models (W^{kD} is defined by (1.32), $\hat{K}_q(\omega) \equiv 1$).

Let $\hat{u}_i(\mathbf{x}, \omega)$ in (6.187) be the low-frequency spectrum of such a wave. From (1.31), (1.32), and (6.104) one can deduce that the quantity $\partial/\partial y_{k_1} \ldots \partial/\partial y_{k_m} \hat{H}_{ij}$ is proportional to ω^m. We choose the time origin so that the source starts at an instant close to $t = 0$, while the origin of coordinates is chosen at a point that is close to the region Ω or belongs to it. Let $L = \max_{x \in \Omega} |\mathbf{x}|$. Then the moment $\dot{g}^{(m,n)}(0, 0)$ does not exceed a value proportional to $L^m t_e^n$ (one recalls that t_e is the source duration). Assuming L and t_e to be proportional quantities for a seismic source, we conclude that, when $\omega t_e < 1$, the terms in (6.187) decay at least as rapidly as $(\omega t_e)^{(m,n)}$. Thus, the infinite sums in (6.186) and (6.187) can, with adequate accuracy for low enough frequencies ω, be replaced by sums involving a few first terms such that $m + n \leqslant M$. Representing in this form displacements (or their spectrum) in surface waves recorded at a number of sites at the free surface, we can derive a set of equations for determining the moments $\dot{g}^{(m,n)}$ of total degree $m + n \leqslant M$. Different types of surface waves and different modes yield independent equations for $\dot{g}^{(m,n)}$, as they contain different kinds of information on the source processes. The moments can be similarly estimated using body wave records (see Pavlov and Gusev, 1980).

We wish to note that the spatio-temporal parameters of an earthquake source cannot be estimated directly from the moments of the equivalent force (without using the moments of the stress glut tensor), because the zero spatial moment of $\dot{g}_i$ is identically zero (see formula (6.170)).

7. SOME RESULTS FROM STUDIES OF REGIONAL LITHOSPHERIC STRUCTURE BY SURFACE WAVES

7.1. THE METHOD OF PATH ELIMINATION IN A STUDY OF THE EURASIAN CRUST

Observations of digital seismological observatories, especially seismic arrays, provide unique opportunities for deriving regional lithospheric structure. This is achieved by using records of surface waves whose epicenter-station paths traverse regions of different tectonic structures. Reliable inference from these observations is facilitated by a high accuracy of raw data ensured by the great dynamic range of digital recording, large signal-to-noise ratios, the possibilities offered by modern methods of space-frequency-time analysis and computer polarization analysis. We are going to demonstrate some of the possibilities offerred by measurements of this type by analysing surface waves recorded at the NORSAR from a variety of seismic sources in Eurasia. Surface wave processing was mainly based on the two techniques described in Chapter 5 implemented as computer programs, namely, space-time and frequency-time analysis.

The vertical component records of Rayleigh surface waves were used for a variety of seismic events within the southern part of the Barents Sea shelf, the Russian plate, East and West Kazakhstan, Soviet Central Asia and China, including a foreshock and the main Gazli earthquake of 1976. Figure 7.1 shows ten surface wave paths from the epicenters to NORSAR, labelled 1 through 10 (the larger numerals).

The dispersion curves of phase and group velocities for fundamental Rayleigh mode are presented in Figure 7.2; they are similar, except along the paths traversing the sub-Caspian depression (paths 7, 8) and the southern part of the Barents Sea Shelf (paths 9, 10). The group velocities there are significantly lower than for the other paths within a wide range of periods. We have used a tectonic map of Eurasia (Yanshin, 1966) to divide the paths into areas (shown by the smaller numerals in the figure) having differing tectonic structures: the Baltic Shield (area 1 for all the paths); the Russian Platform (area 2 for paths 2 through 8); the Barents Sea shelf (area 2 for paths 9, 10); the Urals, Western Siberian plate, and the Kazakh Folded Country (area 3 for paths 3, 4); the sub-Caspian depression (area 3 for paths 6, 7); the Tarbagatai and Tien Shan mountain areas (area 4 for path 4); the Turanian plate (area 4 for paths 6, 7); the area boundaries are shown in Figure 7.1 by bars crossing the paths.

Using paths traversing comparatively homogeneous regions as a reference, we found differential dispersion curves for individual areas. The first step was to find

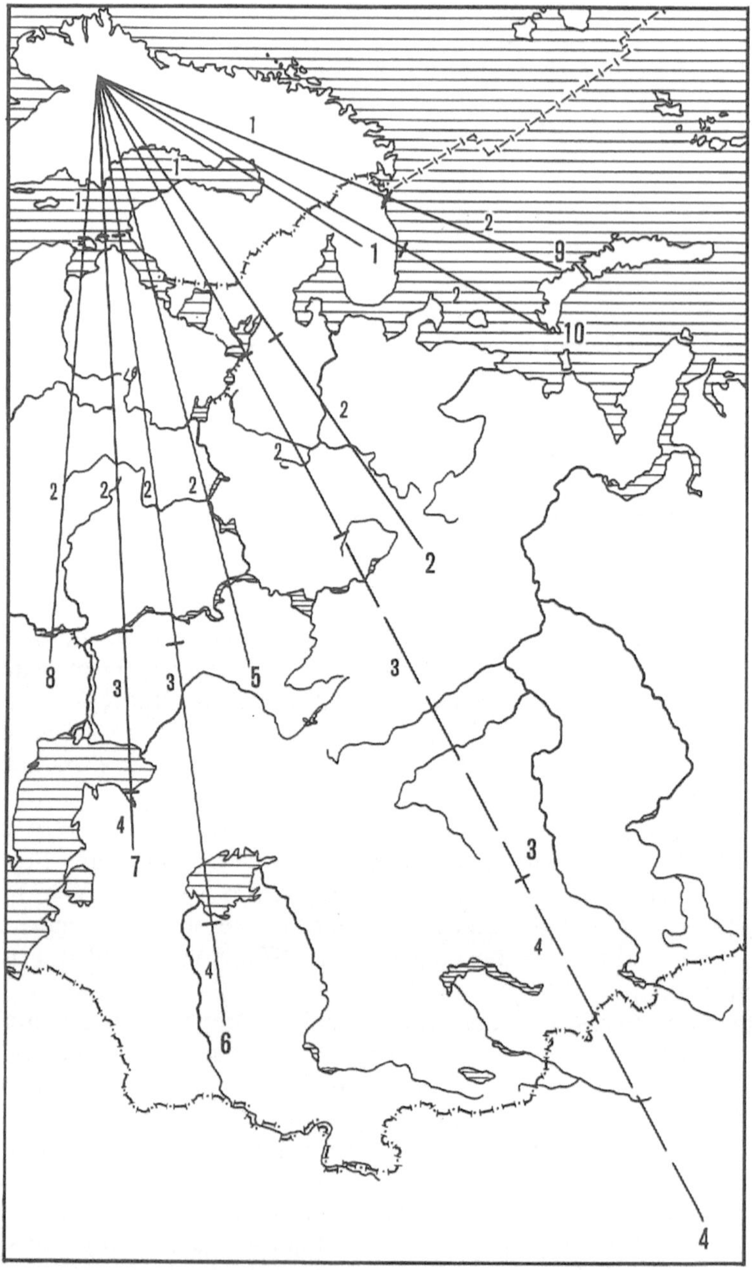

Fig. 7.1. Paths of surface waves from epicenters to NORSAR. Larger Arabic numerals are path numbers, smaller ones denote areas belonging to individual regions.

models that fitted the observed data satisfactorily by the trial-and-error method using computer programs for Rayleigh wave calculations in a vertically varying medium, subsequently refining the model by a computer search according to

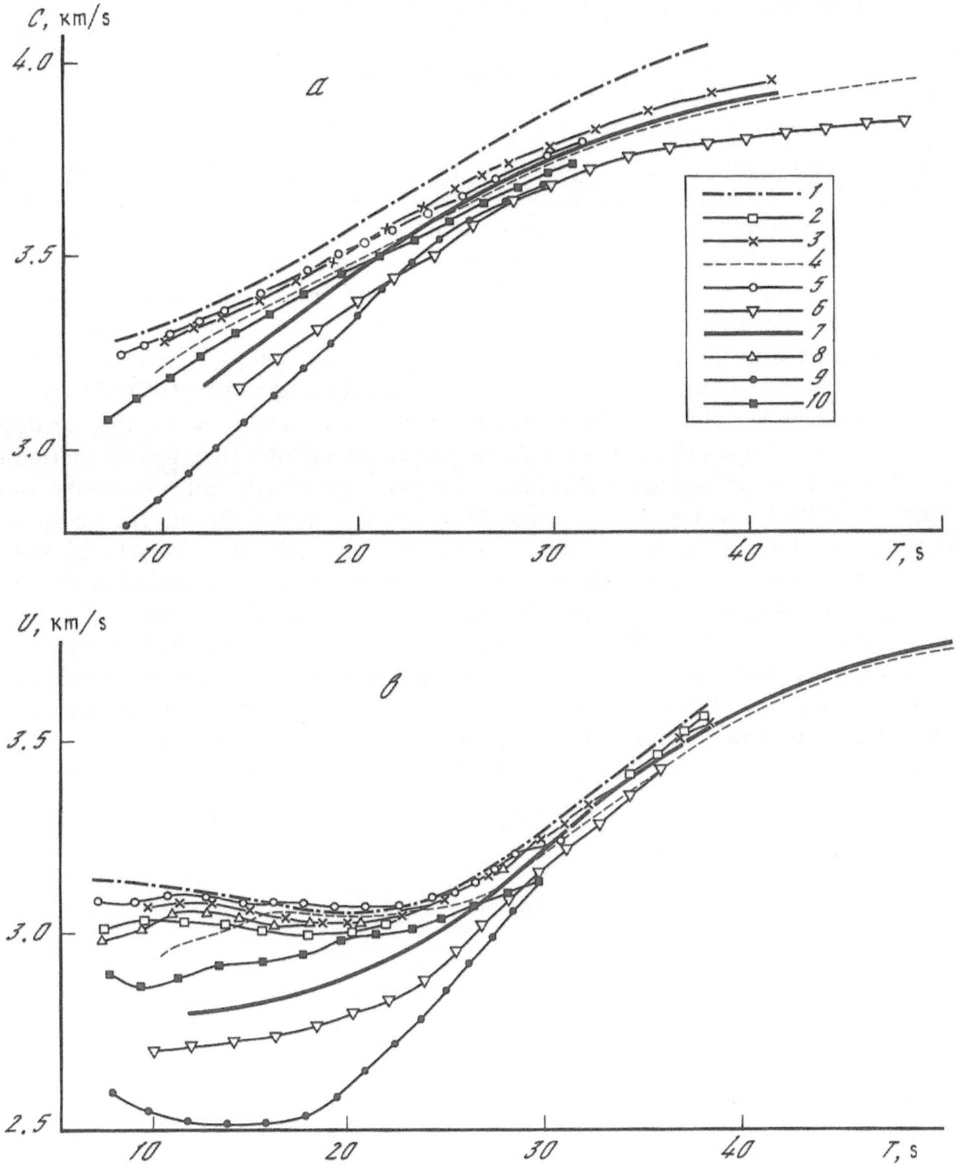

Fig. 7.2. Observed dispersion curves of fundamental mode Rayleigh waves. (a) phase velocities, (b) group velocities, 1 to 10 are path numbers.

Wiggins' (1972) technique. The refinement has not led to significant changes in the vertical velocity profile, but has allowed us to estimate the extreme possible errors and the resolution of the data. It has been established that the most reliably determined are shear velocities in the crust and uppermost mantle, the density and compressional velocities in the crust and upper mantle being less accurate.

Below we discuss these analyses for individual regions.

OBSERVATIONS AND THEIR INTERPRETATION

1. The Baltic Shield

Observations. Path 1 about 1300 km long lies wholly within a single tectonic region, the Baltic Shield. The results of space-time-frequency analysis show the arrival azimuth of the main signal at the array to be very close to the theoretical value (45.2°); no secondary waves above the noise background have been detected. Group and phase velocity curves for the fundamental mode of Rayleigh waves are reliably determined by FTAN in the range 6 to 45 sec. Higher Rayleigh modes have not been identified.

It is of interest to compare our single-station data to other observations on the Baltic Shield (Calcagnile and Panza, 1978; Neunhofer and Guth, 1976; Noponen, 1966) (Figure 7.3). These authors determined phase velocities by the straightforward two-station method; it is only in Noponen (1966) that group velocities were obtained by differentiating phase velocity curves (the observations were conducted in the Finnish part of the Shield in a north—south direction). The phase velocities are close to ours, the misfit is within 0.06 km/sec everywhere, but is systematic, the curve having different slopes. This leads to appreciable discrepancies in group velocity (as large as 0.15 km/sec), especially noticeable at periods around 20—30 sec. The phase velocities determined by Neunhofer and Guth (1976) for the period range 17—35 sec slightly exceed our velocities in the range 20—30 sec (the path, Copenhagen—Pulkovo, traverses the southern portion of the Shield). Calcagnile and Panza (1978) quote data for a number of paths on the Shield, in particular, for the Konsberg—Oulu path, which practically coincides with

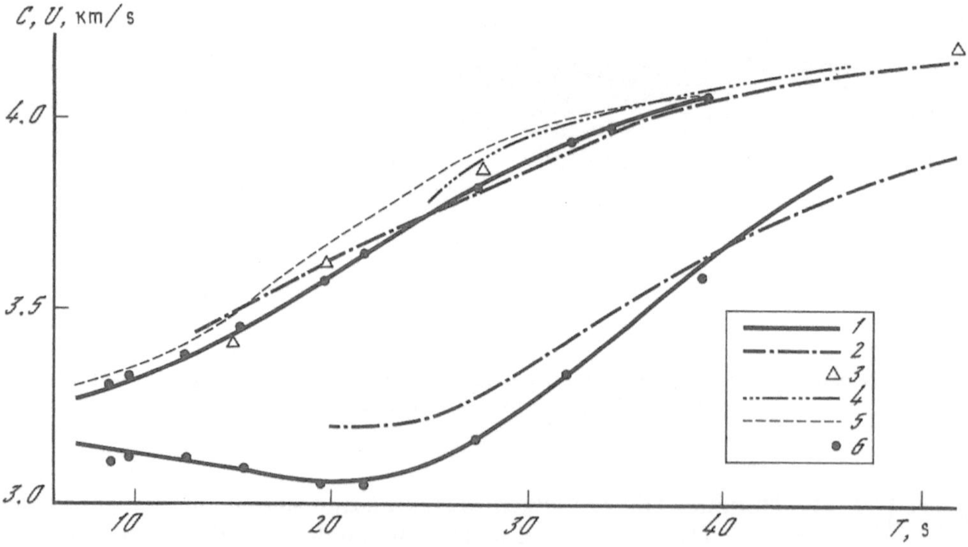

Fig. 7.3. Observed and theoretical curves of phase and group velocity for the Baltic Shield. (1) observations for path 1; (2) data from Noponen (1966); (3) Neunhofer and Guth (1976); (4) Calcagnile and Panza (1978); (5) data for the Canadian Shield; (6) theoretical values for the model.

part of our path. The observed phase velocities are larger that ours by 0.08—0.03 km/sec within an overlapping range of periods.

It is also of interest to compare our observations to the Brune and Dorman (1963) data for the Canadian Shield (see Figure 7.3). Within the entire period range observed, velocities on the Canadian Shield are larger than ours by 0.03—0.1 km/sec, the greatest difference occurring around 20—30 sec.

Interpretation. The similarity of the dispersion curves for different parts, the closeness of the observed arrival azimuth to the theoretical, and the absence of strong later arrivals indicate a high degree of homogeneity throughout the area of study. Since our observations possess a number of advantages over the data of other investigators (digital recording, a high signal-to-noise ratio, absence of possible refraction of rays along epicenter—station paths), we are inclined to consider our observations to be the more accurate. This conclusion justifies an attempt to refine the velocity model for the Shield put forward by Calcagnile and Panza (1978), Der and Landisman (1972), Noponen (1966). Figure 7.4 presents the resulting model, as well as models due to other investigators, and a model for the Canadian Shield. The details are provided in Table 7.I. Our model has slightly lower velocities for the uppermost crust and slightly higher mantle velocities than those in Noponen's (1966) model UL 056 and in other models.

2. The Russian Plate

Observations. We have three paths (2, 5, 8) wholly lying within the Baltic Shield and the Russian plate.

Path 2 about 2400 km long traverses the central Baltic Shield and the northern Moscow Syneclise, some portion of the path going across the southern end of the Timan Range. The length of the 'shield' part is about 1500 km, the 'platform' part (including the Timan rise) is 900 km. Space-time analysis gives an arrival azimuth that is slightly but significantly different from the theoretical value (68.7°), deviating by about 2.5° southward in the range of maximum amplitudes. No later arrivals of noticeable amplitude have been detected. Frequency-time analysis definitely identifies, apart from the fundamental mode of Rayleigh waves in the period range 8—42 s, the first higher mode with periods of 6—30 s (Figure 7.5). The level of maximum amplitude in the fundamental mode is about 25 times that in the first higher mode.

Path 5 traverses the southeastern Baltic Shield and the adjacent part of the Russian plate, the Moscow Syneclise, the Volga—Urals anteclise, terminating in the junction region of the sub-Urals depression, the sub-Caspian depression, and the southeasternmost Russian plate. The lengths of the 'shield' and 'platform' parts of paths 5_1 and 5_2 are equal to 840 km and 1990 km, respectively. The signal is rather weak, but the azimuthal deviations do not exceed 1°—1.5° (southward). The dispersion curves of the fundamental mode can be reliably determined within the period range 7—30 s only.

Path 8 about 2700 km long traverses the southern Baltic Shield, the western Russian plate, follows the trend of the Voronezh anteclise, and traverses the Don—Medveditsa ridge, terminating at the junction of the sub-Caspian depression and the Stavropol rise. The lengths of 'shield' and 'platform' parts are 505 km and

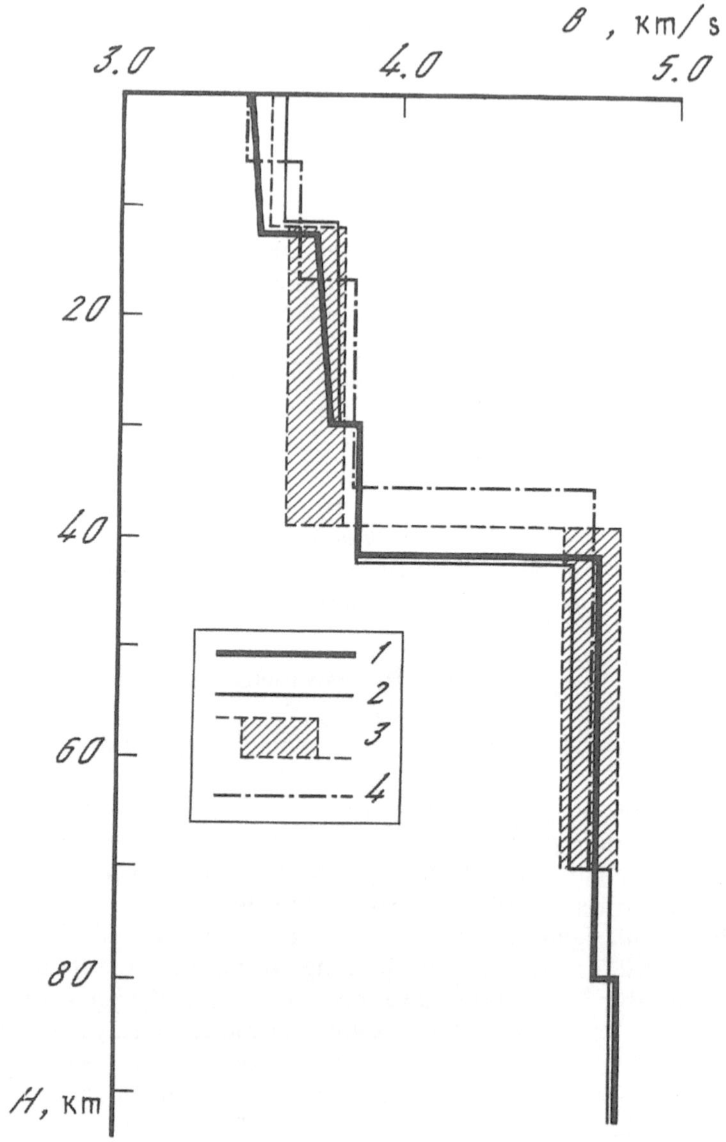

Fig. 7.4. Models of crustal structure for the Baltic Shield. (1) model for path 1; (2) model UL056 from Noponen (1966); (3) region of solutions from Calcagnile and Panza (1978); (4) Canadian Shield model (Brune and Dorman, 1963).

2175 km, respectively. It can be seen from the space—time diagram in Figure 7.6 that no azimuthal departures of the main signal from the theoretical value 110° are observed. There is a wave in the later arrivals having an arrival azimuth of 120° that lags behind the main signal by 100—120 sec. The dispersion curves of the fundamental Rayleigh mode were determined in the interval 8—30 sec.

Interpretation. In determining the dispersion curves for the platform parts of paths 2, 5, 8 we made the assumption that the dispersion curves for paths 1 and

TABLE 7.I
Crustal models for some Eurasian regions based on surface wave dispersion data.

Region	Number of path and segment	Number of layer	Range of z km	Range of b km/sec
Baltic Shield	1	1	0 — 12	3.45—3.50
		2	12 — 30	3.70—3.76
		3	30 — 42	3.85—3.85
		4	42 — 80	4.72—4.72
		5	80 —	4.80
Russian plate	2_2	1	0 — 2	2.30—2.30
		2	2 — 12	3.35—3.50
		3	12 — 22.5	3.50—3.70
		4	22.5— 45	3.70—4.00
		5	45 — 65	4.40—4.40
		6	65 —140	4.20—4.20
		7	140 —	4.60
	5_2	1	0 — 2	2.60—2.60
		2	2 — 11.5	3.35—3.55
		3	12 — 22.5	3.65—3.75
		4	22.5— 43	3.80—3.90
		5	43 —	4.55
	8_2	1	0 — 2	2.60—2.60
		2	2 — 10	3.50—3.50
		3	10 — 22.5	3.70—3.75
		4	22.5— 43	3.80—3.90
		5	43 —	4.45
Western-Siberian plate	3_3	1	0 — 1	2.90—2.90
		2	1 — 12	3.50—3.65
		3	12 — 25	3.65—3.65
		4	25 — 43	3.75—3.75
		5	43 —	4.45
Sub-Caspian syneclise	7_3	1	0 — 4	2.00—2.00
		2	4 — 16	3.00—3.00
		3	16 — 23	3.60—3.60
		4	23 — 38	3.90—3.90
		5	38 —	4.60

2_1, 5_1, 8_1 are identical. That the assumption is reasonable is confirmed by the stability of dispersion for different paths within the shield. The resulting curves for paths 2_2, 5_2, and 8_2 are presented in Figure 7.7.

They lie appreciably below the typical platform curves reported by Brune (1969) and Knopoff (1972). The crustal models fitting the observed dispersion are presented in Figure 7.8 along with a model for the shield. A sedimentary layer 2 km thick is held fixed in all these models. Shear velocity in the layer was 2.6 km/sec for paths 5 and 8 and 2.3 km/sec for path 2. Since the observed periods are longer that 8 sec, the parameters of the layer are estimated rather inaccurately. The upper portion of the consolidated crust (down to 12 km depth)

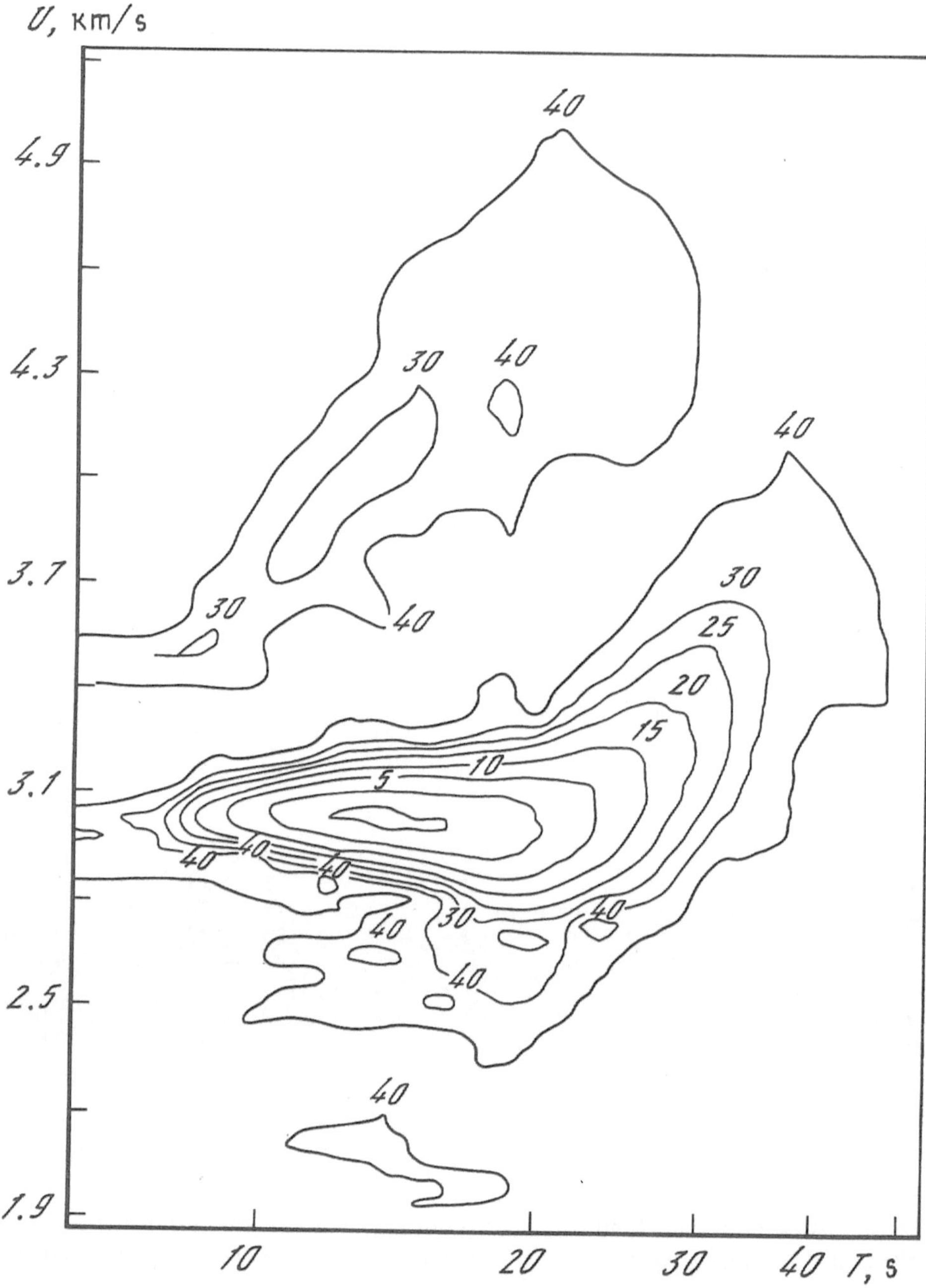

Fig. 7.5. FTAN diagram for path 2. Numerals at the curves denote the decrease in intensity (in dB).

has comparatively low shear velocities for paths 5 and 8 (less than those for the shield at the same depths by 0.07—0.10 km/sec). Path 2 has typically lower velocities in the deeper crust too, a great total crustal thickness (45 km), a low velocity in the uppermost mantle (4.45 km/sec). There is evidence for a mantle

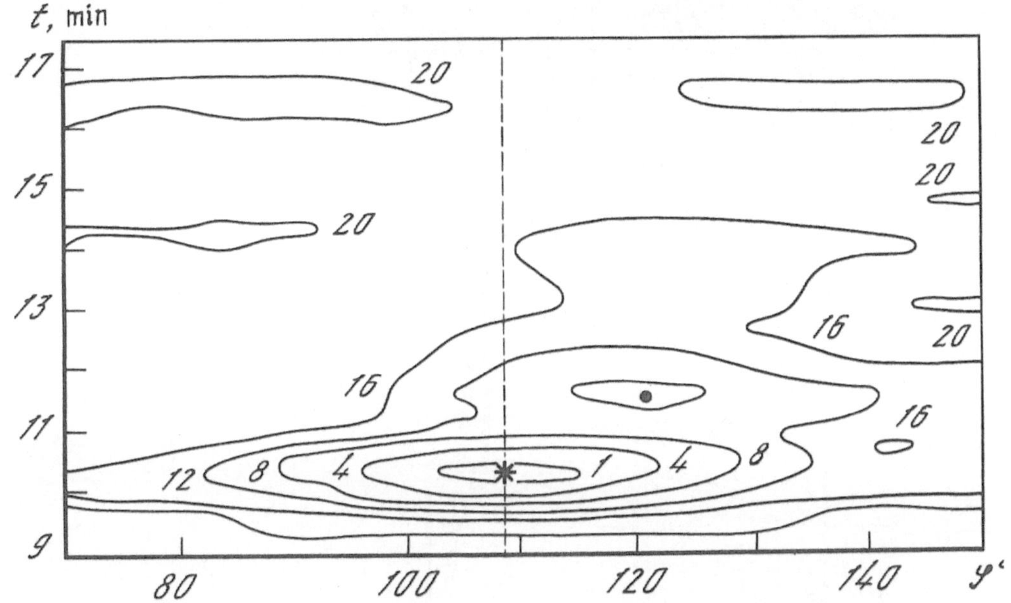

Fig. 7.6. Space-time analysis diagram, a record for path 8 (C = 3.47 km/sec). Numerals by the curves denote decrease in P_c (in dB). The star and the dot mark the locations of maximum amplitude for the main signal and a later arrival.

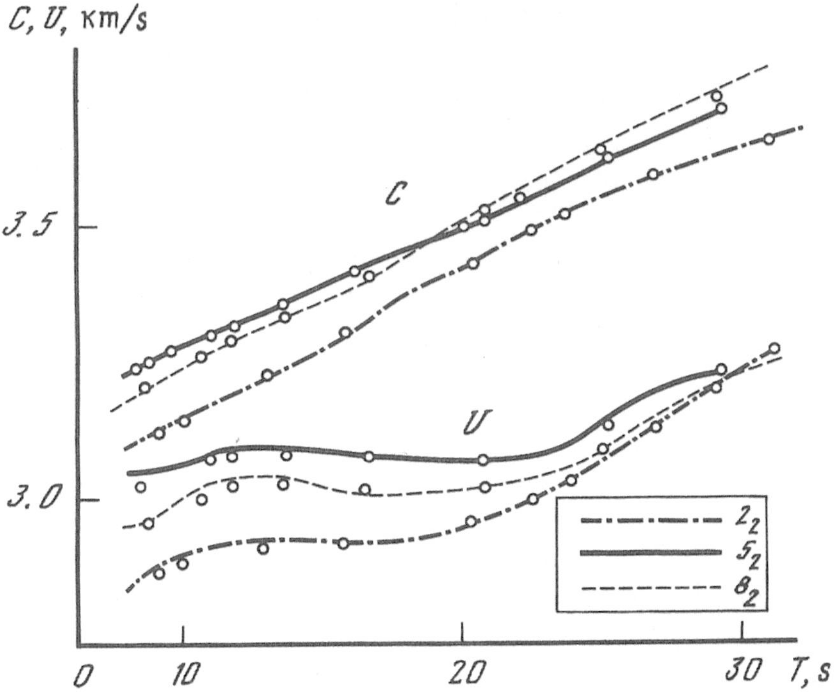

Fig. 7.7. Comparison of observed dispersion curves for areas 2_2, 5_2, 8_2 within the Russian plate and the calculated curves. Open circles are the theoretical velocities for the models in Figure 7.8.

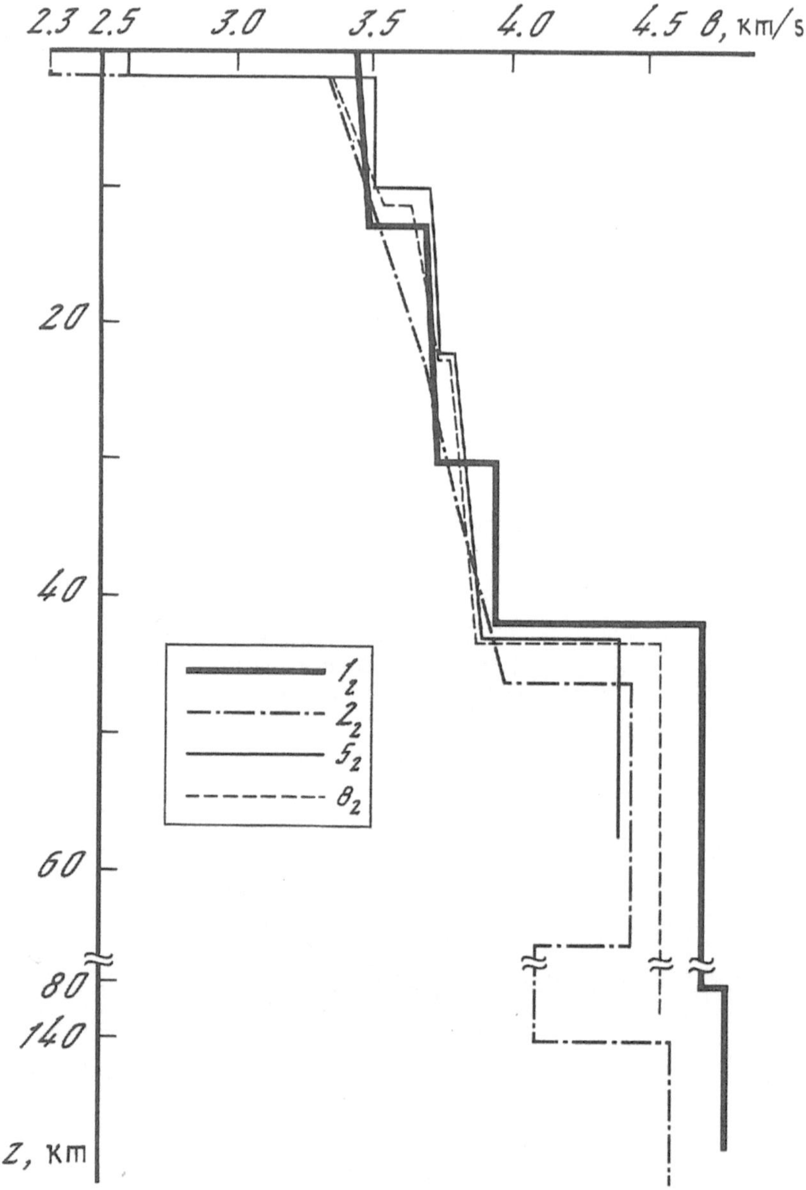

Fig. 7.8. Models of crustal structure for areas in the Russian plate.

layer of low velocity at depths greater than 65 km. Belyaevsky *et al.* (1977) present body-wave evidence for a layer of low velocity in the mantle of the north-eastern Russian plate, Levshin and Berteussen (1979) report lower upper-mantle velocities in the seas north of the plate.

3. Western Siberian Plate and Kazakh Folded Country

Observations. Path 3 (based on three observed records) traverses the Baltic Shield, the Russian plate in its northeastern part, somewhat southwards of path 2, the Middle Urals, and the southern portion of the Western Siberian plate, and then farther along the boundary between this plate and the Kazakh Folded Country. The total length of path 3 is 4260 km, with its separate segments having lengths of 1500 km, 1070 km, 1690 km for 3_1 (Baltic Shield), 3_2 (Russian plate), 3_3 (Urals, Western Siberian plate, and Kazakh Folded Country), respectively.

STAN diagrams show a small (2—2.5° southwards), but significant departure of the observed arrival azimuth from the theoretical (74.7°) for the main signal. No stable (consistent between the records) later arrivals have been identified. The diagrams provide good resolution for the fundamental mode of Rayleigh waves in the range of periods 10—50 sec, the first higher mode being less reliably determined for the periods 11—13 sec and group velocities 3.75—4.1 km/sec. The amplitude level at signal maximum is lower than that of the fundamental mode by a factor of 10.

Interpretation. Measurements for different events yield practically identical dispersion curves. Since path 2 and path $3_1 + 3_2$ are similar, we excluded segments $3_1 + 3_2$ from the observed path, assuming the mean phase and group velocities of the fundamental mode to be the same as for path 2. The resulting curves are presented in Figure 7.9. The phase velocities are rather close to the standard Brune curves (Brune, 1969), the model derived being different from the Brune model in having slightly lower velocities in the deeper crust (Figure 7.10).

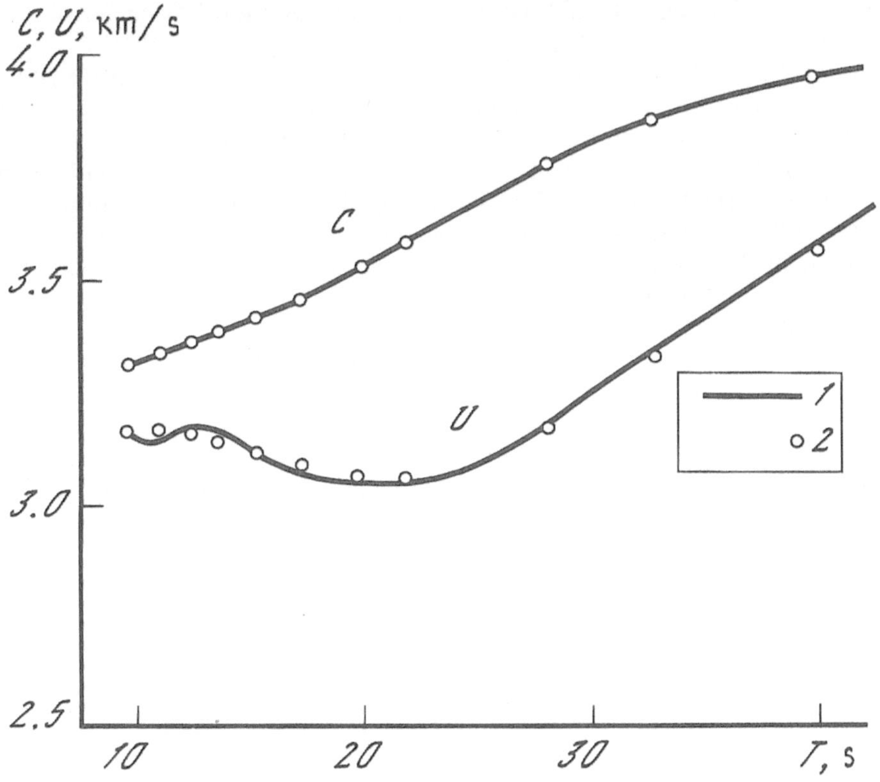

Fig. 7.9. Observed (1) and theoretical (2) dispersion curves for path 3_3 (Western Siberian plate and the Kazakh Folded Country).

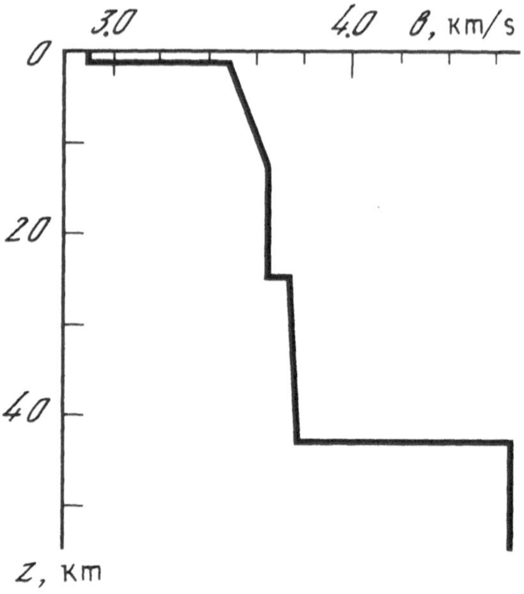

Fig. 7.10. Crustal structure for path 3_3.

4. Tarbagatai and Eastern Tien Shan Mountain Areas

Observations. Path 4 is practically identical with path 3 as far as the latter goes, but continues further to the southeast, traversing the eastern Kazakh Folded Country, the Tarbagatai meganticlinorium, the eastern continuation of Tien Shan, and the eastern end of the Tarim massif. The total path length is 5600 km, with segment 4_4, different from path 3, being 1350 km. Similarly to path 3, space-time analysis reveals rather small lateral deviations of the signal (about 2.5° south of the theoretical azimuth), as well as a number of clear later arrivals with varying arrival azimuths, from ones with azimuths close to that of the main signal (70°—80°) during the first 2 to 5 minutes after the arrival of the main signal to 105° after 10 minutes of the main signal. The maximum amplitude of the later arrivals is 0.2 to 0.4 of the main signal maximum.

FTAN diagrams show the fundamental mode of Rayleigh waves in the period range 15—60 sec and several later arrivals, the most prominent among these lagging some 220 sec behind the main signal maximum with an arrival azimuth of 70° (that is, northward of the theoretical value); the range of periods in this arrival is from 10 to 30 sec.

Interpretation. The dispersion for segment 4_4 of path 4 was estimated by eliminating the effects of segments $4_{1,2,3}$ which coincided with path 3. The resulting curves of group and phase velocity are shown in Figure 7.11; they may have been contaminated by interference with later arrivals in the period range 15—20 sec. Using the fact that our curves look like Brune's curves (1969) for the Alpine belt (Figure 7.11), we selected a model that was in satisfactory agreement with the observations in the period range 20—50 sec. The model (Figure 7.12) has a considerably thicker crust and lower mantle velocities than the platform models.

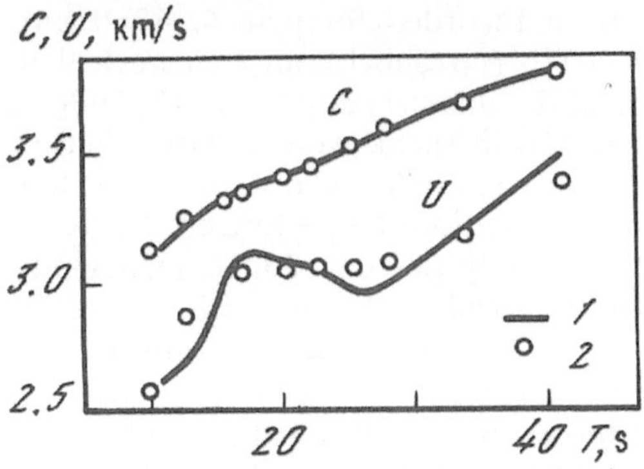

Fig. 7.11. Observed (1) and theoretical (2) dispersion curves of group and phase velocity for path 4_4 (Tien Shan and Tarbagatai mountain areas).

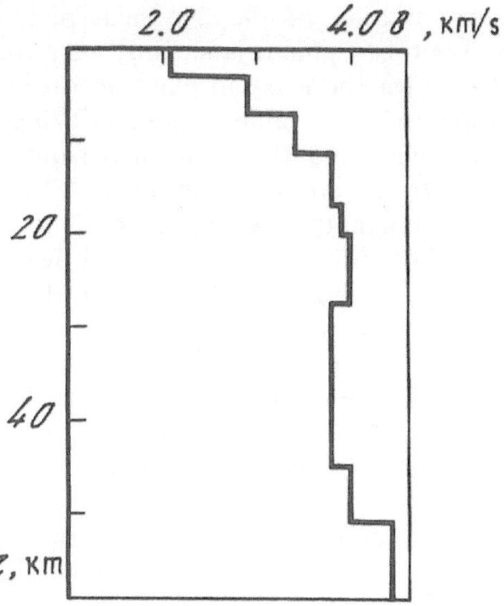

Fig. 7.12. Crustal structure for path 4_4.

5. Sub-Caspian Depression

Observations. Paths 6 and 7 traverse, in addition to the Russian platform and the Baltic Shield, the central sub-Caspian syneclise, terminating either at the northern margin (path 7) or in the central portion of the Turanian plate (path 6). According to the space-time analysis, the signal maximum in path 7 propagates nearly in the theoretical direction, but later arrivals contain waves with azimuths of 110° and 120° (that is, much more to the south) and time lags of 60 sec and 120 sec.

The dispersion of the fundamental mode is reliably determined in the period range 8—50 sec, subsequent arrivals involving a much more restricted range of periods, from 8 to 15 sec.

Three events have been recorded for path 6, 4260 km in length: the Gazli earthquake of April 8, 1976 (foreshock), an aftershock that occurred two hours later on the same day, and the main shock of May 17, 1976. The signal amplitudes for the foreshock and the main shock were so large that all NORSAR channels were overloaded except one specially adjusted to have low sensitivity. For this reason space-time analysis could not be performed; the records of the single low-sensitivity channel were the only raw data for frequency-time analysis. The aftershock signal amplitude is small, only the main signal Airy phase has been identified, the observed arrival azimuth being identical with the theoretical value (98°). The resulting curves of phase and group velocity (see Figure 7.2) are less stable and regular that those for the other paths because of the low signal-to-noise ratio. The interval of observed periods is 10—50 sec.

Interpretation. The combined length of the 'shield' and 'platform' segments of path segments $7_{1,2}$ is about 2500 km, the length of segment 7_3 belonging to the sub-Caspian syneclise being 750 km. The southeastern segment 7_4 which lies within the northwestern margin of the Turanian plate is 180 km long. Since independent evidence for this segment is lacking, we have assumed the dispersion along it to be the same as for the Russian plate; accordingly, segments 7_1, 7_2, and 7_4 have been lumped together, the same dispersion being ascribed to it as for part 8, which traverses the shield and the plate at a similar azimuth. The resulting dispersion curves for path 7_3 are given in Figure 7.13. The group and phase velocities are seen to be significantly lower than for all the other segments.

The model that fits the observations best is shown in Figure 7.14 (for the relevant theoretical curves see Figure 7.13). We see that the syneclise has a large mean thickness of the sediments (4 km of loose and 12 km of compact sediments with a shear velocity of 3 km/sec), a thin granite layer (7 km), a moderate total crustal thickness (38 km). These figures are in general agreement with the model obtained by averaging over the velocity structures along deep seismic sounding lines that traversed the syneclise (Bronguleev and Komarov, 1978; Egorkin *et al.*, 1979; Tsimmer, 1977).

Path 6 contains a long segment that belongs to the Turanian plate in a place where a significantly thicker crust may be expected (Volvovsky, 1973). Appreciably lower velocities are found for segment $6_3 + 6_4$ compared with 7_3. This shows that the path under discussion has lower velocities compared with 7_3. Increasing sediment thickness eastwards, in the sub-Caspian syneclise, is consistent with the deep seismic sounding evidence; the observations on segment 6_3 have not been quantitatively interpreted, because independent data on the Turanian plate are lacking.

6. Southern Part of the Barents Sea Shelf

Observations. Paths 9 and 10 traverse the northern Baltic Shield and the southern Barents shelf; path 9_2 traverses the shelf along the line Rybachii Peninsula—Matochkin Shar Bay, path 10_2 extends from the middle of the Kola coast to the Karskie Vorota Strait. Paths 9 and 10 being 2400 and 2100 km in length, the lengths of the shelf segments are 900 and 700 km, respectively. There are six independent, highly consistent velocity measurements for path 9 and three

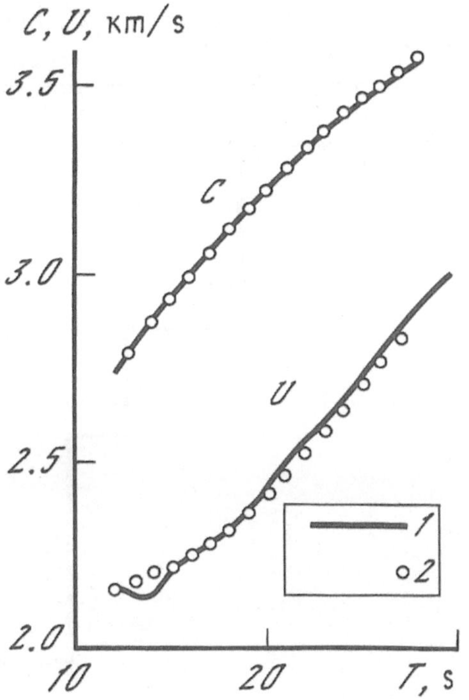

Fig. 7.13. Observed (1) and theoretical (2) curves of group and phase velocity for path 7_3 (sub-Caspian syneclise).

independent ones for path 10 showing somewhat greater scatter. The dispersion curves for the shelf segments 9_2 and 10_2 are shown in Figure 7.15 along with curves for SCAN model of the Baltic Shield (Der and Landisman, 1972). Typically, both paths have abnormally low Rayleigh wave velocities within a broad range of periods, such curves being unusual for platform areas. It should be noted that there is a local extremum in the group velocity curves for path 10 in the period range 17—22 sec which appears to be caused by some interference occurrences unresolved in our analysis. For this reason the measurements for path 10 are the less reliable and present difficulties for interpretation.

Interpretation. The models that fit the structure for paths 9 and 10 are presented in Figure 7.16. The structure for path 9_2 (Figure 7.16a) has a 2-km layer of unconsolidated sediments (a = 3 km/sec), a transition zone at 15 km depth between less dense sediments (a = 4 km/sec) and denser ones (a = 5 km/sec), a transition zone to the 'granite' layer (a = 6.2 km/sec at 25 km depth), the ordinary 'basalt' layer with velocity a = 6.8 km/sec at 25—40 km depth, and a comparatively low-velocity upper mantle (a = 7.9 km/sec below the M discontinuity).

The structure for path 10_2 (Figure 7.16b) consists of a 1-km layer of sediments with a = 3 km/sec; a zone of dense sediments with a rapid increase in velocity from 4 to 5.4 km/sec at 2—8 km depth; a layer whose parameters are similar to those of 'granite', with a = 5.4—6.0 km/sec in the depth interval 8—25 km; a 'basalt' layer; and an upper mantle, as for path 9.

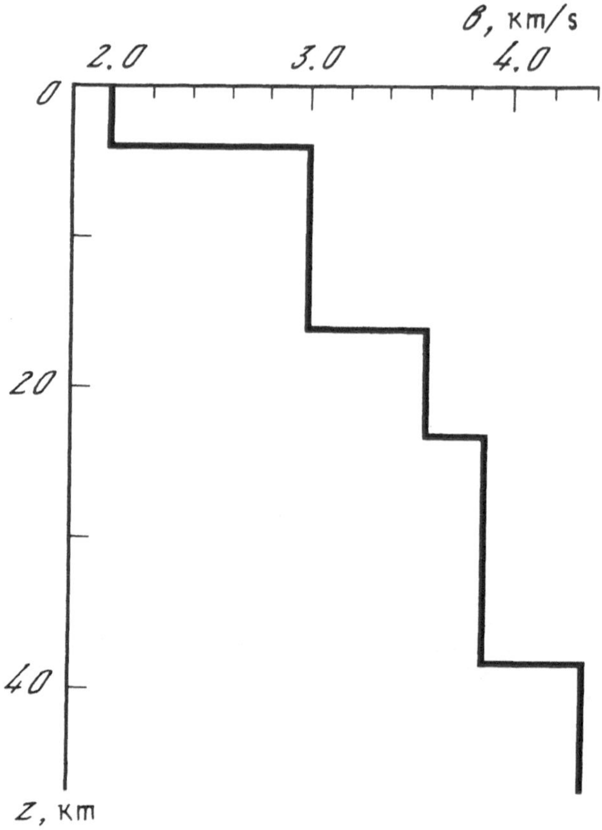

Fig. 7.14. Crustal structure for path 7_3.

In trying to interpret these velocity structures, we should remember that they were derived assuming, firstly, an approximately constant velocity structure along each path segment and, secondly, great circlepaths of actual wave propagation. Neither of these assumptions is quite flawless, but they cannot invalidate the principal result from our analysis of dispersion curves, namely, identification of abnormally low velocities in the upper crust. There are grounds to believe that a fuller accommodation of various distorting factors can merely increase the thickness or diminish the velocity in the upper anomalous part of the crust.

The above evidence for a thick sequence of sediments in the southern Barents Shelf is in good agreement with the results of recent, more detailed geophysical research (Volk *et al.*, 1984; Kijko and Mitchell, 1983).

ON THE ORIGIN OF LATER ARRIVALS

Space-time and time-frequency analyses of surface waves recorded at the NORSAR reveal a number of clear arrivals in the wave train following the fundamental mode of Rayleigh waves, whose arrival azimuths are markedly different from that of the Rayleigh wave and from the array-source line. On the

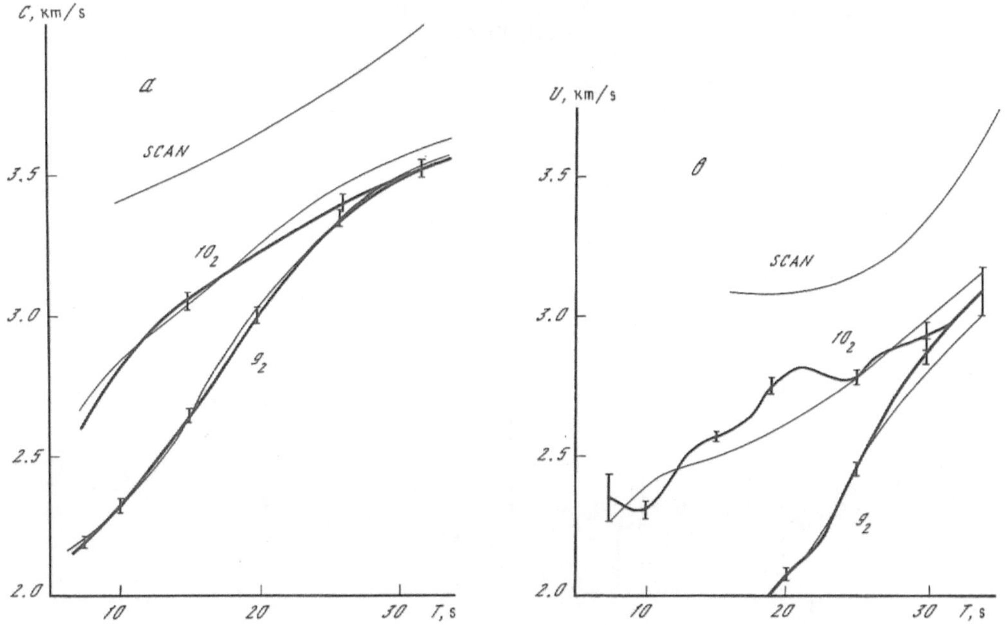

Fig. 7.15. Dispersion curves of phase (a) and group (b) Rayleigh wave velocity for paths 9 and 10. Vertical bars are rms errors.

strength of combined evidence from several kinds of data these arrivals were interpreted as Rayleigh surface waves generated as reflections of a direct Rayleigh wave from near-vertical discontinuities between crustal blocks. We have succeeded in determining the locations of the reflectors for a number of arrivals. Thus, for example, clear later signals recorded for paths 7 and 8 having arrival azimuths of 120°—130° at the NORSAR and lagging 120—160 sec behind the main signal (see Figure 7.6) have been interpreted as surface waves reflected from the northeastern margin of the Dnieper—Donetsk basin where the physical properties of upper crustal rocks change dramatically (Volvovsky, 1973).

One reflecting discontinuity identified from observations for path 10 is located in the middle part of path 9 and trends along the path; there is evidence (see Volk *et al.*, 1984) of a fault zone in that area. Another reflector of less contrast and depth is located around 74 °N, 30 °E, being possibly associated with the southern termination of the Swalbard platform. Lastly, a well-defined reflector is confined to the very steep continental slope west of the Narvik—Tromsø coast (Talwani and Eldholm, 1973). However, taking possible effects of lateral deviations in surface waves into consideration, one is not entitled to regard these reflector locations as absolutely certain.

The research reported here demonstrates the considerable potential of the single-station surface wave method in studying regional crustal structure. Observations at digital seismic arrays are the most convenient for these purposes, possessing as they do high signal-to-noise ratios, high measurement precision and azimuthal resolution which allows lateral deviations of signals to be estimated.

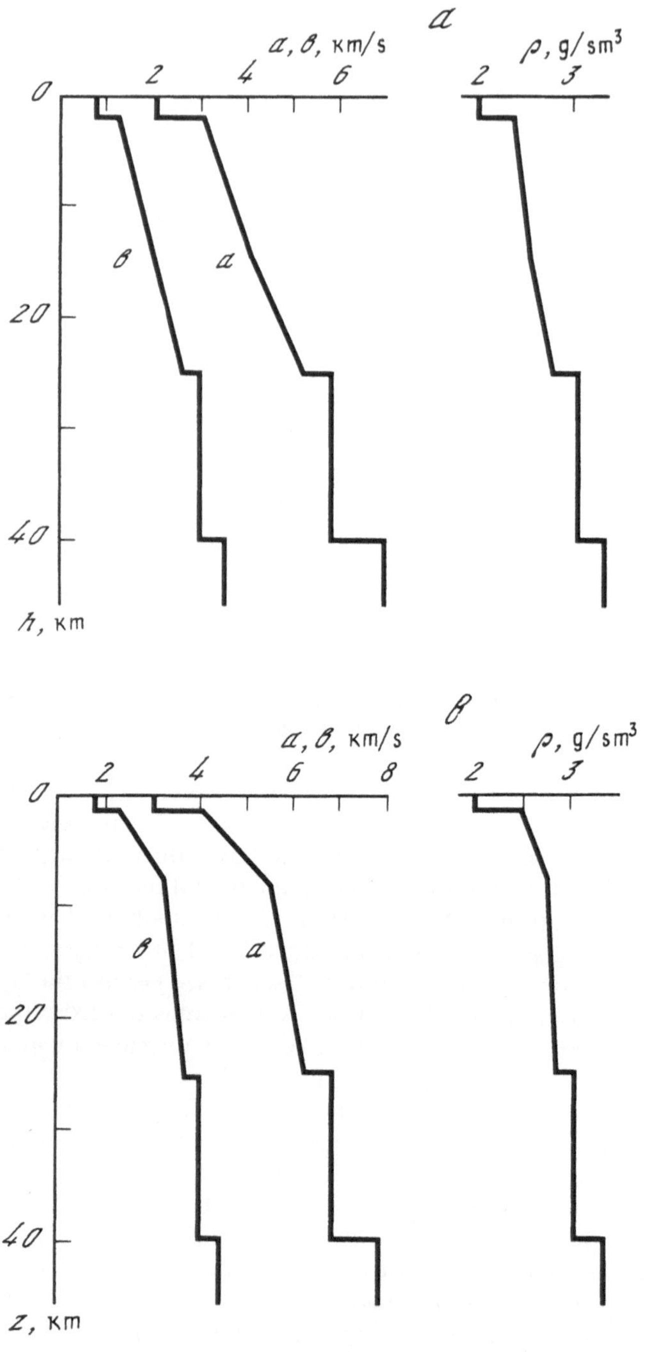

Fig. 7.16. Velocity and density distribution for paths 9_2(a) and 10_2(b).

Results for areas that had been previously studied by deep seismic sounding corroborate the average crustal models derived from surface wave dispersion data.

7.2. RESULTS OBTAINED BY THE BACKUS–GILBERT METHOD

The methods for determining lateral variations of surface wave velocities described in Sections 6.5, 6.6 have been applied to interpret group and phase velocities measured over paths across regions characterized by marked lateral crustal and upper mantle heterogeneity (Dmitrieva *et al.*, 1986, Yanovskaya, 1982, Yanovskaya and Nikolova, 1984, Sabitova and Yanovskaya, 1986).

Black Sea basin. Rayleigh wave group velocities over the paths across the Black Sea basin were obtained by S. A. Kapitanova for the periods 10—20 sec at stations Simferopol, Kishinev, Anapa, Sochi, Tbilisi and Sofia from sources in Turkey, the Carpathians, and the Caucasus (Dmitrieva, Kapitanova and Yanovskaya, 1986). This set of data was supplemented with the group velocities obtained by S. Nikolova (Nikolova and Yanovskaya, 1984) over paths across East Carpathians and the northern Balkan Peninsula. The 60 paths used in this study are shown in Figure 7.17.

The group velocity distributions for the periods 10, 15 and 20 sec obtained by the method described in Section 6.4 are shown in Figure 7.18. The effective radius of the averaging area is about 150—200 km everywhere, except in the southeastern part of the region, where it is increased to reach 300 km on account of the poor resolving power of the data.

The reality of a solution is verified by a good correlation of the velocity isolines with the coast line, especially for the period 10 sec. The Rayleigh wave group velocity for this period mainly depends on the thickness of the sedimentary layer, as well as the water depth. According to the deep seismic sounding data, the thickness of sediments in the Black Sea basin varies from 4—6 km in the western part to 12—14 km in the area to the south of the Crimea, just where the lowest group velocities occur.

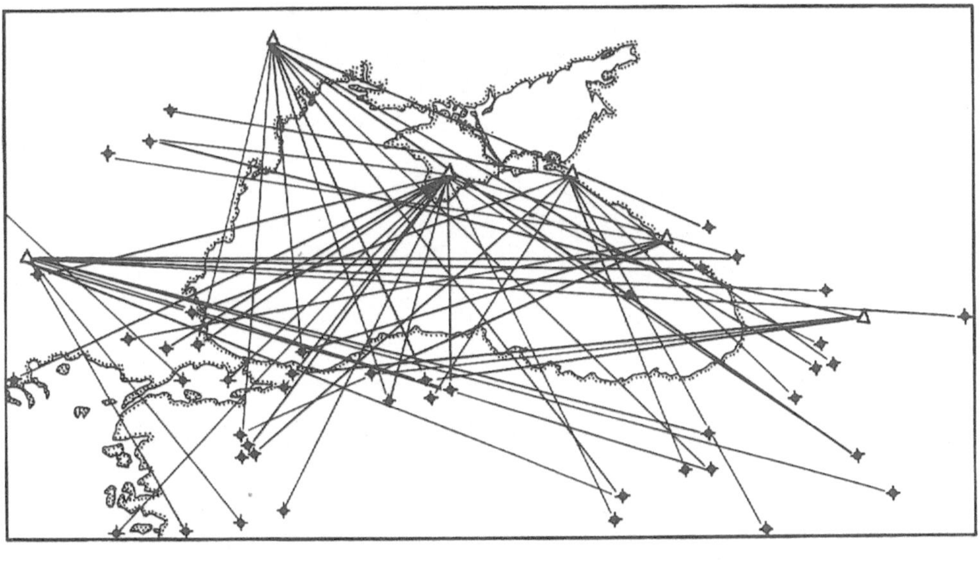

Fig. 7.17. The paths of Rayleigh waves across Black Sea basin.

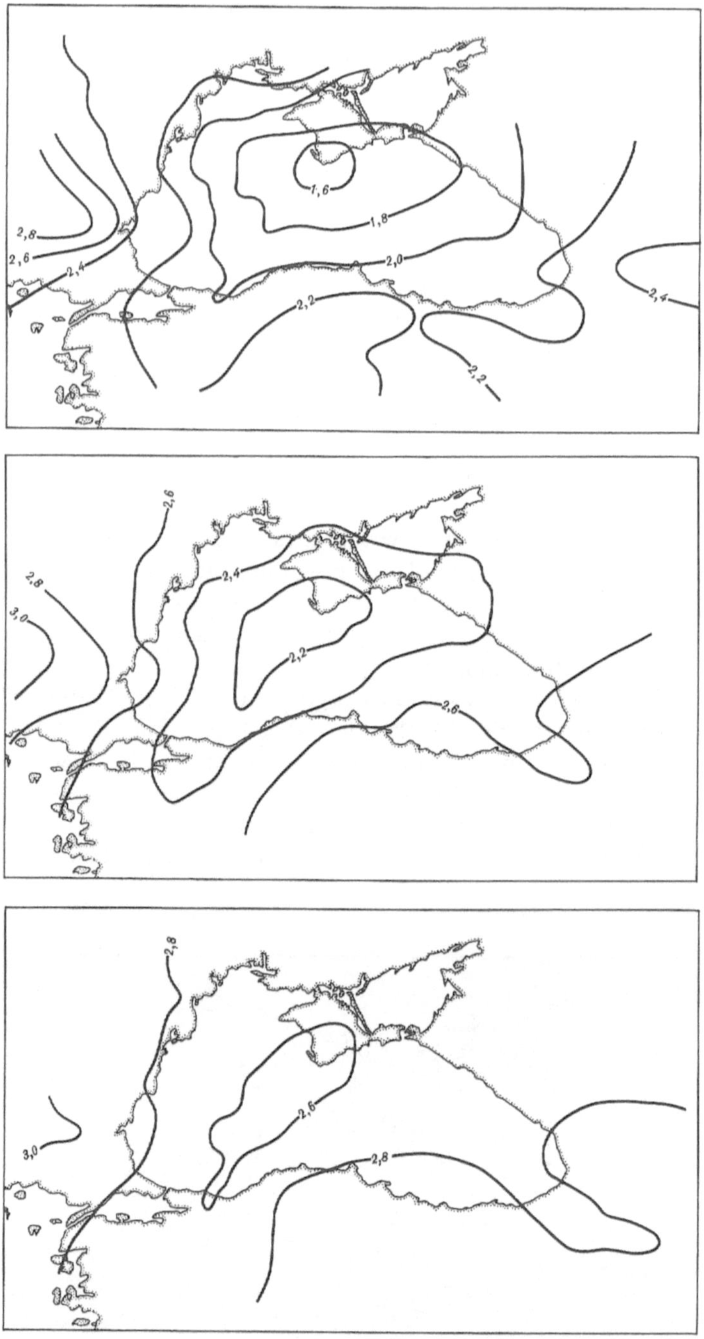

Fig. 7.18. Group velocity distributions for periods: (a) $T = 10$ sec, (b) $T = 15$ sec, (c) $T = 20$ sec.

For larger periods the velocity is affected by the structure of the lower layers. It is interesting to note that the area of minimum velocity values moves to the southwestern part of the Black Sea with increasing period. This points to a similarity in deep structure for the Black and Aegean Sea areas.

North Atlantic region (*Yanovskaya, 1982*). The data on Rayleigh wave group velocities for the periods 20, 30, 40, 50 and 60 sec over paths across the North Atlantic were obtained by Christensen *et al.* (1980), who determined the group velocities using the frequency-time analysis. The method which we used to determine group velocity distributions differs from that described in 6.4 in that the unknown function $m(x, y)$ must now satisfy an extra restriction: it must be represented as a sum of two functions, each depending on a single coordinate. As was shown by Yanovskaya (1984), if the axes are chosen so that they are directed along and across the strike of the main tectonic structures in the study area, the solution will adequately represent the real velocity distribution. In the North Atlantic the strike is parallel to the Mid-Atlantic ridge, therefore one of the coordinate axes was taken along the ridge.

The paths are shown in Figure 7.19. The dashed lines enclose the study area; they are parallel to the coordinate axes. The velocity distributions are shown in Figure 7.20. The averaging area is about 1500 km across.

Since the depth of penetration for a surface wave increases with the period, examination of the velocity $U(\theta, \varphi)$ for different periods enables one to determine the depth behaviour of the lateral variation in velocity structure. There is a pronounced difference for the shortest period $T = 20$ s between the areas to the

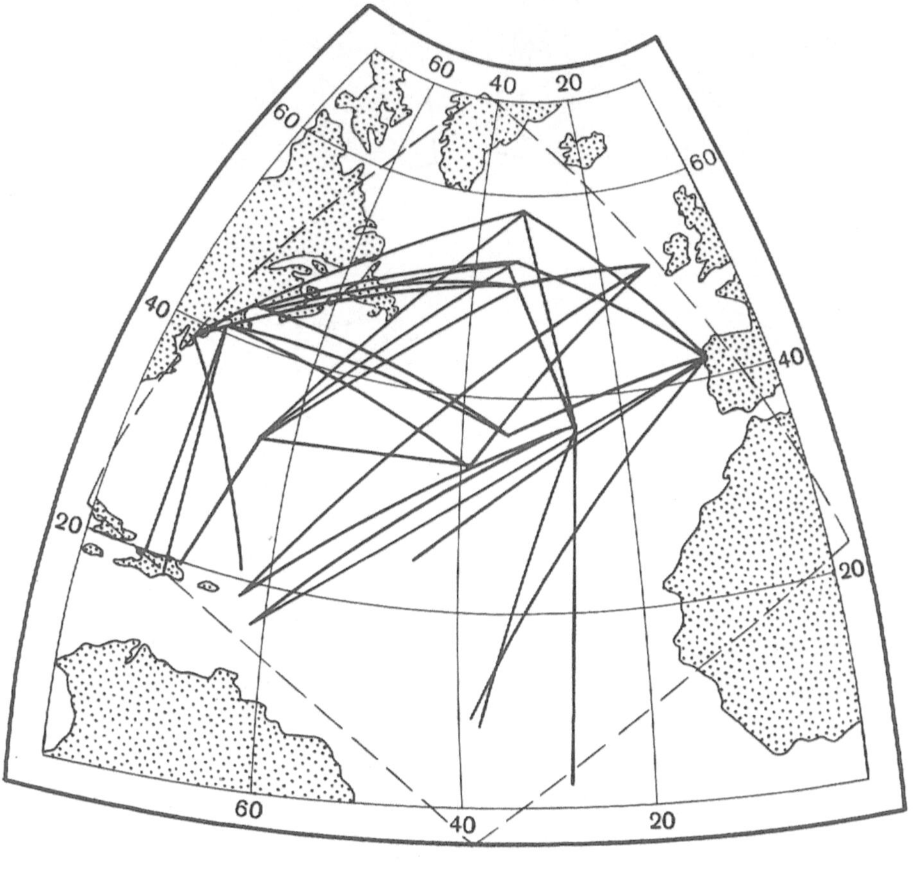

Fig. 7.19. Pattern of paths across North Atlantic region.

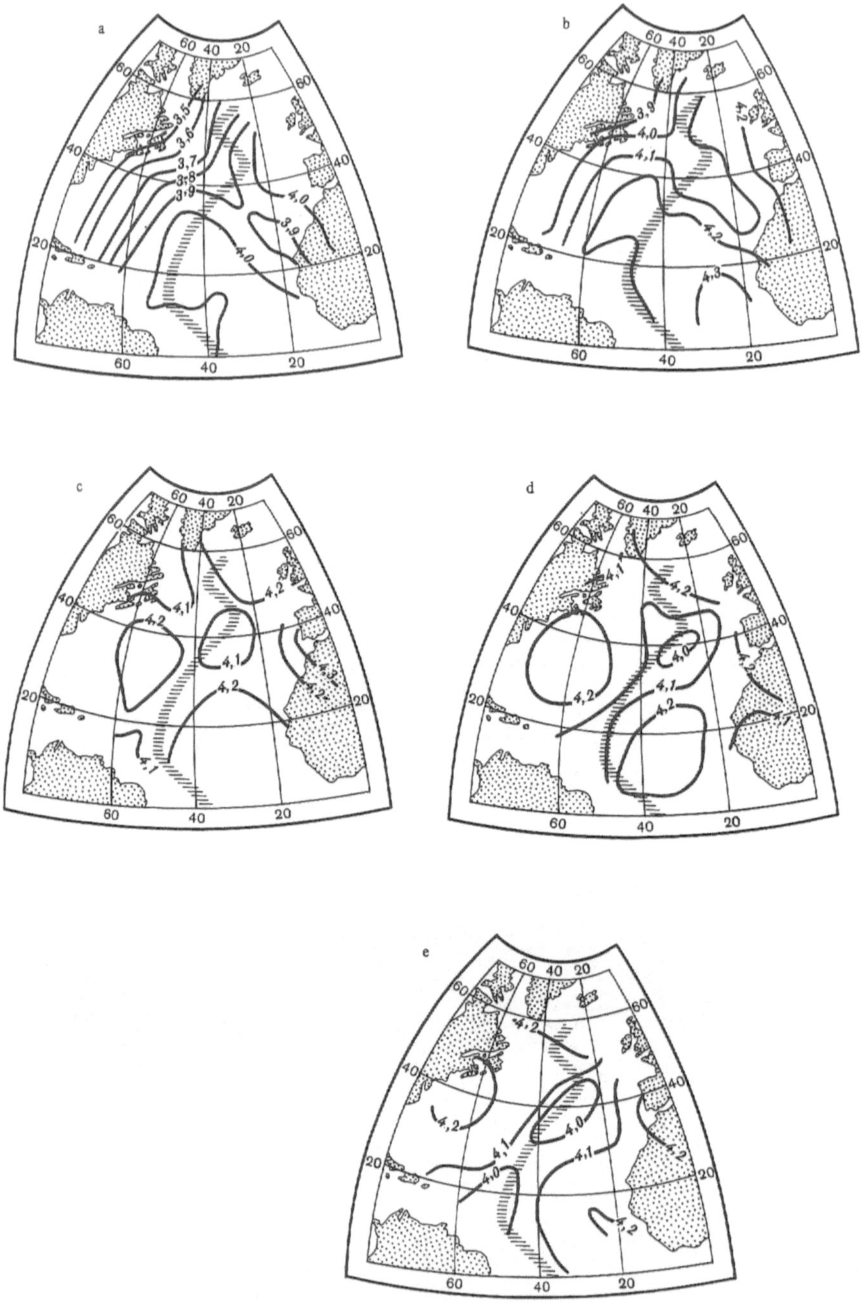

Fig. 7.20. Group velocity distributions for periods: (a) $T = 20$ sec, (b) $T = 30$ sec, (c) $T = 40$ sec, (d) $T = 50$ sec, (e) $T = 60$ sec. Hatched strip indicates the North-Atlantic ridge.

west and east of the Mid-Atlantic ridge. Surface wave velocity rapidly decreases westward from 4.0 to 3.5 km/sec, remaining nearly constant eastwards. The rift zone thus manifests itself at depths of several tens of kilometers merely as a boundary between two heterogeneoous structures. With increasing period the rift behaves like a narrow zone of lower surface wave velocity, differences in structure between the westward and eastward block disappearing.

The lower velocities of surface waves in the rift zone reflect lower upper-mantle shear velocities beneath the Mid-Atlantic ridge. This corroborates the hypothesis of light mantle material ascending there. When mantle material rises towards the surface and flows away from the rift on both sides of it, a strongly asymmetrical structure forms. Lower surface wave velocity to the west can be accounted for, in the first place, by crustal thickening towards the American continent and, secondly, by lower crustal rock density in the westward block compared with the eastward. A crustal density asymmetry on both sides of the rift axis has also been discovered from gravity observations.

It is of interest to note that the velocities vary along the rift too, a minimum occurring about 40° latitude (in the Azores area). This result is consistent with the hot spot concept, which asserts that convection from the mantle takes place in the form of strongly localized flows; one of these spots happens to be located in the Azores area.

Kirgizia. The data on Rayleigh wave group velocity dispersion in the range of periods 6—12 sec along paths across Kirgizia and adjacent areas were obtained by T. M. Sabitova. The paths are shown in Figure 7.21. Lateral distributions of group velocities for different periods obtained by the method described in Section 6.4 are shown in Figure 7.22a—d, where high velocity zones are hatched and low velocity zones are dotted for clarity. The main deep faults in the region are also shown in these figures.

Waves of periods 6—12 sec penetrate to depths of the sedimentary layer and

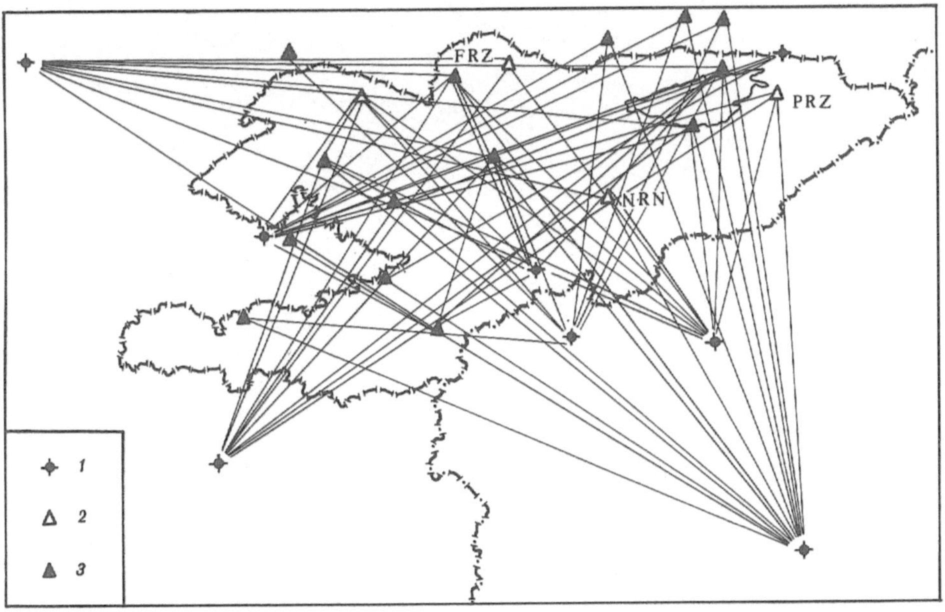

Fig. 7.21. Pattern of paths across Kirgizia Tien-Shan, (1) epicenter, (2) seismological station of the base network, (3) regional seismological station.

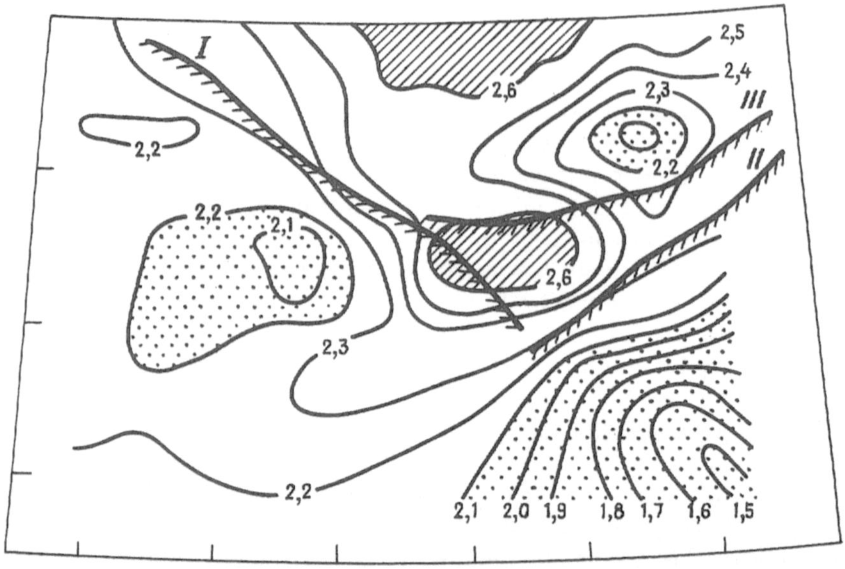
I
2,5
2,4
2,6
2,3
III
II
2,2
2,2
2,2
2,1
2,6
2,3
2,2
2,1
2,0
1,9
1,8
1,7
1,6
1,5

a

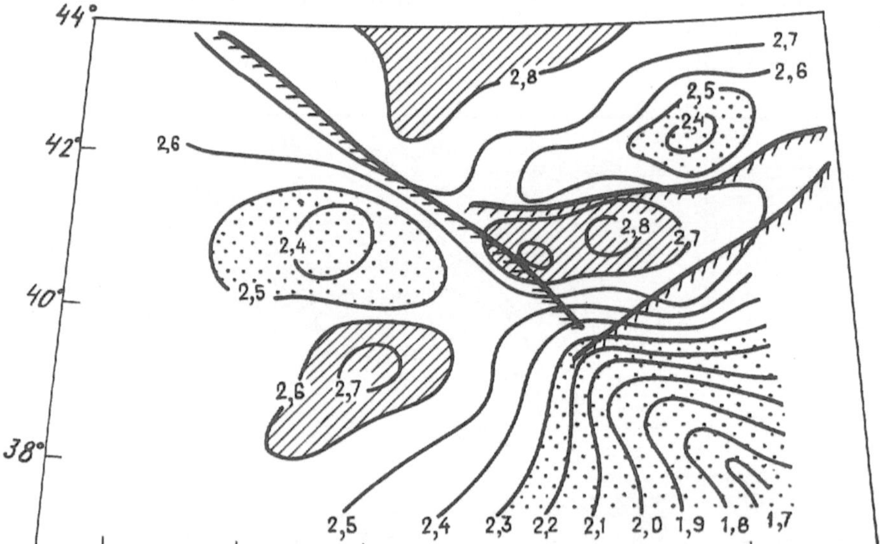
44°
42°
40°
38°
2,7
2,6
2,8
2,5
2,4
2,6
2,8
2,7
2,4
2,5
2,6
2,7
2,5
2,4
2,3
2,2
2,1
2,0
1,9
1,8
1,7
68°
70°
72°
74°
76°
78°

в

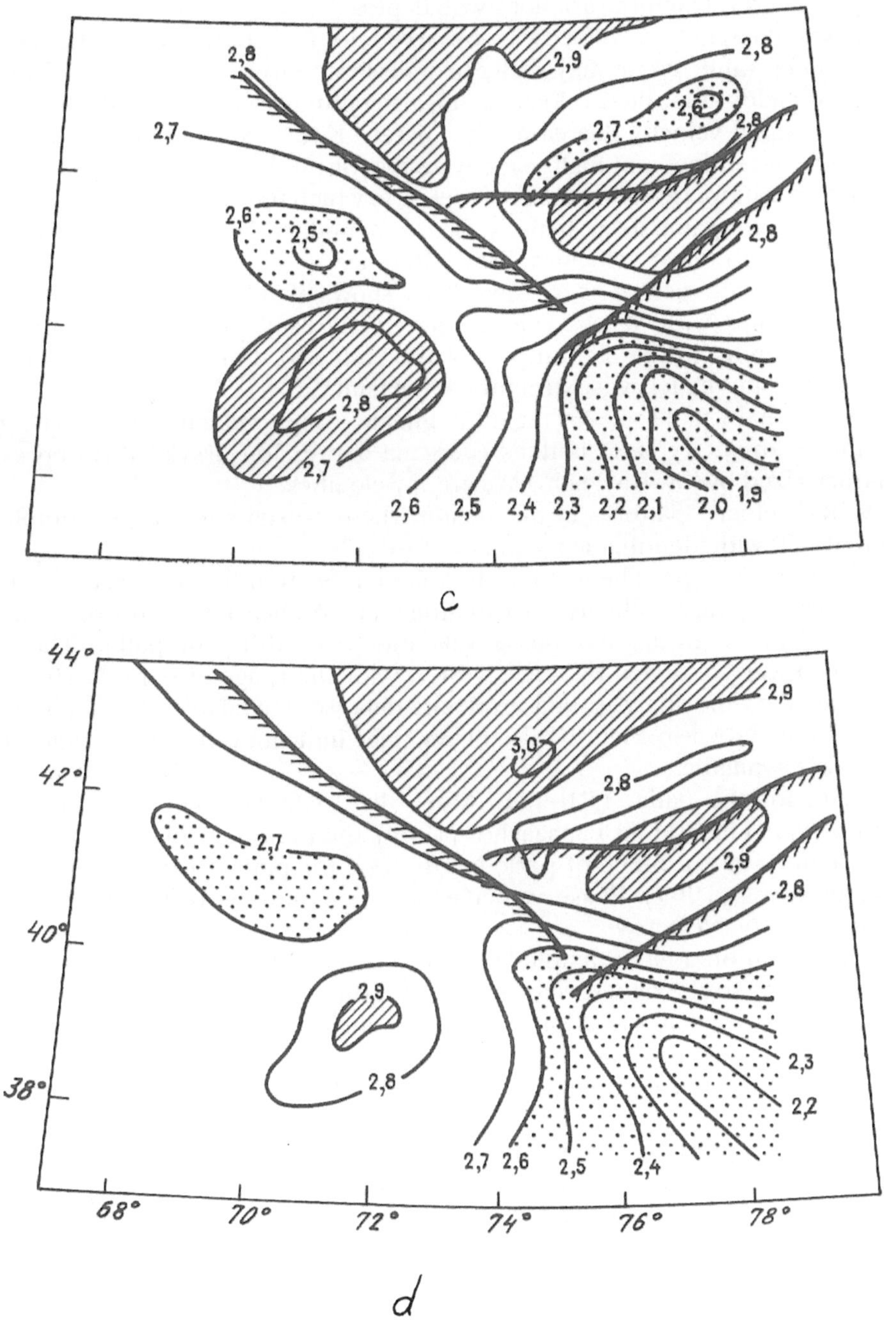

Fig. 7.22. Group velocity distributions over the area under investigation corresponding to the periods $T = 6$ sec (a), $T = 8$ sec (b), $T = 10$ sec (c), $T = 12$ sec (d). I — Gissar–Kokshaal deep fault, II — Talas–Fergana deep fault, III — Nikolaev line.

the upper crust. A comparative analysis of the results for increasing period reveals a change in lateral structure variations with depth.

A strong decrease in velocity is observed to the southeast of the Gissar—Kokshaal deep fault. There may really be a discontinuity across the fault, but since the method yields smoothed velocities, this discontinuity emerges in the solution as a zone of gradual velocity decrease. The Gissar—Kokshaal deep fault behaves as a boundary between low and high velocity zones for all the periods within the range under consideration, revealing the great depth to which this fault penetrates.

The Talas—Fergana deep fault is also a boundary between two media with different Rayleigh wave velocities. The difference is the largest for short periods and decreases with period increasing. This shows that the difference in crustal structure vanishes with depth. The group velocity gradient in the vicinity of the fault is a maximum in the central part of the fault and diminishes in the northern part. This indicates a piecewise structure of the fault.

The three main faults in the region delineate a high velocity block. The zones where the sedimentary layer is thick (Fergana depression, Issyk—Kul depression) are characterized by low Rayleigh wave group velocities.

Thus, the velocity patterns in this region where crustal structure is complicated correlated well with identifiable tectonic features.

Southeastern Europe. The method described in Section 6.5 was used to estimate Rayleigh wave phase velocity distributions in southeastern Europe from the combined data of group and phase velocities over different paths. The group velocity data were obtained by A. N. Nesterov for the range of periods 10—30 sec using frequency-time analysis. Phase velocity data were available over 5 paths only. The paths are shown in Figure 7.23, heavy lines indicating those for which phase velocities are available.

The data for the paths TRI—IST, AQU—IST, AQU—ATH were taken from Calcagnile *et al.* (1984) and Papazachos (1969), the periods being 14—50 sec. The phase velocities in the range of periods 18—38 s for path ATH—IST were taken from Papazachos (1969). Phase velocities between stations KOS and LVV were measured by A. N. Nesterov.

As was shown on a test example in section 6.5, the phase velocity solution turns out to be similar to the actual velocity distribution, even when phase velocity data are scarce. A lack of information on phase velocities is compensated by a large amount of group velocity data.

Lateral velocity distributions for the periods 10, 20 and 30 sec are shown in Figure 7.24a—c.

Rayleigh wave phase velocities are the lowest in the area of the Pannonian and Transylvanian basins. This result is in agreement with other geophysical data which indicate a large thickness of the sedimentary layer in this area. For large periods, another velocity minimum is revealed in central Greece. High velocities are observed in the northern part of the Aegean Sea, as well as in the Pelagonian massif, where the sedimentary layer is practically absent.

Thus, a joint interpretation of phase and group velocities yields results which show an overall agreement with those from other geophysical studies, so that we can expect that some other unknown features of crustal structure can be detected from surface wave data.

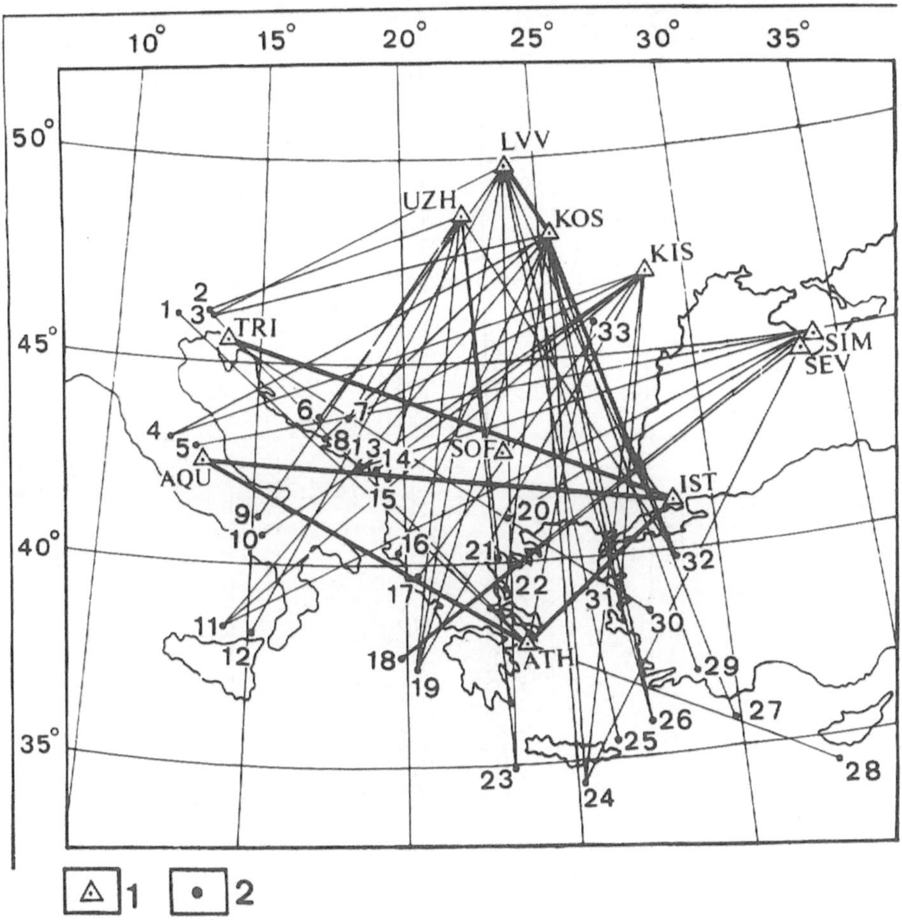

Fig. 7.23. Ray paths across southeastern Europe. (1) Seismological station. (2) Epicenter. Solid lines show the paths along which the phase velocities were available.

7.3. ANOMALOUS RAYLEIGH SURFACE WAVES IN NORTHEAST EURASIA

The tectonic activity of North Eurasia seems to be controlled by the interaction between the North American and Eurasian plate (Savostin, Karasik and Zonenshain, 1984). In northwestern Siberia the boundary between the plates approaches the pole of relative plate rotation, coming to the continent around the Verkhoyansk–Kolyma Tectonic Region. There are some considerations that tend to distinguish among its structures the main one which expresses the principal trend in the tectonic evolution of the Region. This is the Momsky continental rift related to the Chersky Range system (Figure 7.25) (Grachev, 1973). The hypothesis of the rift's existence mainly rests on the results of geological and geomorphic research. There is practically no evidence on the deep structure. The available maps of crustal structure (Truskov, 1975) largely based on gravity

Fig. 7.24. Phase velocity distributions corresponding to the period $T = 10$ sec (a), $T = 20$ sec (b), $T = 30$ sec (c). Numerals within circles show the main tectonic structures in the region: 1 — Eastern Alps; 2 — Carpatian arc; 3 — Dinarides; 4 — Rhodope massiff; 5 — Ukrainian shield; 6 — Pannonian basin; 7 — Transylvanian basin.

observations do not indicate the rift. Indeed, they show a crustal thickening beneath the Chersky Range, that is, an opposite structure to what one should have expected for a rift. Thus, even summary parameters of deep structure, for instance, the typical crustal thickness in major areas of the Verkhoyansk—Kolyma Region, are subject to large uncertainties, while it is just those parameters which largely determine our conclusions as to the general tectonic evolution of the Region.

In trying to derive information on the average seismogeologic model of the Verkhoyansk—Kolyma Region, in the first place the Momsky rift, we undertook a surface wave study (Lander, 1984). We found out that the polarization of Rayleigh waves after traversing the Verkhoyansk—Kolyma Region is essentially different from what one could expect, thus posing a more specific seismological problem associated with phenomena having no obvious theoretical explanation.

We used the fundamental mode of Rayleigh and Love waves recorded at base seismic stations. There are four stations in and around the Verkhoyansk—Kolyma

Region: Tiksi, Yakutsk, Magadan, Seymchan, equipped with SK or SKD seismographs which can record surface waves at periods up to 50 sec (sometimes as long as 60 sec). The observations divide into two data sets by source location (epicentral zones of earthquakes). One set includes teleseismic records from sources in Indonesia and the southwest Pacific. Clear surface waves from these earthquakes have been recorded at all the four stations. This allows wave velocities to be measured for interstation paths. Our study mainly comprised two sets of paths along the Verkhoyansk Range (path I) and the Chersky Range (see Figure 7.25). It has seemed convenient to divide the latter set into two smaller ones, paths II and III. The Verkhoyansk Range was thus treated as a 'standard' for comparison in determining the main features of the Chersky Range structure.

The other set comprised records of 'near' earthquakes occurring in the Kuril–Kamchatka and Aleutian arc (see Figure 7.25) made at station Tiksi. The paths scan a sector that includes the Chersky Range and some considerable adjacent area. This makes it possible to investigate, with respect to Tiksi, the 'azimuthal dependences' of surface waves with a view to finding in the hypothetical rift area some properties that would be anomalous compared with the adjacent area.

Apart from the velocities, we have measured the polarization of Love and Rayleigh waves in all records. Below, the polarization azimuth of a Rayleigh wave is understood to be the line of intersection made by the wave polarization plane with the horizontal plane. We now turn to the results of the measurements.

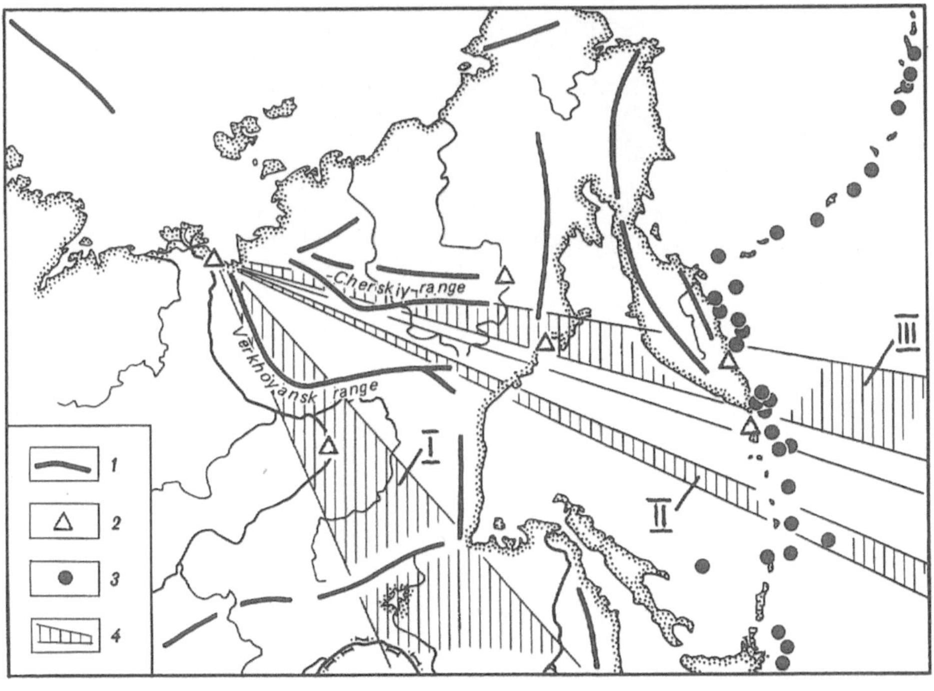

Fig. 7.25. Area of study. (1) major morphological features in the Verkhoyansk–Kolyma Region and adjacent area; (2) seismic station; (3) epicenter of a near event; (4) part of teleseismic wave paths (the hatched sectors include many paths); I, II, III – path numbers.

Polarization anomalies of Rayleigh waves (observed at Tiksi alone). The polarization azimuths for Rayleigh waves coming towards Tiksi from the southeast are markedly different from the directions towards the epicenters. The anomalies are as large as 40—50°. They are clearly observed on all Tiksi records having the southeastern arrival directions during the entire period of observation (1962 to 1978). The anomalies have a well-defined frequency dependence. Figure 7.26 shows the contours of Rayleigh wave polarization anomalies mapped on a grid of wave period and the Tiksi-epicenter direction. The results have been averaged using a window of width 4° in azimuth and 0.07 in log T. The maximum anomalies occur at periods around 20 sec. They disappear both towards shorter and longer periods. The main anomalous region lies within the period range 15—35 sec. Note that this range corresponds to wavelengths (60—150 km) that are the most structure-sensitive at the Moho depths. Short-period 'crustal' and long-period

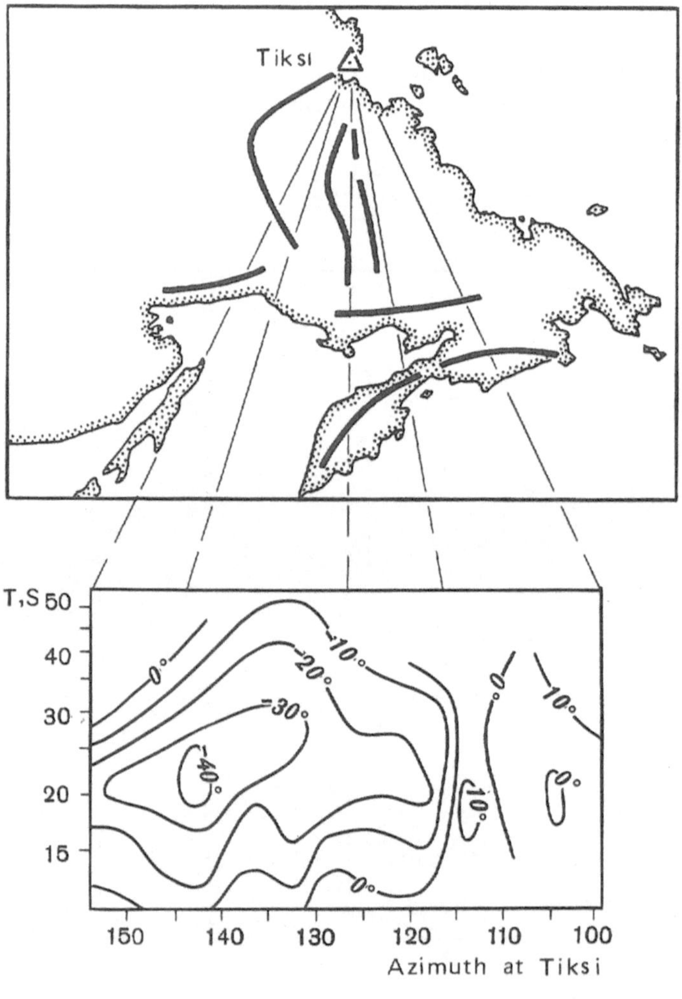

Fig. 7.26. Map of polarization anomalies for Rayleigh waves of different periods along paths crossing North-Eastern Eurasia.

'mantle' Rayleigh waves do not experience distortions when travelling across the Verkhoyansk—Kolyma Region.

Note that the absolute polarization azimuth within the main anomaly region is practically independent of the Tiksi-epicenter direction, being around 105° (a region of 'stable' polarization azimuths).

Azimuthal dependence of polarization anomalies. Anomalously polarized waves are only recorded within a relatively narrow range of Tiksi-epicenter directions. Thus, no such effects occur on records of waves coming towards Tiksi from the north. The boundary of the anomalous region coresponding to 115° azimuth is clearly visible in Figure 7.26. It is at about the same azimuth that the boundary between the high-rise Chersky Range and the lowlands lying farther eastwards as viewed from Tiksi. Although the polarization anomalies for azimuths greater than 145° apparently persist, the absolute azimuths rapidly increase compared with the typical 'stable' azimuth value. This suggests that the processes producing the polarization anomalies for paths with directions greater that 145° are different from those for the paths corresponding to the 'stable' azimuths. A careful analysis of the recorded signals demonstrates that the spatial components of Rayleigh wave records with 150° to 190° directions do correlate much worse than the records in the main anomaly region. It may be that two interfering waves come from 150° to 190° directions: the main Rayleigh wave having a polarization around the theoretical azimuth and a secondary, anomalously polarized signal generated by the main wave in the period range 15—35 sec.

Very remarkably, both boundaries of the 'stable' azimuth region correspond to the directions bounding the Chersky Range. We observe, however, that the range of directions identified by the 'stable' azimuth region corresponds to the entire high-rise Chersky Range rather than to the rift valley proper. We hypothesize then that the anomalous polarization is caused by an inhomogeneity traversed by the waves that is located relative to Tiksi within about the same range of directions as the Chersky Range.

Normal polarization of Love waves at Tiksi. This statement would be of no interest whatsoever, but for the fact that the records referred to are those in which the anomalously polarized Rayleigh waves were observed. The two types of surface waves propagating along similar paths seem to respond to inhomogeneities in different ways. The bulk of polarization azimuths of Love waves lie within 10 percent of the theoretical values, well within the measurement uncertainty.

Differences between Rayleigh wave velocities along paths I, II, III. Figure 7.27 shows, for each path, boundaries between the regions occupied by the measured dispersion curves of phase and group velocity. The scatter is large enough, but roughly equals the uncertainty of a single measurement. The following qualitative features can be inferred from this figure.

(1) The dispersion for path I is a typical one for a mountain area of moderate altitude. The II and III group velocities for periods of 20 to 40 s are comparatively higher. At longer periods (up to 50 s) group velocities show no significant differences between the three paths. The II and III group velocities are closer to the typical average continental ones than to those for mountain areas.

(2) The II and III phase velocity curves have smaller slopes than those for path I.

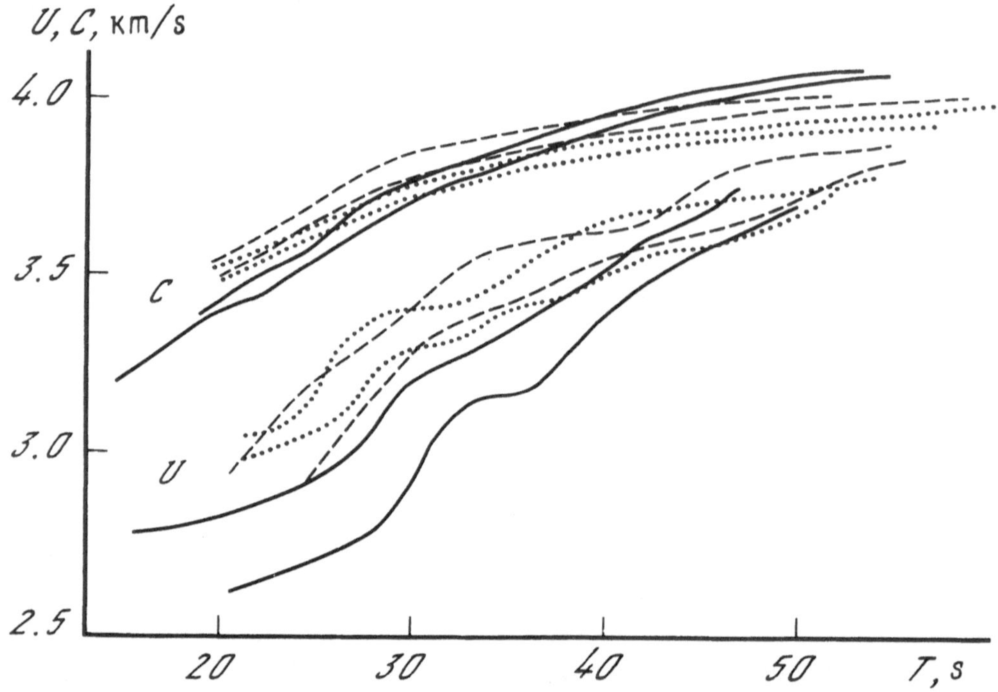

Fig. 7.27. Experimental dispersion curves of phase (C) and group (U) velocities for parts of the Verkhoyansk Range (solid line) and Chersky Range: path II (dashed line), path III (dotted lines).

This fact is closely associated with the difference in group velocity mentioned above. However, the important thing for interpretation is that the phase velocity curves intersect one another. Path I has lower velocities at shorter periods (20—30 s) and higher velocities at longer periods (around 50 s).

(3) Paths II and III are different in phase velocities alone, the group velocities being the same to within the accuracy bounds.

Note that the principal discrepancies in dispersion curves for data sets I and II occur for the periods 20—40 sec, corresponding to the frequency range of the polarization anomaly. Therefore the hypothesis of a connection between the polarization anomaly and the velocity inhomogeneity seems fairly reasonable.

We now discuss the results. It has been presumed in the foregoing discussion that most of the measured surface wave parameters are relevant for the structures of interest in the Verkhoyansk—Kolyma Region. This could be true, provided the waves propagate close enough to their theoretical paths, namely, great circles. Doubts of the validity of this assumption are prompted by the considerable polarization anomalies observed.

In the first place we notice that the polarization of Rayleigh waves that are abnormally polarized at Tiksi is close to the theoretical at the other stations, that is, at the 'entry' into the Verkhoyansk—Kolyma Region. The azimuth deviations are within ± 10 percent, in agreement with the measurement uncertainty. Thus, all the anomalous effects have their origin in the Verkhoyansk—Kolyma Region.

The normal wave propagation at the 'entry' into the Verkhoyansk—Kolyma Region provides no evidence as to the geometry of rays within it. With large deviations from great circle paths, one cannot well predict what particular ray, of those which have entered the Region, would reach Tiksi. The desired correspondence can be established by comparing similar parameters derived from records of teleseismic and near events. The azimuth patterns of Rayleigh wave polarization anomalies for both types of events are in good agreement. Both types of data are available within the range 121° to 154° of Tiksi—epicenter directions. In Figure 7.28 the range is divided into four subranges with the scatter of maximum anomalies separately for teleseismic and near events given for each subrange. One can see that a shift along the axis of directions (for example, for teleseismic events) by 10° impairs the consistency significantly. However, we have noted that the direction of wave propagation too is close to the theoretical at the 'entry' into the Region. Forgetting about the uncertainties for the moment, one can assert that the true rays which coincide with the theoretical ones at the 'entry' are just those later recorded at Tiksi. Since this conclusion refers to fairly a large range of directions relative to Tiksi (about 30°), significant departure from the theoretical ray scheme within the Verkhoyansk—Kolyma Region seems unlikely.

Can a polarization anomaly be associated with Rayleigh wave reflections at tectonic boundaries? When a surface wave propagates in a laterally homogeneous, nondissipative medium, the amplitude spectrum persists along a ray. On the other hand, theoretical calculations and interpretation of actual records (Sikharulidze, 1978; Levshin and Berteussen, 1979) show that reflection coefficients have a much more restricted range of frequency than the incident wave that generates them. Figure 7.29 presents smoothed amplitude spectra of Rayleigh waves

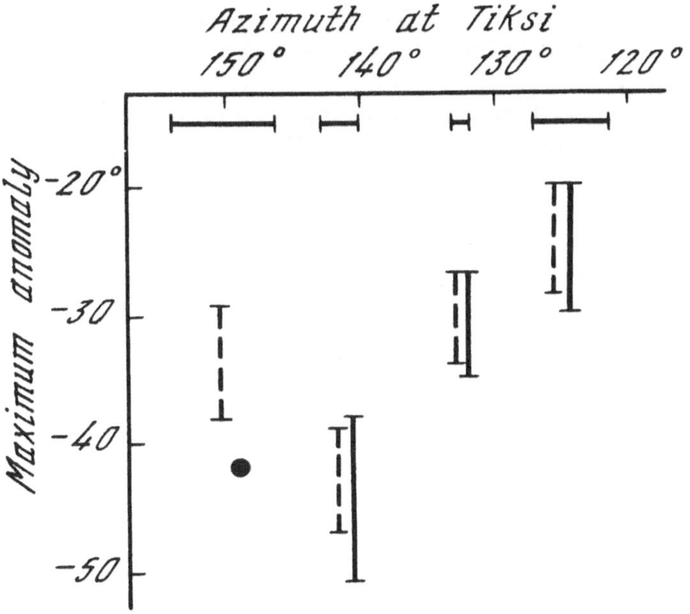

Fig. 7.28. Scatter in maximum polarization anomalies of Rayleigh waves. Solid bars are for teleseismic data, dashed bars are for near events.

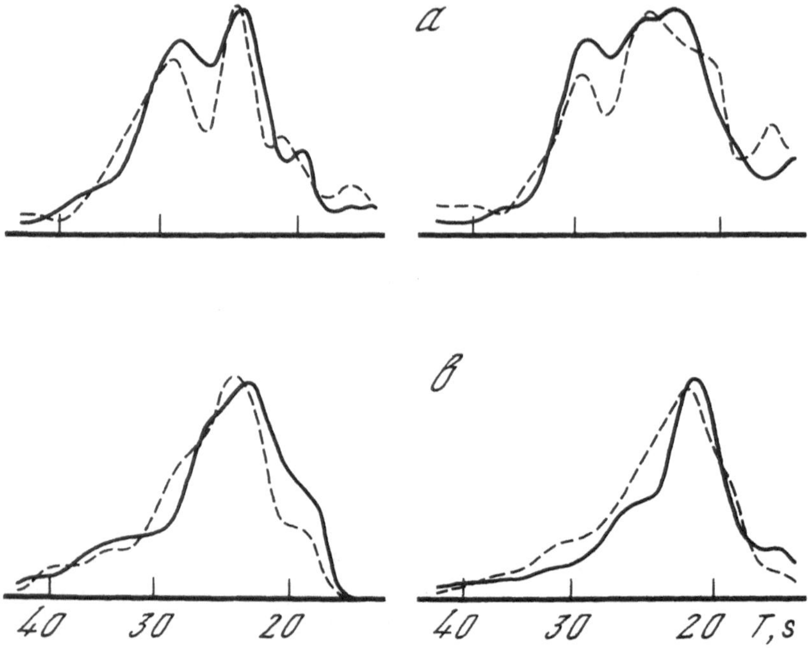

Fig. 7.29. Rayleigh wave spectra at the entry into the Verkhoyansk—Kolyma Region (dashed line) and at Tiksi (solid line). (a) Path across the Verkhoyansk Range. (b) The same for the Chersky Range.

obtained at two stations each. The Tiksi spectra preserve the overall features and period band recorded at the 'entry'. The observed departures are not consistent from event to event.

One should pay special attention to the fact that differences in spectral shape do not correlate with the behavior of polarization anomalies. The latter show a simple and well-pronounced frequency pattern, reaching maximum values around a 20 sec period. Essential changes in spectral shape for waves that have traversed the Verkhoyansk—Kolyma Region are lacking. This enables one to assert that the Rayleigh waves recorded at Tiksi have not experienced significant reflections at lithospheric geologic units.

To sum up, the simple ray schemes are unable to account, if only in qualitative terms, for the significant polarization anomalies observed on Tiksi records. This phenomenon is more likely to be associated with some diffraction effects that occur near the boundary of a major linear unit when a wave is travelling along it. The Momsky rift may well be the unit in question.

REFERENCES

Abo-Zena, A. M., 1979. 'Dispersion function computations for unlimited frequency values'. *Geophys. J.*, **58**, 91–105.

Agnew, D., Berger, J., Buland, R. *et al.*, 1976. 'International deployment of accelerometers: A network for very long period seismology'. *Trans. Amer. Geophys. Union*, **57**, 180–188.

Aki, K. and Richards, P. G., 1980. *Quantitative Seismology*. W. H. Freeman and Co., San Francisco, vols. I, II.

Akopyan, S. Ts., Zharkov, V. N. and Lyubimov, V. M., 1975. 'On the dynamical shear modulus in the Earth'. *Dokl. AN SSSR*, **223**(1), 87–90.

Alekseev, A. S., Babich, V. M. and Gelchinsky, B. Ya., 1961. 'A ray method for calculating wave front intensity'. In: *Questions of Dynamical Seismic Wave Propagation*. Leningrad Univ. Press, Issue 5. pp. 3–24 (in Russian).

Alsop, L. E., 1966. 'Transmission and reflection of Love waves at the vertical discontinuity'. *J. Geophys. Res.*, **71**, 3969–3984.

Alsop, L. E., Goodman, A. S. and Gregersen, S., 1974. 'Reflection and transmission of inhomogeneous waves with particular application to Rayleigh waves'. *Bull. Seismol. Soc. Amer.*, **64**, 1635–1652.

Anderson, D. L. and Dziewonski, A. M., 1982. 'Upper mantle anisotropy: Evidence from free oscillations'. *Geophys. J. Roy. Astron. Soc.*, **69**, 383–404.

Anderson, D. L., 1984. 'Surface wave tomography'. *Trans. Amer. Geophys. Union*, **65**, 147–148.

Avetisyan, R. A. and Yanovskaya, T. B., 1973. 'Expansion of Rayleigh wave group velocity into spherical harmonics'. *Izv. AN SSSR, Fizika Zemli* (*Solid Earth*), No. 11, 27–33.

Azbel, I. Ya., Dmitrieva, L. A. and Yanovskaya, T. B., 1980. 'Calculation of geometrical spreading in a medium varying in three dimensions'. In: *Methods and Algorithms for Interpretation of Seismological Data*. Moscow, Nauka, pp. 113–126 (Computational Seismology, 13) (in Russian).

Babich, V. M. and Buldyrev, V. S., 1972. *Asymptotic Methods in Short-Wave Diffraction Problems*. Moscow, Nauka. (in Russian).

Babich, V. M. and Chikhachev, B. A., 1975. 'Propagation of Love and Rayleigh waves in weakly inhomogeneous media'. *Vestnik LGU*, No. 1, 32–38.

Babich, V. M., Chikhachev, B. A. and Yanovskaya, T. B., 1976. 'Surface waves in a vertically inhomogeneous half-space with weak horizontal inhomogeneity'. *Izv. AN SSSR, Fizika Zemli* (*Solid Earth*), No. 4, 24–31.

Babich, V. M. and Grigoryeva, N. S., 1980. 'Space-time ray method for wave field calculations in a weakly inhomogeneous layered medium'. In: *Interference Waves in Layered Media*. Leningrad, Nauka, pp. 5–18 (in Russian).

Babich, V. M., Buldyrev, V. S. and Molotkov, I. A. 1985. *Space-Time Ray Method*. Leningrad, Leningrad University Press (in Russian).

Backus, G., 1964. 'Geographical interpretation of measurements of average phase velocities of surface waves over great circular and great semi-circular paths'. *Bull. Seismol. Soc. Amer.*, **57**(2), 571–610.

Backus, G. and Gilbert, F., 1968. 'The resolving power of gross Earth data'. *Geophys. J. Roy. Astron. Soc.*, **16**, 169–205.

Backus, G. and Gilbert, F., 1969. 'Constructing *P*-velocity models to fit restricted sets of travel-time data'. *Bull. Seismol. Soc. Am.*, **59**, 1407–1420.

Backus, G. and Gilbert, F., 1970. 'Uniqueness in the inversion of inaccurate gross Earth data'. *Phils. Trans. Roy. Soc. London*, **226**, 123—192.

Backus, G. and Mulcahy, M., 1976. 'Moment tensors and other phenomenological descriptions of seismic sources. 1. Continuous displacements'. *Geophys. J. Roy. Astron. Soc.*, **46**, 341—362.

Backus, G. and Mulcahy, M., 1976a. 'Moment tensors and other phenomenological descriptions of seismic sources. 2. Discontinuous displacements'. *Geophys. J. Roy. Astron. Soc.*, **47**, 301—330.

Backus, G., 1977. 'Interpreting the seismic glut moments of total degree two or less'. *Geophys. J. Roy. Astron. Soc.*, **51**, 1—25.

Backus, G., 1977a. 'Seismic sources with observable glut moments of spatial degree two'. *Geophys. J. Roy. Astron. Soc.*, **51**, 27—45.

Bagramyan, A. Kh., Savarensky, E. F., Sikharulidze, D. I., and Yanovskaya, T. B., 1978. 'The effect of deep faults on surface wave spectra'. *Izv. AN SSSR, Fizika Zemli* (*Solid Earth*), No. 3. 86—89.

Barmin, M. P., 1983. 'Algorithms for regional division of geodesics on a sphere'. Moscow, *Mat. MTs D-B MGK AN SSSR*. (in Russian).

Barmin, M. P., Levshin, A. L. and Starovoit, O. E., 1984. 'Iterative frequency-time analysis of surface waves'. In: *Mathematical Modelling and Interpretation of Geophysical Observations*. Moscow, Nauka, pp. 11—128 (Vychislitelnaya seismologiya, 16), (in Russian).

Bazaraa, M. S. and Shetty, C. M., 1979. *Nonlinear Programming. Theory and Algorithms*. New York, Wiley.

Belyaevsky, N. A., Volvovsky, B. S., Volvovsky, I. S. *et al.*, 1977. 'Seismic crustal structure of Eastern Europe'. In: *Crustal and Upper Mantle Structure from Seismic Observations*. Kiev, Naukova dumka. pp. 4—19, (in Russian).

Berteussen, K. A., Levshin, A. L. and Ratnikova, L. I., 1982. 'Regional studies of Eurasian crust using surface waves'. In: *Mathematical Earth Models and Earthquake Prediction*. Moscow, Nauka. pp. 105—117 (Computational Seismology, 14) (in Russian).

Blum, P. A. and Gaulon, R., 1971. 'Detection et traitment des ondes sismiques de tres basses frequences'. *Ann. Geophys.*, **27**, 123—140.

Boore, D. M., 1970. 'Love waves in nonuniform wave guides: Finite difference calculations'. *J. Geophys. Res.*, **75**, 1512—1527.

Boore, D. M., 1972. 'Finite difference methods for seismic wave propagation in heterogeneous materials'. In: *Methods in Computational Physics*. New York: Academic press, pp. 1—137.

Bronguleev, V. V. and Komarov, V. P., 1978. 'Structural plans of the basement top and subsalt complex in the sub-Caspian syneclise'. *Byul. MOIP. Otd. geol.*, **53**(3), 8—23. (in Russian).

Brune, J. N., 1969. 'Surface waves and crustal structure'. In: *The Earth's Crust and Upper Mantle*. Washington. pp. 320—242.

Brune, J. N. and Dorman, J., 1963. 'Seismic waves and Earth structure in the Canadian Shield'. *Bull. Seismol. Soc. Am.*, **53**, 167—210.

Bukchin, B. G., 1979, 'Propagation of Love waves across a vertical contact between two quarter-spaces'. In: *Theory and Analysis of Seismological Observations*. Moscow, Nauka, pp. 70—79 (Computational Seismology, 12) (in Russian).

Bukchin, B. G., 1980. 'Reflection and transmission of Love waves at the ocean-continent boundary: theory'. In: *Methods and Algorithms for Interpretation of Seismological Data*. Moscow, Nauka, pp. 109—113. (Computational Seismology, 13) (in Russian).

Bukchin, B. G. and Levshin, A. L., 1980. 'Propagation of Love waves across a vertical discontinuity'. *Wave Motion*, **2**, 293—302.

Bukchin, B. G., Levshin, A. L., and Ratnikova, L. I., 1984. 'On apparent anisotropy observed in inhomogeneous media'. In: *Logical and Computational Methods in Seismology*. Moscow, Nauka, pp. 116—123 (Computational Seismology, 17) (in Russian).

Bulaevsky, V. A., Zvyagina, R. A., and Yakovleva, M. A., 1977. Numerical Methods of Linear Programming. Special Problems. Moscow, Nauka. (in Russian).

Bungum, H., Husebye, E. S. and Ringdal, F., 1971. 'The NORSAR array and preliminary results of data analysis'. *Geophys. J. Roy. Astron. Soc.*, **25**, 115—126.

Bungum, H. and Capon, J., 1974. 'Coda pattern and multipath propagation of Rayleigh waves at NORSAR'. *Phys. Earth and Planet. Inter.*, *9*, 111—127.

Burridge, R. and Weinberg, H., 1977. 'Horizontal rays and vertical modes'. In: *Wave Propagation and Underwater Acoustics* (Lecture Notes in Physics, 70). Ed. F. B. Keller and J. S. Papadakis. Berlin, Heidelberg: Springer-Verlag. pp. 86—152.

Burton, P. W., 1973. 'Estimation of Q_γ from seismic Rayleigh waves'. *Geophys. J. Roy. Astron. Soc.*, **36**, 167—189.

Calcagnile, G. and Panza, G. F., 1978. 'Crust and upper mantle structure under the Baltic shield and Barents Sea from dispersion of Rayleigh waves'. *Tectonophysics*, **47**, 59—71.

Calcagnile, G., Mascia, U., Del Gaudiio, V. and Panza, G. F., 1984. 'Deep structure of Southeastern Europe from Rayleigh waves'. *Tectonophysics*, **110**, 189—200.

Capon, J., 1969. 'High-resolution frequency-wavenumber spectrum analysis'. *Proc. IEEE*, **57**, 1408—1418.

Capon, J., 1970. 'Analysis of Rayleigh-wave mutlipath propagation at LASA.' *Bull Seismol. Soc. Amer.*, **60**(5), 1701—1731.

Cara, M., 1973. 'Filtering of dispersed wave trains'. *Geophys. J. Roy. Astron. Soc.*, **33**, 65—80.

Cerveny, V., Molotkov, I. A., and Psenčik, I., 1977. *Ray Method in Seismology*. Prague.

Chesnokov, E. M., 1977. *Seismic Anisotropy of the Upper Mantle*. Moscow, Nauka (in Russian).

Chou, C. W. and Booker, J. R., 1979. 'A Backus—Gilbert approach to the inversion of travel-time data for three-dimensional velocity structure'. *Geophys. J. Roy. Astr. Soc.*, **59**, 325—344.

Christensen, D. H., Kimball, J. K. and Mauk, F. J., 1980. 'Rayleigh wave group velocity dispersion in the North and South Atlantic oceans'. *Bull. Seisml. Soc. Amer.*, **70**, 1187—1809.

Copson, E. T., 1965. *Asymptotic Expansions*. Cambridge University Press.

Crampin, S., 1981. 'A review of wave motion in anisotropic and cracked elastic media'. *Wave Motion*, **3**(4), 343—391.

Der, Z. A. and Landisman, M., 1972. 'Theory for error resolution and separation of unknown variables in inverse problems with application to the mantle and crust in Southern Africa and Scandinavia'. *Geophys. J. Roy. Astron. Soc.*, **27**, 137—178.

Ditmar, P. G. and Yanovskaya, T. B. 1987. 'A generalization of the Backus—Gilbert method for estimation of lateral variations of surface wave velocity'. *Izv. AN SSSR. Fizika Zemli (Solid Earth)*, No. 6, 30—40.

Dmitrieva, L. A., Kapitanova, S. A. and Yanovskaya, T. B., 1985. 'The use of the Backus—Gilbert technique to determine the two-dimensional distribution of surface wave velocity in the Black Sea basin'. In: *Derivation of Earth Structure from Seismological Evidence*. Kiev, Naukova Dumka, pp. 35—67 (in Russian).

Dorofeev, V. M. and Zharkov, V. N., 1978, 'On determining the mechanic Q of the Earth's mantle'. *Izv. AN SSSR, Fizika Zemli (Solid Earth)*, No. 9, 55—73, (in Russian) . . .

Dost, B., Van Wettum, A. and Nolet, G., 1984. 'The NARS array'. *Geologie en Mijnbow*, **63**, 381—386.

Drake, L. A., 1972. 'Rayleigh waves at the continental boundary by finite element method'. *Bull. Seismol. Soc. Amer.*, **62**, 1259—1268.

Dunkin, J. W., 1965. 'Computation of modal solution in layered elastic media at high frequencies'. *Bull. Seismol. Soc. Amer.*, **55**, 335—358.

Dziewonski, A., Bloch, S. and Landisman, M., 1969. 'A technique for the analysis of transient seismic signals'. *Bull. Seismol. Soc. Amer.*, **59**, 427—444.

Dziewonski, A., 1971. 'On regional differences in dispersion of mantle Rayleigh waves'. *Geophys. J. Roy. Astron. Soc.*, **22**, 289—325.

Dziewonski, A., Mills, J. and Bloch, S., 1972. 'Residual dispersion measurement: A new method of surface wave analysis'. *Bull. Seismol. Soc. Amer.*, **62**, 129—139.

Dziewonski, A. M. and Anderson, D. L., 1981. 'Preliminary reference Earth model'. *Phys. Earth and Planet. Inter.*, **3**, 1157—1180.

Egorkin, A. V., Chernyshev, N. M., Belokopytov, V. N. *et al.*, 1979. 'Structure of sub-salt deposits in the sub-Caspian syneclise based on the results of regional seismic studies'. *Izv. AN SSSR, Ser. Geol.*, No. 10, 105—114 (in Russian).

Ewing, M., Mueller, S., Landisman, M. and Sato, Y., 1959. 'Transient analysis of earthquake and explosion arrivals'. *Pure and Appl. Geophys.*, **44**, 83—118.

Feofilaktov, V. D., 1977. *Noise in Long-Period Seismometry*. Moscow, Nauka (in Russian).

Flinn, E. A., 1965. 'Signal analysis using rectiliniarity and direction of particle motion'. *Proc. IEEE*, **53**(12), 1874—1876.

Forsythe, G. E., Malcolm, M. A. and Moler, C. B., 1977. *Computer Methods for Mathematical Computations*. Englewood Cliffs (New Jersey).

Frez, J. and Schwab, F. 1976. 'Structural dependence of the apparent initial phase of Rayleigh waves'. *Geophys. J. Roy. Astron. Soc.*, **44**(2), 311—331.

Gabrielov, A. M., 1984. 'Diffraction of Love waves at a vertical discontinuity'. In: *Mathematical Modelling and Interpretation of Geophysical Data*. Moscow, Nauka, pp. 165—170 (Computational Seismology, 16) (in Russian).

Galmakov, B. G. and Sitnikov, A. V., 1984. 'Polarization and azimuth analysis of seismic records for studying secondary Rayleigh waves'. *Izv. AN SSSR, Fizika Zemli* (*Solid Earth*), No. 9, 40—49.

Galperin, E. I., 1984. *The Polarization Method of Seismic Exploration*. D. Reidel Publ. Co.

Gegechkori, T. Sh. and Yanovskaya, T. B., 1979. 'Reflection of Rayleigh waves at a contact of two media with a strong velocity contrast'. *Soobshch. AN GSSR*, **95**(3), 537—576, (in Russian).

Gjevik, B., 1974. 'Ray tracing for seismic surface waves'. *Geophys J.*, **39**, 29—39.

Grachev, A. F., 1973. 'The Momsky continental rift (Northeastern USSR)'. In: *Geophysical Exploration Methods in Use in the Arctic*. 8. Leningrad, Nauka, pp. 56—75 (in Russian).

Gregersen, S., 1974. 'Surface waves in isotropic, laterally inhomogeneous media'. *Pure and Appl. Geophys.*, **114**, 821—832.

Gregersen, S. and Alsop, L. E., 1974. 'Amplitudes of horizontally refracted waves'. *Bull. Seismol. Soc. Amer.*, **64**, 535—553.

Gregersen, S. and Alsop, L. E., 1976. 'Mode conversion of Love waves at the continental margin'. *Bull. Seismol. Soc. Amer.*, **66**, 1855—1872.

Gregersen, S., 1978. 'Possible mode conversion between Love and Rayleigh waves at the continental margin'. *Geophys. J. Roy. Astron. Soc.*, **54**, 121—127.

Grudeva, N. P., Levshin, A. L., Malinovskaya, L. N. and Rosenknop, L. M., 1973. 'Spectra of seismic waves from two Kamchatka earthquakes'. In: *Pattern Recognition and Spectral Analysis in Seismology*. Moscow, Nauka, pp. 107—117 (Computational Seismology, 10) (in Russian).

Haskell, N. A., 1953. 'The dispersion of surface waves in multilayered solid media'. *Bull. Seismol. Soc. Amer.*, **43**, 17—34.

Herrera, I., 1964. 'On a method to obtain a Green function for multilayered half-space'. *Bull. Seismol. Soc. Amer.*, **54**, 1087.

Hudson, J. A. and Knopoff, L., 1964. 'Transmission and reflection of surface waves at a corner'. *J. Geophys. Res.*, **69**, 281.

Its, E. N. and Yanovskaya, T. B., 1977. 'Reflection and transmission of Rayleigh waves at a vertical interface'. In: *Pattern Recognition and Spectral Analysis in Seismology*. Moscow, Nauka, pp. 214—222 (Computational Seismology, 10) (in Russian).

Its, E. N. and Yanovskaya, T. B., 1979a. 'Reflection and transmission of surface waves at oblique incidence on a vertical discontinuity'. In: *Theory and Analysis of Seismological Observations*. Moscow, Nauka, pp. 86—92 (Computational Seismology, 12) (in Russian).

Its, E. N. and Yanovskaya, T. B., 1979b. 'Determination of reflection and transmission coefficients for surface waves incident on a vertical contact using the Green function technique'. *Izv. AN SSSR, Fizika Zemli* (*Solid Earth*), No. 6, 11—21.

Its, E. N. and Yanovskaya, T. B., 1983. 'Reflection and transmission of surface waves at an inclined discontinuity'. In: *Earthquake Prediction and the Study of Earth Structure*. Moscow, Nauka, pp. 87—92 (Computational Seismology, 15) (in Russian).

Its, E. N. and Yanovskaya, T. B., 1985. 'Propagation of surface waves in a half-space with vertical, inclined or curved interfaces'. *Wave Motion*, **7**, 79—84.

Its, E. N. and Malischewski, P., 1987. 'Propagation of Rayleigh waves across a vertical unwelded contact of elastic media'. *Izv. AN SSSR, Fizika Zemli* (*Solid Earth*), No. 6, 66—72, (in Russian).

Jackson, D. D., 1972. 'Interpretation of inaccurate, insufficient and inconsistent data'. *Geophys. J. Roy. Astron. Soc.*, **28**, 97—110.

Jobert, N. and Jobert, G., 1983. 'An application of ray theory to the propagation of waves along a laterally heterogeneous spherical surface'. *Geophys. Res. Lett.*, **10**(12), 1148—1151.

Kanareikin, D. B., Pavlov, N. F. and Potekhin, V. A., 1966. *Polarization of Radar Signals*. Moscow, Sovetskoe Radio (in Russian).

Kats, S. A. and Mikhailova, N. G., 1977. 'Non-particle-path polarization analysis of linearly polarized waves recorded by three-component instruments'. In: *Pattern Recognition and Spectral Analysis in Seismology*. Moscow, Nauka, pp. 223—232. (Computational Seismology, 10) (in Russian).

Kazi, M. H., 1976. 'Spectral representation of Love wave operator'. *Geophys. J. Roy. Astron. Soc.*, **47**, 225—249.

Kennett, B. L. N. and Clarke, T. J., 1983. 'Rapid calculation of surface wave dispersion'. *Geophys. J. Roy. Astron. Soc.*, **72**, 619—631.

Kijko, A. and Mitchell, B. J., 1983. 'Multi-mode Rayleigh wave attenuation and Q in the crust of the Barents shelf'. *J. Geophys. Res.*, **8**, 315—328.

Knopoff, L. and Hudson, J. A., 1964. 'Transmission of Love waves past a continental margin'. *J. Geophys. Res.*, **69**, 1649—1653.

Knopoff, L., 1964. 'A matrix method for elastic wave problems'. *Bull. Seismol. Soc. Amer.*, **54**, 431—438.

Knopoff, L., 1972. 'Observation and inversion of surface wave dispersion'. *Tectonophysics*, **13**, 497—519.

Kogan, S. Ya., 1966. 'A review of seismic wave absorption theory'. *Izv. AN SSSR, Fizika Zemli* (*Solid Earth*), No. 11, 1—28; No. 12, 1—16.

Kolesnikov, Yu. A. and Toksoz, M. N., 1982. 'Lowering the sensitivity of long-period vertical seismometers to air pressure fluctuations'. In: *Mathematical Models of Earth Structure and Earthquake Prediction.* Moscow, Nauka, pp. 183—189 (Computational Seismology, 14) (in Russian).

Kolesnikov, Yu. A. and Toksoz, M. N., 1984. 'The use of summation for suppression of seismic noise due to wind'. In: *Logical and Computational Methods in Seismology.* Moscow, Nauka, pp. 177—188 (Computational Seismology, 17) (in Russian).

Kostrov, B. V., 1970. 'Theory of tectonic earthquake sources'. *Izv. AN SSSR, Fizika Zemli* (*Solid Earth*), No. 4, 84—101.

Kushnir, A. F., Levshin, A. L. and Lokshtanov, D. E., 1988. 'Determination of velocity cross-section by surface wave spectra using nonlinear programming methods'. In: *Problems of Seismological Informatics.* Moscow, Nauka, (Computational Seismology, 21) (in Russian).

Kuznetsov, D. S., 1965. *Special Functions.* Moscow, Vysshaya shkola (in Russian).

Lander, A. V., 1974. 'On interpretation of frequency-time analysis results'. In: *Computer Analysis of Digital Seismic Data.* Moscow, Nauka, pp. 279—315 (Computational Seismology, 7) (in Russian).

Lander, A. V., 1975. 'Frequency-time representation of a linearly dispersed signal with a Gaussian spectrum'. In: *Interpretation of Seismological and Neotectonic Evidence.* Moscow, Nauka, pp. 122—128 (Computational Seismology, 8) (in Russian).

Lander, A. V., 1978. 'Some methodological problems in the measurement of spectral characteristics and interpretation of surface wave data'. In: *Problems in Earthquake Prediction and Earth Structure.* Moscow, Nauka, pp. 93—110 (Computational Seismology, 11) (in Russian).

Lander, A. V., 1984. 'Surface wave anomalous occurrences in Northeast Eurasia and their relation to the Momsky rift area'. In: *Mathematical Modelling and Interpretation of Geophysical Data.* Moscow, Nauka, pp. 127—155 (Computational Seismology, 16) (in Russian).

Lander, A. V., Levshin, A. L., Pisarenko, V. F. and Pogrebinsky, G. A., 1973. 'On the frequency-time analysis of oscillations'. In: *Computational and Statistical Methods in Interpretation of Seismic Evidence.* Moscow, Nauka, pp. 236—249 (Computational Seismology, 6) (in Russian).

Lander, A. V. and Levshin A. L., 1982. 'Azimuth-polarization anomalies of surface waves and the methods for studying them'. In: *Development of G. A. Gamburtsev's Ideas in Geophysics.* Moscow, Nauka, pp. 248—260 (in Russian).

Lay, T. and Kanamori, H., 1985. 'Geometric effects of global lateral heterogeneity on long-period surface wave propagation'. *J. Geophys. Res.*, **90**(B1), 605—621.

Lazareva, A. P. and Yanovskaya, T. B., 1975. 'Effect of lateral velocity variation on surface wave amplitudes'. In: *Problems of Geophysics*, **25**, 197—207. Leningrad University Press (in Russian).

Leveque, J. J., 1980. 'Regional upper mantle *S*-velocity models from phase velocities of great-circle Rayleigh waves'. *Geophys. J. Roy. Astron. Soc.*, **63**, 23—43.

Levshin, A. L., Pisarenko, V. F. and Pogrebinsky, G. A., 1972. 'On a frequency — time analysis of oscillations'. *Ann. Geophys.*, **28**, 211—218.

Levshin, A. L., 1973. *Surface and Channel Seismic Waves.* Moscow, Nauka (in Russian).

Levshin, A. L. and Yanovskaya, T. B., 1976. 'Reflection and transmission of Love waves at a vertical discontinuity'. In: *Investigations of Seismicity and Earth Models.* Moscow, Nauka, pp. 160—173 (Computational Seismology, 9) (in Russian).

Levshin, A. L. and Berteussen, K. A., 1979. 'Anomalous propagation of surface waves in the Barents Sea as inferred from NORSAR recordings'. *Geophys. J. Roy. Astron. Soc.*, **56**, 97—118.

Levshin, A. L., 1980. 'The method of seismic surface waves: state-of-the-art and prospects'. *Izv. vuzov. Geologiya i razvedka*, No. 8, 92—111, (in Russian).

Levshin, A. L. and Berteussen, K. A., 1980. 'Crustal structure in the southern Barents Shelf area based on surface wave data'. In: *Methods and Algorithms in Interpretation of Seismological Data.* Moscow, Nauka, pp. 142—145 (Computational Seismology, 13) (in Russian).

Levshin, A. L., Ratnikova, L. I. and Saks, M. V., 1980. 'On dispersion and attenuation of elastic waves in rocks'. In: *Methods and Algorithms in Interpretation of Seismological Data.* Moscow, Nauka, pp. 134—141 (Computational Seismology, 13) (in Russian).

Levshin, A. L., 1984. 'On the effect of lateral inhomogeneities on surface waves measurements'. In: *Mathematical Modelling and Interpretation of Geophysical Observations.* Moscow, Nauka, pp. 71—83 (Computational Seismology, 16) (in Russian).

Lidsky, V. B. and Neigauz, M. G., 1962. 'Fast recurrence method for a self-conjugate second-order system'. *Zhurn. vychisl. matematiki i mat. fiziki*, **2**, 161—165.

Lin'kov, E. M., Tipisev, S. Ya. and Butsenko, V. V., 1980. 'Noise-protected long-period seismograph'. In: *Geophysical Instrumentation*, **80**, 78—87. Leningrad, Nedra.

Lutikov, A. I., 1979. 'The effect of an inclined discontinuity on the reflection and transmission of Rayleigh waves'. *Izv. AN SSSR, Fizika Zemli* (*Solid Earth*), No. 10, 58—66.

Lysmer, J., 1970. 'Lumped mass method for Rayleigh waves'. *Bull. Seismol. Soc. Amer.*, **60**, 89—104.

Lysmer, J. and Drake, L. A., 1981. 'The propagation of Love waves across nonhorizontally layered structures'. *Bull. Seismol. Soc. Amer.*, **61**, 1233—1251.

Mal, A. K. and Knopoff, L., 1965. 'Transmission of Rayleigh waves at a corner'. *Bull. Seismol. Soc. Amer.*, **55**, 455—466.

Malischewski, P., 1974. 'The influence of curved discontinuities on the propagation of surface waves'. *Gerlands Beitr. Geophys.*, **83**, 355—362.

Malischewski, P., 1976. 'Surface waves in media having lateral inhomogeneities'. *Pure and Appl. Geophys.*, **114**, 833—843.

Matsievsky, S. A., 1980. 'Electronic seismograph with a capacitance transducer'. In: *Methods and Algorithms for Interpretation of Seismological Data.* Moscow, Nauka, pp. 173—184. (Computational Seismology, 13) (in Russian).

McGarr, A. and Alsop, L. E., 1967. 'Transmission and reflection of Rayleigh waves at vertical boundaries'. *J. Geophys. Res.*, **72**, 2169—2180.

Means, J. D., 1972. 'The use of the three dimensional covariance matrix in analysing the polarization properties of plane waves'. *J. Geophys. Res.*, **77**, 5551—5559.

Mendiguren, J., 1977. 'Inversion of surface wave data. 1. Source mechanism studies'. *J. Geophys. Res.*, **82**, 889—894.

Mills, J. M., 1978. 'Great circle Rayleigh wave attenuation and group velocity. Regionalization and pure path models for shear velocity and attenuation'. *Phys. Earth and Planet. Inter.*, **17**, 323—252.

Molotkov, L. A., 1961. 'On propagation of elastic waves in media involving thin plane-parallel layers'. In: *Problems in the Dynamical Theory of Seismic Wave Propagation*, **5**, 240—280. Leningrad, Nauka.

Molotkov, L. A., 1972. 'On matrix representations of the dispersion relation for stratified elastic media'. *Zap. nauch. sem. LOMI*, **25**, 116—131.

Molotkov, L. A., 1984. *Matrix Method in the Theory of Wave Propagation in Stratified Solids and Fluids.* Leningrad, Nauka. (in Russian).

Nakanishi, I. and Kanamori, H., 1982. 'Effects of lateral heterogeneity and source process time on the linear moment tensor inversion of long-period Rayleigh waves'. *Bull. Seismol. Soc. Amer.*, **72**, 2063—2080.

Nakanishi, I. and Anderson, D. L., 1982. 'Worldwide distribution of group velocity of mantle Rayleigh waves as determined by spherical inversion'. *Bull. Seismol. Soc. Amer.*, **72**, 1185—1194.

Nakanishi, I. and Anderson, D. L., 1983. Measurements of mantle velocities and inversion for lateral heterogeneity and anisotropy. 1. Analysis of great-circle phase velocities'. *J. Geophys. Res.*, **88**, 10267—10283.

Nakanishi, I. and Anderson, D. L., 1984. 'Measurements of mantle wave velocities and inversion for lateral heterogeneity and anisotropy. Part 2: Analysis by single station method'. *Geophys. J. Roy. Astr. Soc.*, **78**, 573—617.

Neigauz, M. G. and Shkadinskaya, G. V., 1966. 'A technique for calculating Rayleigh surface waves in a vertically varying half-space'. In; *Computer Interpretation of Seismic Waves.* Moscow, Nauka, pp. 121—129 (Computational Seismology, 2) (in Russian).

Neunhofer, H. and Guth, D., 1976. 'Dispersion of seismic surface waves in middle and northern Europe'. In: *Seismology and solid Earth Physics. B.: Zentralinst. Phys. Erde*, **31**, 399—403.

Nikolaev, A. V. (ed), 1978. *Seismic Wave Fields in Fault Zones*, Moscow, Nauka (in Russian).

Nolet, G. and Panza, G., 1974. 'Array analysis of seismic surface waves: Limits and possibilities'. *Pure and Appl. Geophys.*, **114**, 775—796.

Nolet, G., 1976. 'Higher Modes and the Determination of Upper Mantle Structure'. Ph.D. Thesis/ Veining-Meinesz Lab. Utrecht Univ. Utrecht.

Nolet, G. (ed), 1987. *Seismic Tomography*. D. Reidel Publ. Co.

Noponen, I., 1966. 'Surface wave velocities in Finland'. *Bull. Seismol. Soc. Amer.*, **56**, 1093—1104.

Okal, E. A., 1977. 'The effect of intrinsic oceanic upper-mantle heterogeneity on regionalization of long-period Rayleigh-wave phase velocities'. *Geophys. J. Roy. Astr. Soc.*, **49**, 357—370.

Oliver, J. and Murphy, L., 1971. 'Seismological global network'. *Science*, **174**, 256—261.

Olver, F. W. J., 1974. *Asymptotics and Special Functions*. New York, Academic Press.

Panza, G., 1976. 'Phase velocity determination of fundamental Love and Rayleigh waves'. *Pure and Appl. Geophys.*, **114**, 753—763.

Papazachos, B. C., 1969. 'Phase velocities of Rayleigh waves in South-Eastern Europe and Eastern Mediterranean sea'. *PAGEOPH*, **75**, 47—55.

Patton, H., 1980a. 'Crust and upper mantle structure of the Eurasian continent from the phase velocity and Q of surface waves'. *Rev. Geophys. and Space Phys.*, **18**(3), 605—625.

Patton, H., 1980b. 'Reference point equalization method for determining the source and path effects of surface waves'. *J. Geophys. Res.*, **85**(B2), 821—848.

Pavlov, V. M. and Gusev, A. A., 1980. 'On recovery of motion at the source of a deep earthquake from the far-field body waves'. *DAN SSSR*, **255**(4), 824—828.

Pekeris, C. L., 1948. 'Theory of propagation of explosive sound in shallow water'. *Geological Society of American Memoirs*, No. 27.

Peterson, J. *et al.*, 1976. 'The seismic research observatory'. *Bull. Seismol. Soc. Amer.*, **66**(6), 2049—2068.

Peterson, J. and Qu Kexin (Project Managers), 1987. 'The China digital seismograph network. A joint report by the Institute of Geophysics', State Seismol. Bureau, China, and the Albuquerque Seismol. Lab., USGS; Beijing, China.

Pisarenko, V. F., 1973. 'On estimation of spectra based on nonlinear functions of the covariance matrix'. In: *Computational and Statistical Methods in Interpretation of Seismic Data*. Moscow, Nauka, pp. 263—285 (Computational Seismology, 6) (in Russian).

Plešinger, A. and Horalek, J., 1976. 'The seismic broadband recording and data processing system FBV/DPS and its seismological applications'. *Geophys. J.*, **42**, 201—217.

Plešinger, A., 1981. 'Acquisition, processing and interpretation of broadband seismic data'. In: *Geophysical Syntheses in Czechoslovakia* A. Zatopek (Ed.). Bratislava, pp. 59—78.

Pod'yapolsky, G. S., 1963. 'Reflection and transmission at a boundary between two elastic media with an unwelded contact'. *Izv. AN SSSR, Ser. Geofis.*, No. 4, 525—531.

Popov, M. M. and Pšencik, I., 1978. 'Computation of ray amplitudes in inhomogeneous media with curved interfaces'. *Stud. Geophys. et Geod.*, **22**(3), 248—258.

Popov, M. M., 1983. 'Summation of Gaussian beams in the isotropic elasticity theory'. *Izv. AN SSSR, Fizika Zemli* (*Solid Earth*), No. 9, 39—50.

Rao, C. R., 1965. *Linear Statistical Inference and Its Applications*. New York, Wiley.

Rodi, W. L., Glover, P., Li, M. T. C., and Alexander, S. S. 1975. 'A fast accurate method for computing group velocity partial derivatives for Rayleigh and Love modes'. *Bull. Seismol. Soc. Amer.*, **65**, 1105—1114.

Roult, G. and Romanowicz, B. 1983. 'GEOSCOPE, a new long period global network — first results'. *Trans. Amer. Geophys. Union*, **65**(58), 220.

Sabitova, T. M. and Yanovskaya, T. B. 1986. 'Lateral heterogeneities of the Earth's crust in Kirgiz Tien Shan as derived from distribution of group surface wave velocities'. *Izv. AN SSSR. Fizika Zemli* (*Solid Earth*), No. 6, 31—38.

Santo, T. A., 1961. 'Division of the southwestern Pacific Area into several regions in each of which Rayleigh waves have the same dispersion characters'. *Bull. Earthquake Res. Inst. Univ. Tokyo*, **39**(4), 603—630.

Santo, T. A. and Sato, Y., 1966. 'World-wide survey of the regional characteristics of group velocity dispersion of Rayleigh waves'. *Bull. Earthquake Res. Inst. Tokyo Univ.*, **44**, 939—964.

Sato, R., 1961. 'Love waves propagated across transitional zone'. *Jap. J. Geophys.*, **2**, 117—134.

Sato, Y. and Santo, T. A., 1969, 'World-wide distribution of the group velocity of Rayleigh waves as determined by dispersion data'. *Bull. Earthquake Res. Inst. Tokyo Univ.*, **47**, 31—41.

Savostin, L. A., Karasik, A. M. and Zonenshain, L. P., 1984. 'The opening history of the Eurasian arctic basin'. *DAN SSSR*, **275**(5), 1156—1161.

Schlue, J. W. and Knopoff, L., 1978. 'Inversion of surface-wave phase velocities for an anisotropic structure'. *Geophys. J. Roy. Astron. Soc.*, **54**, 697—702.

Schlue, J. W., 1981. 'Seismic surface wave propagation in three dimensional finite-element structures'. *Bull. Seismol. Soc. Amer.*, **71**, 1003—1010.

Schwab, F. and Knopoff, L., 1970. 'Surface wave dispersion computations'. *Bull. Seismol. Soc., Amer.*, **60**, 321—344.

Schwab, F., Nakanishi, K., Cuscito, M. *et al.*, 1984. 'Surface wave computations and the synthesis theoretical seismograms at the high frequencies'. *Bull. Seismol. Soc. Amer.*, **74**, 1555—1578.

Shkadinskaya, G. V., 1971. 'Calculation of Rayleigh surface waves in a radially varying sphere'. In: *Algorithms for Interpretation of Seismic Data.* Moscow, Nauka. pp. 178—188 (Computational Seismology, 5) (in Russian).

Sikharulidze, D. I. and Mandzhgaladze, P. V., 1977. 'Modelling the horizontal reflection of Rayleigh waves at a vertical interface between two quarter-spaces'. *Soobshch. AN GSSR*, **85**(2), 345—347.

Sikharulidze, D. I., 1978. *Earth Structure as Deduced from Surface Wave Evidence.* Tbilisi, Metsnireba. (in Russian).

Sobel, P. A. and Seggern, D. H., von, 1978. 'Application of surface-wave ray tracing'. *Bull. Seismol. Soc. Amer.*, **68**(5), 1359—1380.

Souriau, A. and Souriau, M., 1983. 'Test of tectonic models by great circle Rayleigh waves'. *Geophys. J. Roy. Astron. Soc.*, **73**, 533—551.

Starovoit, O. E., Feofilaktov, V. D., Shulpin, L. A. and Yaroshevich, M. I., 1971. 'Quartz strainmeter at the Central Seismological Observatory "Obninsk"'. *Izv. AN SSSR, Fizika Zemli* (*Solid Earth*), No. 11, 85—94.

Starovoit, O. E., Zakharova, A. I., Mishatkin, V. N. and Barmin, M. P., 1983. 'A system for acquisition, processing and maintenanance of seismological information'. In: *Computerized Acquisition and Processing of Seismological Information.* Moscow, Radio i svyaz, pp. 53—61 (in Russian).

Talwani, M. and Eldholm, O., 1973. 'Boundary between continental and oceanic crust at the margin of rifted continents'. *Nature*, **241**(5388), 325—330.

Tanimoto, T., 1985. 'The Backus—Gilbert approach to the three-dimensional structure in the upper mantle. I. Lateral variation of surface wave phase velocity with its error and resolution'. *Geophys. J. Roy. Astr. Soc.*, **82**, 105—123.

Tatham, R., 1975. 'Surface-wave dispersion applied to the detection of sedimentary basins'. *Geophysics*, **40**(1), 40—55.

Thomson, W. T., 1950. 'Transmission of elastic waves through a stratified solid'. *J. Appl. Phys.*, **21**, 89—93.

Tikhonov, A. N. and Arsenin, V. Ya., 1974. *Methods for Solution of Incorrectly Stated Problems.* Moscow, Nauka. 223 pp. (in Russian).

Toksoz, M. N. and Ben-Menahem, A., 1963. 'Velocities of mantle Love and Rayleigh waves over multiple paths'. *Bull. Seismol. Soc. Amer.*, **53**, 741—764.

Truskov, Yu. N. (ed.). *Tectonics of Yakutia, 1975.* Novosibirsk, Nauka. (in Russian).

Tsai, Y. B. and Aki, K., 1969. 'Simultaneous determination of the seismic moment and attenuation of the seismic surface waves'. *Bull. Seismol. Soc. Amer.*, **59**, 275—287.

Tsimmer V. A., 1977. 'On the nature of deep interfaces in the sub-Caspian basin'. In: *Crustal and Upper Mantle Structure Based on Seismic Evidence.* Kiev, Naukova Dumka, pp. 302—306 (in Russian).

Valyus, V. P., 1968. 'Determination of velocity structure based on combined data'. In: *Some Forward and Inverse Problems in Seismology.* Moscow, Nauka, pp. 3—14 (Computational Seismology, 4) (in Russian).

Vilkovich, E. V., Levshin, A. L. and Neigauz, M. G., 1966. 'Love waves in a vertically varying medium: sphericity, variation of parameters, attenuation'. In: *Computer Interpretation of Seismic Waves.* Moscow, Nauka, pp. 130—149 (Computational Seismology, 2) (in Russian).

Volk, V. E., Gaponenko, G. I., Zatsepin, E. N. *et al.*, 1984. 'Crustal structure of Arctica based on geophysical evidence'. In: *Geology of Arctica: Reports at the 27th International Geological Congress.* Moscow, Nauka, vol. 4, pp. 26—36 (in Russian).

Volvovsky, I. S., 1973. *Seismic Studies of Crustal Structure in the USSR*. Moscow, Nedra (in Russian).
Weidner, D. J. and Aki, K., 1973. 'Focal depth and mechanism of mid-ocean ridge earthquakes'. *J. Geophys. Res.*, **78**(11), 1818—1831.
Whitham, G. B., 1974. *Linear and Nonlinear Waves*. New York, Wiley-Interscience.
Wielandt, E. and Streckeisen, G., 1982. 'The leaf-spring seismometer: Design and performance'. *Bull. Seismol. Soc. Amer.*, **72**, 2349—2367.
Wiggins, R. A., 1972. 'The general linear inverse problem: Implication of surface waves and free oscilations for Earth structure'. *Rev. Geophys.*, **10**, 251—285.
Wiggins, R. A., 1976. 'A fast, new computational algorithm for free oscillations and surface waves'. *Geophys. J. Roy. Astron. Soc.*, **47**, 135—150.
Woodhouse, J. H., 1974. 'Surface waves in the laterally varying layered structure'. *Geophys. J. Roy. Astron. Soc.*, **37**, 461—490.
Woodhouse, J. H., 1976. 'On Rayleigh's principle'. *Geophys. J. Roy. Astron. Soc*, **46**(1), 11—22.
Wu, F. T., 1972. 'Mantle Rayleigh wave dispersion and tectonic provinces'. *J. Geophys. Res.*, **77**, 6445—6453.
Yanovskaya, T. B., 1980. 'A technique for inversion of seismic arrival times in a laterally varying medium'. In: *Methods and Algorithms for Interpretation of Seismological Data*. Moscow, Nauka, pp. 96—101. (Computational Seismology, 13) (in Russian).
Yanovskaya, T. B., 1982. 'Distribution of surface wave group velocities in the North Atlantic'. *Izv. AN SSSR, Fizika Zemli* (*Solid Earth*), No. 2, 3—11.
Yanovskaya, T. B. and Porokhova, L. N., 1983. *Inverse Problems in Geophysics*. Leningrad University Press (in Russian).
Yanovskaya, T. B., 1984. 'Solution of the inverse problem of seismology for laterally inhomogeneous media'. *Geophys. J. Roy. Astron. Soc.*, **79**, 293—304.
Yanovskaya, T. B. and Nikolova, S. B., 1984. 'Distribution of Rayleigh and Love surface wave group velocities in Northeast Europe and Asia Minor'. *Geofiz. zhurn. Bolg. AN*, **10**(4), 83—92 (in Russian).
Yanovskaya, T. B., Maaz, R. and Neunhofer, H. 1987. 'A technique for combining phase and group velocities of surface waves in estimating the lateral variation of the Earth's structure'. *Izv. AN SSSR. Fizika Zemli* (*Solid Earth*), No. 6, 41—47.
Yanshin, A. L. (Ed.), 1966. *Tectonic Map of Eurasia*. Moscow, GIN AN SSSR, GUGK MINGEO SSSR (in Russian).
Yudakhin, F. N., 1978. *Geophysical Fields, Deep Structure and Seismicity of Tien-Shan*. Frunze, Ilim. (in Russian).

INDEX

MODERN APPROACHES IN GEOPHYSICS

formerly *Seismology and Exploration Geophysics*

1. E. I. Galperin, Vertical Seismic Profiling and Its Exploration Potential. 1985. ISBN 90–277–1450–9.
2. E. I. Galperin, I. L. Nersesov and R. M. Galperina, Borehole Seismology. 1986. ISBN 90–277–1967–5.
3. Jean-Pierre Cordier, Velocities in Reflection Seismology. 1985. ISBN 90–277–2024–X.
4. Gregg Parkes and Les Hatton, The Marine Seismic Source. 1986. ISBN 90–277–2228–5.
5. Guust Nolet (ed.), Seismic Tomography. 1987. ISBN 90–277–2521–7.
6. N. J. Vlaar, G. Nolet, M. J. R. Wortel and S. A. P. L. Cloetingh (eds.), Mathematical Geophysics. 1988. ISBN 90–277–2620–5.
7. J. Bonnin, M. Cara, A. Cisternas and R. Fantechi (eds.), Seismic Hazard in Mediterranean Regions. 1988. ISBN 90–277–2779–1.
8. Paul L. Stoffa (ed.), Tau-p: A Plane Wave Approach to the Analysis of Seismic Data. 1989. ISBN 0–7923–0038–6.
9. V. I. Keilis-Borok (ed.), Seismic Surface Waves in a Laterally Inhomogeneous Earth. 1989. ISBN 0–7923–0044–0.